Time and Time Again

Determination of longitude at sea in the 17th Century

Time and Time Again

Determination of longitude at sea in the 17th Century

Richard de Grijs 何锐思

Kavli Institute for Astronomy and Astrophysics, Peking University, China
中国北京大学科维理天文与天体物理研究所

IOP Publishing, Bristol, UK

ISBN   978-0-7503-1194-6 (ebook)
ISBN   978-0-7503-1195-3 (print)
ISBN   978-0-7503-1196-0 (mobi)

DOI   10.1088/978-0-7503-1194-6

Version: 20171101

IOP Expanding Physics
ISSN 2053-2563 (online)
ISSN 2054-7315 (print)

British Library Cataloguing-in-Publication Data: A catalogue record for this book is available from the British Library.

Published by IOP Publishing, wholly owned by The Institute of Physics, London

IOP Publishing, Temple Circus, Temple Way, Bristol, BS1 6HG, UK

US Office: IOP Publishing, Inc., 190 North Independence Mall West, Suite 601, Philadelphia, PA 19106, USA

To my parents,

for their continuous support of all my choices in life,

in the hope that this manuscript will finally put to rest
a standing joke in the family triggered by my exasperated exclamation
as a high-school pupil that

'History is merely excess baggage, surplus to requirements.'

Contents

Preface

As so often happens, ideas and project developments may take unexpected turns. I never intended to write a book about the determination of longitude at sea in the 17th Century, yet here we are. Upon completion of a scientific monograph on the physics of astronomical distance determination in late 2010, I set out to use that book as the basis of a 30-hour lecture course, *An Introduction to Distance Measurement in Astronomy*, aimed at senior undergraduate students and first-year PhD candidates at Peking University. While preparing my lecture notes, I came across an outreach video [1] produced by the *Spitzer Space Telescope* Science Center which discussed the application of 'light echoes' to measure distances to objects like supernovae—massive stars that explode, and hence become significantly brighter, at the end of their relatively short lifetimes. That video made me digress momentarily from the hard-core physics I was due to teach, since it caught my attention with a discussion of a number of historical supernovae, including the supernova now known as 'Cassiopeia A.'

The presenter pointed out that this supernova should have been seen in the 17th Century, but that there are no clear historical references to it, at least not in European records. Therefore, the exact date of the explosion has largely remained a mystery. A cursory search of the relevant literature showed me that European records were circumstantial at best, although the first 'Astronomer Royal' John Flamsteed, in England, and the founding director of the Observatoire de Paris, Giovanni Domenico Cassini, may have seen an object that *could* have been the supernova [2,3]. Being based in China, I wondered whether the Chinese Imperial records might be helpful in this regard, so I approached Professor Xiaochun Sun (孙小淳) at the Institute for the History of the Natural Sciences (Chinese Academy of Sciences) in Beijing. That line of research is still continuing, but since I don't read Chinese I undertook an extensive Internet search of databases that might contain relevant, late-17th Century materials.

At that time, I came across a news item highlighting that the Dutch Huygens ING institute had launched an online, virtual research environment, *ePistolarium* [4], under the umbrella of the 'Circulation of Knowledge and Learned Practices in the Seventeenth-Century Dutch Republic' project. The *ePistolarium* database contains the full text and metadata of some 20 000 'geleerdenbrieven' (scholarly letters) received and sent by nine leading 17th Century intellectuals who were based in the Dutch Republic during some or all of their life, including the French philosopher René Descartes and the father-and-son team Constantijn and Christiaan Huygens. With my interest piqued, I explored whether this treasure trove of historical letters might be helpful for projects related to the history of astronomy. Xiaochun advised and encouraged me to look into the European efforts to develop methods to accurately determine one's longitude at sea, since his research group was engaged in addressing similar questions, although more in the context of land measurements in East Asia. To spur me on to properly engage with the historical records, he offered me a lecture slot as invited speaker at the 'International Symposium on the Sino–

French Geodetic Survey of the Qing Empire in the 18th Century,' held in November 2014. That invitation forced me to engage with the materials more seriously, leading to a 10-page conference contribution, *The search for longitude: Preliminary insights from a 17th Century Dutch perspective* [5]. This latter article forms the basis of the book in front of you.

Having perused the *ePistolarium* database and the relevant literature, I realized that historical accounts of efforts to assist in the determination of longitude at sea often focus almost exclusively on John Harrison's role in 18th Century Britain, which eventually led to him being awarded approximately half of the British Government's Longitude Prize by 1776. In this book, instead, I decided to start from Galileo Galilei's late-16th Century suggestions—although Galileo himself may have been inspired by Leonardo da Vinci—for the manufacturing of an operational pendulum clock, a feat which was first achieved by Christiaan Huygens in the Dutch Republic. This is a fascinating period in the history of science that is not well covered elsewhere and which, in particular, allows us to also explore the characters of the scientists involved. The latter is an angle that is often bypassed, yet personalities play an important role in how science progresses, a role that is not always recognized.

The 17th Century is regarded as the 'Golden Age' in the history of the Netherlands. The open, tolerant, and transparent conditions in the 17th Century Dutch Republic allowed the nation to play a pivotal role in the international network of humanists, scholars, and 'natural philosophers' before and during the 'Scientific Revolution.' Intellectuals from all over Europe visited the liberal country on the North Sea, which was then considered the 'storehouse of the intellectual world.' However, in order for this dynamic intellectual atmosphere to thrive, scientific knowledge had to be passed on somehow to the educated members of society. Prior to the launch of the first scientific journals, the primary means of communication among intellectuals consisted of a prolific exchange of letters. This is the context of the scientific advances explored in the manuscript in front of you; its main chapters are based on collections of letters that have never been combined into a single volume before, but which represent a well-defined body of work on one of the key scientific and practical issues of the time, from the late-16th to the early-18th Centuries.

Now that the book has been completed, this is indeed an opportune time to look back and reflect on the contributions from numerous people that have allowed me to progress to this stage. First and foremost, I want to acknowledge the moral support I received from my wife Jie (那洁), particularly during those times I may have inadvertently distracted her from her own work with yet another interesting factoid I had uncovered over the course of my research. Although writing a book can be a lonely endeavour, I have had the pleasure to interact with a large number of colleagues whose help and discussions have been invaluable. These scientists and scholars include, in alphabetical order, Alexi Baker (previously University of Cambridge, UK), Rob van Gent (Utrecht University, Netherlands), Rajesh Kochhar (Indian Institute of Science Education and Research, History of Science, India), Andrea Palmieri (Master Horologer, Industrial Technical Institute 'Leonardo da Vinci,' Florence, Italy), Paolo Palmieri (University of Pittsburgh,

USA), Keith Piggott (Horological Antiquarian, UK), Simon Schaffer (Oxford University, UK), Dava Sobel (journalist and author of the 1995 best-seller *Longitude: The True Story of a Lone Genius Who Solved the Greatest Scientific Problem of His Time*, USA), Virginia Trimble (University of California at Irvine, USA), @trinityhouse_uk, and Daniel Vuillermin (Peking University, China). My good friend Giuseppe Bono (Rome Observatory and Università Tor Vergata, Rome, Italy) has been a great font of information and advice in relation to some of the early chapters he read and critiqued. Thank you! These days, most pertinent historical records are available on the Internet, with the exception of a few of the rarest articles. I thank Pieter C van der Kruit (University of Groningen, Netherlands), Maurizio Falanga and Irmela Schweizer (International Space Science Institute, Switzerland), and Jon Earle (Royal Museums Greenwich, UK) for their help in providing access to materials and information that was inaccessible online, but which instead needed personal intervention. I would also like to thank Chengyuan Li (李程远; Macquarie University, Australia) and Habib Khosroshahi (حبیب خسروشاهی; IPM Institute for Research in the Fundamental Sciences, Tehran, Iran) for their help with translations of, respectively, Chinese and Persian (Farsi) titles and names.

Over the course of my writing, the luxury of spending significant lengths of time in environments that were conducive to the writing process has been very beneficial. A three-week writing residency at *Stiwdio Maelor* in Corris, Wales, helped me to make a major dent in writing the narrative. I would like to thank Veronica Calarco for giving me the opportunity to use one of her studios for this purpose in the summer of 2016, and I appreciated the nightly banter with the regulars at the *Slater's Arms* in Corris, which made me feel really welcome in their midst. I also want to thank David Joyner, Lina Davitt, and the team at the Confucius Institute at Bangor University (Wales) for their fantastic hospitality during this same period.

A second major push was possible thanks to the generous award of an Erskine fellowship at the University of Canterbury in Christchurch, New Zealand, from late April until early June 2017. Thanks are due to Karen Pollard, Simone Scaringi, and the staff at the Department of Physics and Astronomy for their hospitality and great company! I managed to finish the compilation of the entire manuscript and deal with the remaining technical challenges during a final, two-week writing residency in August 2017 at the *Fondation Jan Michalski* in Montricher, Switzerland, where I particularly enjoyed the wonderful assistance from Guillaume Dollmann, the tremendously welcoming atmosphere, and the daily chats with the *Fondation*'s staff members and with my fellow *écrivains*.

Last but not least, I would like to acknowledge the valuable input into the book's development from Leigh Jenkins, my Institute of Physics Publishing commissioning editor. It was great meeting her in person at the January 2017 winter meeting of the American Astronomical Society, allowing both of us to put a face to the name. I also thank Daniel Heatley for his assistance with many aspects related to copyright issues.

Thank you to everyone whose contributions and assistance have allowed me to complete this project successfully. I hope that it was worth your time and effort!

Beijing ǀ 北京, August 2017   Richard de Grijs

References

[1] http://www.spitzer.caltech.edu/video-audio/318-hiddenuniverse020-Echoes-of-a-Supernova-Showcase-

[2] Hughes D W 1980 Did Flamsteed see the Cassiopeia supernova? *Nature* **285** 132–3

[3] Soria R, Balestrieri R and Ohtsuka Y 2013 On Cas A, Cassini, Comets, and King Charles *Publ. Astron. Soc. Aus.* **30** e028

[4] http://ckcc.huygens.knaw.nl/epistolarium/

[5] https://arxiv.org/abs/1501.00627

Foreword

The solution to the Longitude Problem that finally emerged in the late 1700s capped several centuries' worth of effort. The answer lay, all along, in the title of this book: *Time and Time Again*—that is, knowledge of the time at one's location and also, at the same instant, the time at a place of known longitude. It sounds straightforward, but the nature of time and locations of places had first to be determined.

My personal encounter with the Longitude Problem occurred in 1993, during the Longitude Symposium organized at Harvard by William J H Andrewes, who was then curator of the university's collection of historical scientific instruments. The invited speakers at the three-day event represented disparate fields such as astronomy, cartography, history, economics, navigation, and of course horology, the science of precision timekeeping. The Longitude Problem, I learned, was rooted in the very turning of the Earth on its axis as it orbited the Sun. Thus the hard-won ability to pinpoint the whereabouts of a ship on the open ocean arose from a great sea of accumulated knowledge.

The book I wrote as a result of the Symposium focused on the story of John 'Longitude' Harrison, the 18th-century English clockmaker credited with the invention of the first successful sea-clocks, now known as chronometers. It took Harrison some forty years to achieve mastery, and decades of struggle to secure the promised financial reward offered by Parliament to '*such Person or Persons as shall Discover the Longitude at Sea*'.

Richard de Grijs, in contrast, has turned his attention to earlier attempts at a Longitude Solution. Before Harrison pitted his time-keepers against bitter rivals for the £20 000 British prize, many other determined individuals put their minds to the problem and competed for other purses. I might call his book a 'prequel' to mine, except that it is so much more than that. Handsomely illustrated and assiduously footnoted, it probes the intricacies of work by Christiaan Huygens, Robert Hooke, Isaac Newton, their several contemporaries, and their numerous predecessors dating back to Antiquity. The intended audience of *Time and Time Again* will be pleased to see that the text does not shy away from physical constants and mathematical formulas. In fact, it revels in them.

The Longitude tale holds enough science and intrigue for re-telling from various perspectives, especially now that so many of us seek guidance from the Global Positioning System.

Each improvement to the technology of way-finding underscores the difficulty of the original challenge and increases the appreciation (or emphasizes the folly) of prior strategies. Meanwhile the Prime Meridian, or zero-degree line of longitude, which came to rest at Greenwich in 1884 after alighting in two dozen other cities, remains restive. It jogged 5.3 arc-seconds (about 300 feet) to the east when authorities began measuring the Earth from a vantage point in space, and it continues to drift an inch or so a year via the motions of the planet's tectonic plates. In 1984, one hundred years after the International Meridian Conference, the countries of the world agreed on a flexible new reference meridian that adjusts to the Earth's internal activity and variable rate of rotation. Each and every one of us knows exactly where we are.

Dava Sobel

Author biography

Richard de Grijs

Richard de Grijs obtained his PhD in Astrophysics from the University of Groningen (the Netherlands) in 1997. He subsequently held postdoctoral research positions at the University of Virginia (USA) and the University of Cambridge (UK), before being appointed to a permanent post at the University of Sheffield (UK) in 2003. He joined the Kavli Institute for Astronomy and Astrophysics at Peking University (China) in September 2009 as a full professor. Richard has been a scientific editor of *The Astrophysical Journal* since 2006 and took on the role of deputy Editor of *The Astrophysical Journal Letters* in September 2012. He was the founding director (2012–2016) of the East Asian Regional Office of Astronomy for Development, which operates under the auspices of the International Astronomical Union, and he has held the role of Discipline Scientist (Astrophysics) at the International Space Science Institute–Beijing since 2015. Richard received the 2012 Selby Award for excellence in science from the Australian Academy of Science, a 2013 Visiting Academy Professorship at Leiden University from the Royal Netherlands Academy of Arts and Sciences, and a 2017 Erskine Award from the University of Canterbury (New Zealand). His research focuses on many aspects of star cluster physics, from their stellar populations to their dynamics and their use as star-formation tracers in distant galaxies. He is currently also engaged in a number of research projects related to the history of astronomy, with special emphasis on the 17th Century. His interests in historical and cultural aspects of life in East Asia resulted in his appointment to the Council of the Royal Asiatic Society China in Beijing, where he is currently responsible for editing the annual issues of the scholarly *Journal of the Royal Asiatic Society China*. His research group at Peking University is popular with both under-graduate and graduate students. He considers interactions with students among the highlights of his professional life.

Chapter 1

Changing times

1.1 Enlightenment in Western Europe: the Dutch Golden Age

The 17th Century is regarded as the 'Golden Age' in the history of the Netherlands. The open, tolerant, and transparent conditions in the 17th Century Dutch Republic (at the time known as the Republic of the Seven United Netherlands) allowed the nation to play a pivotal role in the international network of humanists and scholars before and during the 'Scientific Renaissance'—the two centuries leading up to the publication of Galileo Galilei's groundbreaking treatise, *Dialogue Concerning the Two Chief World Systems* (1632). The young nation had just declared its independence from the declining Spanish empire, in 1581, and its citizens had fully embraced the spirit of the European Enlightenment; more fully so than any other European nation at the time. The economist Dasgupta [1] has pointed out eloquently that this period was revolutionary precisely because …

> *'it created institutions that enabled the production, dissemination, and use of knowledge … to be transferred from small elites to the public at large.'*

Indeed, the global trade network established by the Dutch, the resulting prosperity, and its political system's relative tolerance made the Dutch Republic a refuge for disenfranchised and persecuted scholars and intellectuals fleeing religious or political censorship elsewhere in Europe. In turn, this facilitated a dynamic intellectual atmosphere and a free exchange of new knowledge in a wide diversity of fields, from philology to natural philosophy and natural history. Intellectuals from all over Europe visited and often made their long-term home in the Dutch Republic; its thriving intellectual environment was then considered the 'storehouse of the intellectual world.'

A truly amazing group of the leading free thinkers of their time flocked to the tolerant environment on the North Sea. Johannes Amon Comenius (1592–1670) faced persecution as a pioneer of Czech Protestantism during the (Catholic) Counter-Reformation; Thomas Hobbes (1588–1679) felt unwelcome in England following the publication of his masterpiece, *Leviathan or The Matter, Forme and Power of a Common Wealth Ecclesiasticall and Civil*, commonly known by its shorter title *Leviathan* (1651); Galileo Galilei (1564–1642) was offered refuge in the Dutch Republic following his censure by the Vatican in 1633 (see chapter 3); and John Locke (1632–1704) was suspected of involvement in the 1683 Rye House Plot to assassinate King Charles II of England, although there was ultimately no evidence to support these allegations.

The intellectual capital gained from offering refuge to persecuted and censured scholars and philosophers soon yielded significant dividends. As a major seafaring nation, the ability to accurately and precisely determine one's position at sea was particularly important for crews on the large fleet of Dutch merchant ships. This was so, because following the Dutch War of Independence—also known as the Eighty Years' War (1568–1648)—Spanish ports were off limits to Dutch Republic-flagged vessels. It is, therefore, no surprise that Dutch and Dutch Republic-based scholars and scientists[1] played a significant role in the development of instruments and technology which would allow suitably trained sailors to navigate the open seas, away from the relative safety of coastal waters. One of the key players among the commercial shipping giants was the Dutch East India Company, the 'Vereenigde Oostindische Compagnie,' which sent its merchants to the far corners of the planet in pursuit of new trade links and rich returns from rare commodities sourced in the East Indies.

It was thus of the utmost importance to have access to reliable maps and charts. However, such navigational tools were, at the time, considered closely guarded state secrets, which were therefore only accessible to the most senior officials. Generally available sources of information were of mixed reliability; they included accounts based on poorly translated texts, mythical tales, and geographic insights that were largely coloured by Biblical allegories. Contemporary information obtained from sailors was generally considered largely unreliable. Government officials and trade departments went to great lengths to secure more reliable accounts; the Portuguese trade department, the 'Casa da India,' even required its navigators to swear on the Bible that their accounts represented the truth. Aimed at simultaneously satisfying the new nation's unbridled intellectual curiosity, careful exploration of new and strange lands, combined with reliable dissemination of new insights thus obtained, became an integral aspect of the voyages undertaken by the Dutch East India Company. In the words of Carl Sagan [2], '*the Middle Ages had ended; the Enlightenment had begun.*'

Leiden University, established in 1575, played a particularly important role in driving the country's scientific renaissance—a pioneering role naturally linked to its fierce resistance to the Spanish occupation during the Eighty Years' War. Many refugee scholars from elsewhere in Europe congregated in its erudite halls. Comenius, René Descartes (1596–1650), Hugo Grotius (1583–1645), Christiaan Huygens (1629–1695), Antonie Philips van Leeuwenhoek (1632–1723), and Jan

---

[1] Refer to chapter 1.2 for a discussion and justification as to the use of this term.

Adriaansz Leeghwater (1575–1650) are just a few of the names of the foreign and domestic intellectual leaders who propelled the young nation to a dominant place in the international pecking order: …

'*[Descartes] lived in Holland for twenty years (1629–1649), except for a few brief visits to France and one to England, all on business. It is impossible to exaggerate the importance of Holland in the seventeenth century, as the one country where there was freedom of speculation. Hobbes had to have his books printed there; Locke took refuge there during the five worst years of reaction in England before 1688; Bayle* [author of the *Historical and Critical Dictionary*] *found it necessary to live there; and Spinoza would hardly have been allowed to do his work in any other country.*' [3]

1.2 Intermezzo: The rise of the scientist

Although I used the term 'scientist' in the previous section, and I will continue to use it as a catch-all term throughout this manuscript to denote those individuals who engaged in the practice of scientific research, I am well aware that this professional designation was coined only in 1834, that is, more than a century after our narrative concludes around the time of the confirmation of the 1714 British Longitude Act. The designation 'scientist' was not commonly used until at least a century later—occasionally and sometimes reluctantly at first, whereas a surge in its uptake was only witnessed in the late-19th Century in the USA and from the turn of the 20th Century in the UK. [4] The early-20th Century scholar of American English, Henry Louis Mencken (1880–1956), expanded upon its use (and the early opposition to it!) in his classic reference work, *The American Language*, …

'*The last-named* [scientist] *was coined by William Whewell, an Englishman, in 1840, but was first adopted in America. Despite the fact that Fitzedward Hall and other eminent philologists used it and defended it, it aroused almost incredible opposition in England. So recently as 1890 it was denounced by the London* Daily News *as "an ignoble Americanism," and according to William Archer it was finally accepted by the English only "at the point of the bayonet."*' [5]

In a footnote he further expanded the argument, stating that …

'*[d]espite this fact an academic and ineffective opposition to it still goes on. On the Style Sheet of the* Century Magazine *it is listed among the 'words and phrases to be avoided.' It was prohibited by the famous* Index Expurgatorius *prepared by William Cullen Bryant for the New York* Evening Post, *and his prohibition is still theoretically in force, but the word is now actually permitted by the* Post. *The Chicago* Daily News *Style Book, dated July 1, 1908, also bans it.*' [6]

The term was, in fact, coined on 24 June 1833 and not in 1840. As is often the case, the first use of the designation 'scientist' seems to have been an off-the-cuff remark by the English philosopher and historian of science William Whewell (1794–1866), at the third annual meeting of the British Association for the Advancement of Science, an organisation of which he was both co-founder and President. Whewell, one of the leading figures of 19th Century British science, was also a Fellow of the Royal Society, President of the Geological Society, and Master of Trinity College at the University of Cambridge. The Association's members had voiced the need for a description of the practitioners of science for a number of years. The designations 'natural philosopher,' 'man of science' (very few, if any, women were included in the early scientific circuits), or 'cultivator of science,' which had been in common use until that time, had gradually lost their meaning.

In fact, the designation 'natural philosopher' had not been used commonly for long itself. In this context, Galileo caused a seismic intellectual shift; before his time, philosophers did not engage with observations but only with moral and logical theory. Galileo instead asserted that ...

'[h]e who looks the higher is the more highly distinguished, and turning over the great book of Nature (which is the proper object of philosophy) is the way to elevate one's gaze,' [7]

so that, in his view, philosophy was just not meant to deal with the words of other men. This led to the notion that those engaging with natural science were referred to as 'natural philosophers.'

The Romantic poet Samuel Taylor Coleridge (1772–1834), who had previously expanded upon the scientific method, attended the meeting. He declared that although he was a true philosopher, the designation 'philosopher' should not be applied to the Association's members without a modifier like 'natural' or 'experimental.' Whewell casually responded, suggesting, ...

'by analogy with artist, we may form scientist.' [8]

'Scientist' first appeared in print in 1834, in an anonymous review of the Scottish science writer Mary Somerville's *On the Connexion of the Physical Sciences*, which was eventually published in the *Quarterly Review*. The reviewer appears to not have taken its use very seriously but instead provides a partly satirical commentary, most likely in response to the changing concept of the nature of science, at a time when natural knowledge was increasingly distinguished from other forms of knowledge: ...

'The tendency of the sciences has long been an increasing proclivity and dismemberment. ... The mathematician turns away from the chemist; the chemist from the naturalist; the mathematician, left to himself, divides himself into a pure mathematician and a mixed mathematician, who soon part company; the chemist is perhaps a chemist of electro-chemistry; if so, he leaves common chemical analysis to others; between the mathematician and the chemist is to be

interpolated a 'physicien' (we have no English name for him*), who studies heat, moisture, and the like. And thus science, even more physical science, loses all traces of unity. A curious illustration of this result may be observed in the want of any name by which we can designate the students of the knowledge of the material world collectively. We are informed that this difficulty was felt very oppressively by the members of the British Association for the Advancement of Science, at their meetings at York, Oxford, and Cambridge, in the last three summers. There was no general term by which these gentlemen could describe themselves with reference to their pursuits.* Philosophers *was felt to be too wide and too lofty a term, and was very properly forbidden them by Mr. Coleridge, both in his capacity as philologer and metaphysician;* savans [savant] *was rather assuming, besides being French instead of English; some ingenious gentleman proposed that, by analogy with artist, they might form scientist, and added that there could be no scruple in making free with this termination when we have such words as* sciolist [someone pretending to be knowledgeable and well-informed], economist, *and* atheist—*but this was not generally palatable; others attempted to translate the term by which the members of similar associations in Germany have described themselves, but it was not found easy to discover an English equivalent for Natur-forscher. The process of examination which it implies might suggest such undignified compounds as nature-poker, or nature-peeper, for these naturae curiosi; but these were indignantly rejected.'* [9]

As a footnote, the reviewer—who we now know was Whewell himself, based on Isaac Todhunter's 1876 biography [10] of the scholar—who referred to himself as '*some ingenious gentleman*,' added, …

'*When the German association met at Berlin, a caricature was circulated there, representing the "collective wisdom" employed in the discussion of their mid-day meal with extraordinary zeal of mastication and dexterity in the use of the requisite instruments, to which was affixed the legend—"wie die Natur-forscher Natur-forschen," which we venture to translate* "the poking of the nature pokers".' [11]

That caricature is reproduced in figure 1.1. Whewell repeated his proposed designation of 'scientist' in print in 1840, in *The Philosophy of the Inductive Sciences*: …

'*As we cannot use physician for a cultivator of physics, I have called him a physicist. We need very much a name to describe a cultivator of science in general. I should incline to call him a Scientist. Thus we might say, that as an Artist is a Musician, Painter, or Poet, a Scientist is a Mathematician, Physicist, or Naturalist.'* [12]

Also in 1840, in *Blackwood's Edinburgh Magazine,* [13] Leonardo da Vinci (1452–1510) was referred to as '*mentally a seeker after truth—a scientist,*' while the Italian Renaissance painter Antonio Allegri da Correggio (1498–1534) was cast as an

Fig. 214. Wie die Naturforscher naturforschen (zirka 1840).

Figure 1.1. Caricature, '*Wie die Natur-forscher Natur-forschen.*' (Holländer E 1905 *Die Karikatur und Satire in der Medizin*, Stuttgart: Verlag von Ferdinand Enke, Fig. 214; not in copyright).

'*assertor of truth—an artist*,' thus clarifying the philosophical distinction between scientists and artists.

Nevertheless, general acceptance of the term 'scientist' took a long time to develop. Scientific luminaries, including the English scientist Michael Faraday (1791–1867), did not easily give in, despite their public appearances: …

> '*As for hailing [the new term] scientist as "good," that was mere politeness: Faraday never used the word, describing himself as a natural philosopher to the end of his career.*' [4]

Separately, William Thomson, First Baron Kelvin (1824–1907), preferred the term 'naturalist' over 'physicist,' following the definition of the former term in Samuel Johnson's *Dictionary of the English Language* of 1755 as 'a person well versed in natural philosophy:' …

> '*Armed with this authority, chemists, electricians, astronomers, and mathematicians may surely claim to be admitted along with mere descriptive investigators of nature to the honourable and convenient title of Naturalist, and refuse to*

accept so un-English, unpleasing, and meaningless a variation from old usage as physicist.' [14]

Faraday was more outspoken in his opposition to being called a physicist: …

'I also perceive another new and good word, the scientist. Now can you give us one for the French physicien? Physicist is both to my mouth and ears so awkward that I think I shall never use it. The equivalent of three separate sounds of i in one word is too much.' [15]

1.3 Scholarly communication and scientific networks in the 17th Century

Returning now to scientific developments in the 17th Century, for the new dynamic, intellectual atmosphere to thrive during the Age of Enlightenment, scientific knowledge had to be disseminated somehow among the educated members of society. Given that the first scientific and academic journals appeared only in 1665, as we will see shortly, the key questions to be addressed include, *How did the 17th Century scientific information system work?* and *How were new scientific information and insights assimilated, distributed, and eventually broadly established in the educated community?*

The Renaissance coincided with a significant increase in scientific activity, a step change from predominantly book-reading philosophers addressing scientific questions in a largely qualitative manner to rapidly increasing numbers of experimental scientists in the modern sense of the word. Initially, an abundance of new knowledge was created by just a few leading intellectuals, 'Renaissance men,' that is, true polymaths who were well versed in disciplines as wide-ranging as biology, art, and engineering. Think of da Vinci, for instance, who developed flying machines, studied weather conditions, was a great artist in his own right, and pursued an in-depth understanding of human anatomy, all at the same time.

As we will see in chapter 3, the scientific reasoning, observation, and experimentation approach developed by luminaries such as Galileo and Francis Bacon (1561–1626) is now known as the modern scientific method. These global developments, often referred to as the 'Scientific Revolution,' fully surfaced in the late-16th Century and changed the scientific landscape completely from the onset of the 17th Century. Knowledge creation became an institutionalized professional occupation in its own right; the numbers of natural philosophers and scholars involved in scientific pursuits increased exponentially. Meanwhile, scientists pursued ever more specialized research questions, forever changing the meaning of the word 'scientist' from the experts in general knowledge the Renaissance had produced.

Whereas scholarly books had been the main vehicle of scientific dissemination until that time, the pace of scientific discovery soon accelerated, indirectly leading to a 17th Century reference to 'information overload.' [16] Specifically, the popular English author and soldier Barnabe (Barnaby) Rich (1580–1617) commented in 1613 that …

'[o]ne of the diseases of this age is the multiplicity of books; they doth so overcharge the world that it is not able to digest the abundance of idle matter that is every day hatched and brought forth.' [17]

Writing, editing, and publishing books was slow and expensive—then as now—while their readership was often limited, and printing a book carried a financial risk. [18]

With the increasing numbers of active scientific researchers came the need for recognition of their achievements by their peers and protection from intellectual piracy. However, in the absence of such safeguards, early-17th Century researchers often treated their discoveries as closely guarded secrets. Whereas in today's scientific environment one would pursue publication of new research insights in some of the most highly ranked journals, without access to such a route to publication many 17th Century researchers resorted to disseminating their discoveries to their fellow learned men in the form of anagrams. The latter could not be deciphered without actual knowledge of the discovery; the key to unscrambling the anagram was usually a single sentence or a phrase. They thus established priority of discovery and would hand the anagram to an official witness for safekeeping. In essence, they would develop their discovery into a profitable undertaking or obtain patronage from a wealthy benefactor before revealing the details, only doing so once they were ready to claim ownership. [19]

Galileo is known to have used anagrams to announce some of his breakthroughs, including his discovery of the rings of the planet Saturn. In 1610, he sent an anagram to Johannes Kepler (1571–1630), the German astronomer who is best known for his laws of planetary motion: …

```
s m a i s m r m i l m e p o e t a l e v m i b u n e n u g t
t a v i r a s
```

Once deciphered—something Kepler was apparently unable to achieve—it read [20]

```
altissimum planetam tergeminum observari.
```
(*'I have observed the uppermost planet triple.'*)

We will encounter additional examples of discovery anagrams, designed by Robert Hooke (1635–1703) as well as by Huygens, in chapters 3, 4, and 5.

This approach would at least allow the inventor to tell the world about their breakthrough, without having to worry about being scooped by their competition. However, this solution was not ideal: the decrypted anagrams could be as vague as the anagrams themselves, it was not necessarily clear *where* the discovery had been made, discoveries may have been made independently by different claimants, and in general the use of anagrams to claim priority of discovery suppressed the dissemination of scientific information. As a case in point, Isaac Newton (1642–1727) and Gottfried Leibniz (1646–1716) both claimed to have developed calculus, Newton's so-called 'fluxional method.' Newton claimed that his priority dated from the 1660s and 1670s, as shown in the following passage from a letter he wrote to Leibniz in 1677. His letter was delivered via Henry Oldenburg (c. 1619–1677) at the Royal Society in London, who is often appropriately referred to as the English 'clearinghouse' of science. …

'The foundation of these operations is evident enough, in fact; but because I cannot proceed with the explanation of it now, I have preferred to conceal it thus: 6accdae13eff7i3l9n4o4qrr4s8t12ux. On this foundation I have also tried to simplify the theories which concern the squaring of curves, and I have arrived at certain general Theorems.' [21]

However, he did not publish his findings openly until 1693. [22] Yet, in the meantime, Leibniz had already published his own treatise on calculus, *Nova Methodus pro Maximis et Minimis*, in 1684. Deciphered, the anagram from Newton's letter reads ...

Data aequatione quotcunque fluentes quantitates involvente, fluxiones invenire; et vice versa
('Given an equation involving any number of fluent quantities to find the fluxions, and vice versa.')

When rearranged alphabetically, it consists of the following numbers of characters, allowing us to retrieve Newton's original reference[2] ...

```
aaaaaa cc d ae eeeeeeeeeeeee ff iiiiiii lll nnnnnnnnn
  (6a)              (13e)           (7i)  (3l)   (9n)
oooo  qqqq  rr  ssss  ttttttttt  uuuuuuuvvvvv  x
(4o)  (4q)  (4s)  (9t)           (12u)
```

Beyond communication through anagrams, the 17th Century witnessed the spread of active correspondence networks across the European continent, reaching as far as the Levant (the Eastern Mediterranean). Scientists would often send their observations and discoveries to a central person who acted as a 'gatekeeper' or conduit for transmission of new developments to the network's other members. The increasing importance of this approach to publicizing their work was greatly facilitated by significantly improved postal systems across Western Europe. Despite the increased use of print, ordinary, handwritten letters still remained a powerful means of communication among the men of science. Exchanging letters was convenient, affordable, as well as fast and not subject to significant censorship. From the beginning of the 17th Century, letters could be exchanged anywhere within Europe on timescales of weeks; by the end of the century, regular letters and diplomatic couriers provided an improved, faster, and more reliable service which covered far-flung locales on the periphery of European science.

Two particularly pivotal correspondents were the London-based German–British polymath Samuel Hartli(e)b (c. 1600–1662), known as the 'Great Intelligencer of Europe,' and the French scientist–priest Marin Mersenne (1588–1648). Despite his role of 'gatekeeper,' Hartlib does not seem to have interfered in the free exchange of knowledge, since he corresponded prolifically with a wide variety of European

---

[2] Here, the diphthong '*ae*' is taken as a separate character, '*u*' and '*v*' are considered the same character, and also note that Newton may have miscounted the number of '*t*'s, or he may have left out a '*t*' in his original Latin text.

scientific and literary movers and shakers, including Comenius, Robert William Boyle (1627–1691), Oliver Cromwell (1599–1658), Descartes, John Milton (1608–1674), Blaise Pascal (1623–1662), and Christopher Wren (1632–1723).

Hartlib operated as one of the era's most active intellectual impresarios. His predominant aim was to record all human knowledge and—in the true 'universalist' spirit—make it available to everyone, irrespective of one's background, so as to facilitate universal dissemination of knowledge. His extensive network of knowledge providers ranged from the great philosophers of the day to gentleman farmers, with many of whom he maintained high levels of correspondence. As a consequence, he turned into a central intellectual of his era by arranging the spread of knowledge and information, examination of patent applications, facilitating learning and informal education, and circulating designs for a range of mechanical instruments and machines, [23] including calculators, double-writing instruments, [24] seed machines, and siege engines for use during wartime.

Hartlib's closest correspondents were Comenius, the philosopher and theologian, and his colleague, the Scottish theologian John Dury (1596–1680). The ultimate goal pursued by Hartlib, Comenius, Dury, and their fellow thinkers—greatly influenced by Bacon's general theory of education—was to facilitate the spread of knowledge at a time when the pursuit of knowledge was scattered and libraries were not commonly accessible. As religious scholars, they considered it their Christian duty to enlighten, educate, and improve society. Comenius considered man 'very suitable for education,' because of the processes of Nature, which he believed to be spontaneously at work in everyone and which could easily be cultivated, because ...

> *'they are all human beings with the prospect of the same future life in the way appointed by heaven yet beset with snares and obstructed by diverse pitfalls.'* [25]

He was an important proponent of the idea of 'universal education,' as is keenly communicated in the opening chapter of his *Pampaedia* (rediscovered in 1935) [26], part four of his unfinished seven-volume pedagogical masterwork, *De rerum humanarum emendatione consultatio catholica* (*'General advice on the improvement of human affairs'*): ...

> *'First, the expressed wish is for full power of development into full humanity not of one particular person, but of every single individual, young and old, rich and poor, noble and ignoble, men and women—in a word, every being born on Earth, with the ultimate aim of providing education to the entire human race regardless of age, class, sex, and nationality.'* [27]

Hence, he says at the end of his first chapter, ...

> *'I had this consideration in mind when I put the symbol of the art of the tree pruner in the frontispiece to this Deliberation, showing gardeners grafting freshly-plucked shoots from the tree of Pansophia into rooted layers in the hope of filling God's whole garden, which is the human race, with saplings of a similar nature.'*

The gardener transplanting shoots onto a young tree, seen in the figure above, is thus clearly analogous to a teacher whose job consists of grafting shoots from the 'tree of knowledge' onto young children and cultivating intellectual growth.

Across the English Channel, meanwhile, Mersenne had become a leading, second-generation 'intelligencer' in the Parisian scholarly community. Although he was not a great philosopher himself, his contributions to the development of theology, music, mathematics, and natural philosophy were certainly competent; he is now best known on his own merits as the 'father of acoustics.' More importantly, however, his facilitation of the scientific enterprise of the early-17th Century was invaluable and gave him the moniker of the 'mailbox of Europe.' He stimulated and corresponded with numerous colleagues, hailing from a wide variety of backgrounds, including all contemporary leading philosophers, mathematicians, musical theorists, anatomists and other medical practitioners, antiquarians, oriental scholars, and theologians—most notably, however, Galileo and Descartes. In 1633, 1634, and 1639 Mersenne translated Galileo's studies on mechanics from Italian into French; it is largely owing to Mersenne's promotion that Galileo's work reached his contemporaries across Europe. Mersenne encouraged discussion of groundbreaking philosophies by posing problems, communicating 'objections' (an early form of peer review), supplying news and information, establishing contacts, and facilitating or pursuing publication.

An ordained Minim friar by vocation, in the early 1620s Mersenne was greatly influenced by Nicolas-Claude Fabri de Peiresc (1580–1637), whose extensive correspondence with scientists and great intellectual contributions to diverse fields—including the determination of longitude across Europe, North Africa, and the Mediterranean—gave him the accolades 'Prince of Erudition' and 'Prince of the Republic of Letters,' a catch-all designation for the largely informal, pre-Renaissance network of prolific scientific correspondents across Europe. De Peiresc's network of nearly 500 correspondents was predominantly concentrated in the major 'centres of learning,' particularly Rome and Paris. Although he was a gentleman scientist in his own right, who actively pursued experiments and dissections in astronomy and optics, his most important contribution and influence was in fostering interdisciplinary and multinational intellectual knowledge exchange, from Italy to France, from humanism to science.

De Peiresc introduced Mersenne to his network of Parisian intellectuals, thus setting the stage for Mersenne's own development as an influential science broker. By 1626, Mersenne had initiated a series of weekly scientific discussions, meanwhile working on establishing a network of correspondents across the continent—Dutch, Flemish, English, Italian, French, German; Protestant as well as Catholic. Indeed, he proved capable of maintaining excellent relationships with people whose scientific and philosophical ideas were sometimes significantly at odds with one another. For instance, in preparation of the publication, in 1641, of his close friend Descartes' *Meditationes*, Mersenne commissioned sets of 'objections' (to which Descartes was meant to respond) from Pierre Gassendi (1592–1655) and Hobbes, two major philosophers who were fundamentally opposed to the Cartesian principles supported by Descartes—which emphasize the use of reason to develop the natural sciences, while separating the mind from the sensations of the body. He was also on good personal terms with Gilles Personne de Roberval (1602–1675), the French mathematician, another vocal anti-Cartesian.

Mersenne was particularly keen to make the latest philosophical insights accessible to a wider public and he strongly believed in science as a collaborative enterprise. To achieve his ideal of establishing a truly global scientific community of like-minded intellectuals, an 'academy,' he reached out to scientists and scholars well beyond his own circle of Parisian mathematicians. Requesting feedback from De Peiresc, in a letter dated 15 July 1635 he explained, …

> *'I would like to have such a peace that we could build an Academy, not just in one city …, but if not of all Europe, at least in the entire France, which would communicate by letters, which would be better than the talks where one often gets too excited.'* [28]

In his pursuit of a global scientific community, he did not care about political, religious, or scientific differences—and apparently he was rather successful, as evidenced by a comment from one of his contemporaries: …

> *'He had become the centre of the world of letters, owing to the contact he maintained with all, and all with him … serving a function in the Republic of Letters similar to that of the heart in the circulation of blood within the human body.'* [29]

Meanwhile, a close friend of De Peiresc and Mersenne, Ismaël Boulliau (1605–1694), built up a network of correspondents that was to become one of the most extensive letter-writing circles of the 17th Century, amassing of order 5000 letters between 1632 and 1693. Although his correspondents were certainly interested in studying the Classics, philology, politics, and diplomacy, most importantly he fostered a more clearly defined scientific focus—particularly on astronomical observations, which he exchanged prolifically with the likes of Gassendi, Johannes Hevelius (1611–1687), and Huygens—and a more extensive geographic reach, beyond the traditional scientific powerhouses of France, the Dutch Republic, Italy, and England, but also encompassing Poland, Scandinavia, and the Levant.

Mersenne's community building eventually led to his establishment of the *Academia Parisiensis* in 1635. Although informal by design, it attracted keen interest from leading scholars and mathematicians, including Blaise and his father Étienne Pascal (1588–1651), Claude Mydorge (1585–1647), Claude Hardy (1604–1678), De Roberval, and Pierre de Fermat (1607–1665).

Similarly, the 'Hartlib circle,' the correspondence network covering Western and Central Europe established by Hartlib and his fellow thinkers around 1630, ...

'was an association of personal friends. Hartlib and Dury were the two key figures; Comenius, despite their best efforts, always remained a cause they were supporting rather than a fellow coordinator. Around them were Hübner, Haak, Pell, Moriaen, Rulise, Hotton, and Appelius, later to be joined by Sadler, Culpeper, Worsley, Boyle and Clodius. But as soon as one looks any further than this from the centre, the lines of communication begin to branch and cross, threading their way into the entire intellectual community of Europe and America. It is a circle with a definable centre but an almost infinitely extendable periphery.' [30]

1.4 Birth of the learned societies and their scientific journals

Hartlib and Mersenne were not alone in their pursuit of knowledge through extensive international networks. The first half of the 17th Century coincided with a rapidly increasing number of private learned academies. [31] Despite our focus so far on developments in France and England, the Italian city states, including Naples, Rome, and Florence, had long also been major centres of scientific inquiry; the foundation of the first scientific society, the *Academia Secretorum Naturae* (the 'Academy of the Mysteries of Nature'), by the Napolitan polymath, scholar, and playwright Giambattista della Porta (1535?–1615), dates from 1560. [32] Subsequently, the *Accademia dei Lincei* (the 'Lincean Academy;' 1603–1630)—whose name paid tribute to the sharp vision of the lynx which symbolizes the keen observational eye that science requires—was established by Frederico Angelo Cesi (1585–1630), the Roman scientist and naturalist; by 1651, following the death of its founder, the Academy's demise was inevitable, however. [33] Another early Italian scientific academy, the short-lived *Accademia del Cimento* (the 'Academy of Experiment;' 1657–1667), [34, 35] was founded by students of Galileo, among others, and funded by Prince Leopoldo (1617–1675) and his brother Ferdinando II de'Medici (1610–1670), Grand Duke of Tuscany. In essence, the society *'under the protection of the Most Serene Prince Leopoldo of Tuscany'* [36] catered to

Leopoldo's personal interests in science while shying away from generating public controversy.

These informal structures contributed to the formal foundation of the Royal Society (formally known as 'The President, Council, and Fellows of the Royal Society of London for the Improvement of Natural Knowledge') on 15 July 1662, the French *Académie des Sciences* on 22 December 1666, and eventually the *Accademia Nazionale Reale dei Lincei*, the Italian Royal National Lincean Academy, in 1874. In France, many academies transferred to government control, sponsorship, or patronage by the middle of the 17th Century, with as notable frontrunner the *Académie française*, which had been founded by Cardinal Armand Jean du Plessis (1585–1642), Duke of Richelieu and Fronsac (better known as Cardinal Richelieu), in 1634. In the mid-17th Century, the vibrant intellectual communities in Paris and London had set their first tentative steps towards establishing permanent literary and scientific academies, on both sides of the Channel under royal patronage.

The Royal Society originated from a combination of different groups of physicians, mathematicians, and natural philosophers, [37] mostly 'gentlemen of means' who worked independently, including the Gresham College 'group of 1645' convened by the German Calvinist scholar Theodore Haak (1605–1690), the Oxford Philosophical Club (1649–1660) around the natural philosopher John Wilkins (1614–1672), and even the Hartlib Circle[3]. However, the relationship between the latter and the Royal Society was fraught with difficulty because of Hartlib's emphasis on utopian universal education—principles which were left out of the Royal Society's charter. The Gresham College group was the dominant contributor of Fellows to the nascent Royal Society; they adhered to Bacon's 'New Science,' proposed in his unfinished utopian novel *New Atlantis* (1627) [38]—an approach that rejected Aristotelian principles (see chapter 3.1.1) and which represented an important step towards the development of the modern scientific method of inquiry.

Following a lecture by Wren on 28 November 1660, Gresham College's '1660 committee of 12' met and agreed to establish a 'College for the Promoting of Physico–Mathematical Experimental Learning.' At their subsequent weekly meeting, Sir Robert Moray (1608/1609–1663) told his colleagues that King Charles II of England (1630–1685; reigned from 1660 until his death) supported the discussion meetings. The King proceeded to sign a royal charter on 15 July 1662. This formally created the 'Royal Society of London,' while a second royal charter dated 23 April 1663 changed the society's name to the 'Royal Society of London for the Improvement of Natural Knowledge.'

However, French scientists at the time claimed that it had instead been *their* influence that led to the establishment of the Royal Society. Leading French thinkers, including Jean-Baptiste du Hamel (1624–1706), Giovanni Domenico Cassini (1625–1712), Bernard le Bovier de Fontenelle (1657–1757), and Melchisédech Thévenot (c. 1620–1692), suggested that their English counterparts, including the Royal Society's first Secretary, Oldenburg, had come up with the idea

---

[3] The group of 1645 and the Hartlib Circle are sometimes confusingly referred to as the 'Invisible College,' although historians disagree about the precise composition of this poorly defined group.

for their learned society after having attended the 'Montmor Academy,' a private academy convened by Henri Louis Habert de Montmor (c. 1600–1679), scholar and 'man of letters,' in 1657. [39] Hooke, never one to mince his words, summarily rejected that notion: …

'[Cassini] makes, then, Mr Oldenburg to have been the instrument, who inspired the English with a desire to imitate the French, in having Philosophical Clubs, or Meetings; and that this was the occasion of founding the Royal Society, and making the French the first. I will not say, that Mr Oldenburg did rather inspire the French to follow the English, or, at least, did help them, and hinder us. But 'tis well known who were the principal men that began and promoted that design, both in this city and in Oxford; and that a long while before Mr Oldenburg came into England. And not only these Philosophic Meetings were before Mr Oldenburg came from Paris; but the Society itself was begun before he came hither; and those who then knew Mr Oldenburg, understood well enough how little he himself knew of philosophic matter.' [40]

The foundation of the French *Académie des Sciences*, meanwhile, was triggered by French finance minister Jean-Baptiste Colbert's (1619-1683) plan to create a general academy. His hand-picked group of leading scholars convened for the first time on 22 December 1666 in King Louis XIV's library, establishing a schedule of twice-weekly meetings thenceforth. It took until 20 January 1699 before the King drafted a set of rules, which conferred upon the organization the designation 'royal,' thus forming the *Académie Royale des Sciences*. As a governmental body, discussions and external communications were expected to remain non-political, while also avoiding religious and social issues of possible contention.

In Germany, meanwhile, the *Academia Naturae Curiosorum* (the 'Academy for the Curious of Nature;' 1652–1693) underwent a transformation from a gathering place for physicians, with meetings held in different towns across the region, into an academy. The emperor of the Holy Roman Empire, Leopold I (1640–1705), formally recognized the society in 1677 and gave it the designation *Leopoldina* in 1687. Independently, Prince-elector Frederick III of Brandenburg (1657–1713) followed Leibniz' suggestion and founded the *Kurfürstlich Brandenburgische Societät der Wissenschaften* ('Electoral Brandenburg Society of Sciences') on 11 July 1700 in Berlin. Once Frederick III had been crowned King *in* Prussia in 1701, thus having become Frederick I of Prussia, the society was renamed the *Königlich-Preußische Societät der Wissenschaften* ('Royal Prussian Society of Sciences'). The Prussian Academy was the first learned society to engage with both the sciences and humanities subjects.

Following the foundation of the Royal Society, the *Académie des Sciences*, and the Prussian Academy [41], the number of scientific societies mushroomed to more than 70 such organisations in the major intellectual centres of Europe by 1793 [42], many of which were officially recognized by their respective governments. Science had become more professional and institutionalized: scientific and literary academies absorbed much of the enterprise of scholarship, including communication and dissemination of news and discoveries among the scientific community. The time was clearly opportune for frequent printed scientific dissemination to take root. In 1663, Hooke had proposed that the Royal Society establish a weekly journal: ...

'And that you may understand what parts of naturall knowledge they are most inquisitive for at this present, they designe to print a paper of advertisements once every week, or fortnight at furthest, wherein will be contained the heads or substance of the inquiries they are most solicitous about, together with the progress they have made and the information they have received from other hands, together with a short account of such philosopicall matters as accidentally occur, a brief discourse of what is new and considerable in their letters from all parts of the world, and what the learned and inquisitive are doing or have done in physick, mathematicks, mechanicks, opticks, astronomy, medicine, chymistry, anatomy, both abroad and at home.' [43]

Hooke's proposal did not come to fruition until after the first journal had been established in France. There, the first issue of the interdisciplinary literary and academic *Journal des Sçavans* launched on 5 January 1665 (see figure 1.2, left). It was run by its founding editor, the Parisian author and lawyer Denis de Sallo, Sieur de la Coudraye (1626–1669), bibliophile and adviser to the *Parlement de Paris*. [44] Just two months later, on 6 March 1665, the first issue of the *Philosophical*

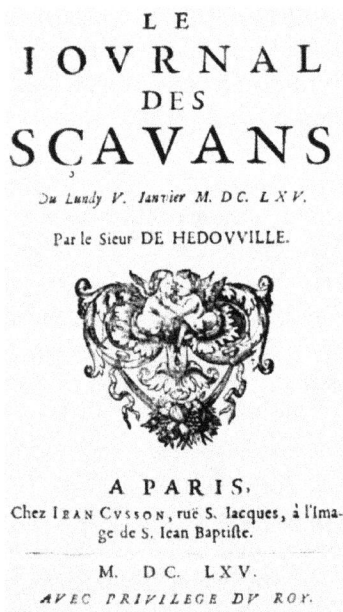

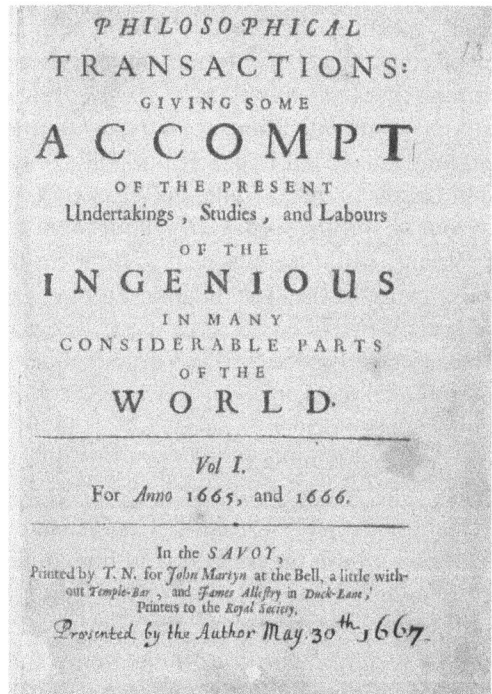

Figure 1.2. (*left*) Frontispiece to Volume **1** of the *Journal des Sçavans* (1665) (Source: Wikimedia Commons; public domain) (*right*) Frontispiece to Volume **1** of the *Philosophical Transactions* (1665) (Source: Royal Society archives; Creative Commons Attribution 4.0 International license).

Transactions, Giving some Account of the present Undertakings, Studies, and Labours of the Ingenious in many considerable parts of the World was published (see figure 1.2, right) upon having been granted a royal charter by King Charles II. [45]

The contents of both journals were rather different from what we expect from scientific journals today. The *Journal des Sçavans* published obituaries of well-known scholars and intellectuals, church history, book reviews, legal and university reports, science news, and scholarly activity, predominantly written by *scriveners* (journalists). It appeared in Paris on Mondays in 12-page *quarto* format—which would soon be followed by unauthorized copies printed in Amsterdam [46]—until 1792, when the French Revolution (1789–1799) intervened. From 1683 to 1686, the journal had assumed the lengthy title *Journal des Savants ou recueil succinct et abrégé de tout ce qui arrive de plus surprenant dans la nature, et de ce qui se fait et se découvre de plus curieux dans les arts et les sciences.* When regular publication resumed under the auspices of the Institut de France in 1816, it became known under its updated title, *Journal des Savants.*

The *Philosophical Transactions,* since 1776 known as the *Philosophical Transactions of the Royal Society,*[4] focussed on disseminating scientific progress from the outset, aiming to inform the Fellows of the Royal Society and other interested readers of the latest scientific news and discoveries. As such, it was the first true scholarly journal and it is generally considered the longest running journal exclusively dedicated to the communication of scientific discoveries. The intended publication was first mentioned, as shown in figure 1.3, in the Royal Society's Council minutes of 1 March 1664 (Julian date; corresponding to 11 March 1665 in the modern, Gregorian system[5]), ordering that ...

'*the* Philosophical Transactions, *to be composed by Mr. Oldenburg, be printed the first Munday [sic] of every Month, if he have sufficient matter for it, and that that Tract be Licensed by the Council of this Society, being first revised by some of the Members of the same. And that the President be desired now the License the first papers thereof, being written in four sheets in folio, to be printed by John Martyn and James Allestree.*' [47]

Note the phrase, '*being first revised by some of the Members of the same,*' which is considered the first reference to the now well-established practice of formal scientific or scholarly peer review preceding publication in a learned journal, an approach that became commonplace from the 1830s. Indeed, the *Philosophical Transactions* were founded on the modern principles of peer review, external validation ('certification'), scientific priority ('registration'), dissemination, and archiving. This is reflected in a

---

[4] For a brief period between 1679 and 1682, Robert Hooke changed the journal's name to *Philosophical Collections.*

[5] Until the 1580s, in England and many European countries the year began on 25 March, so that January 1664 followed December 1664. From 1582, this practice changed across most of Europe but not in England, where the modern Gregorian calendar was adopted as late as in January 1752. Most of continental Europe relevant to the present narrative adopted the Gregorian calendar some time before 1600, although in some cases there may be some date confusion (see, e.g., chapter 5, reference [36]).

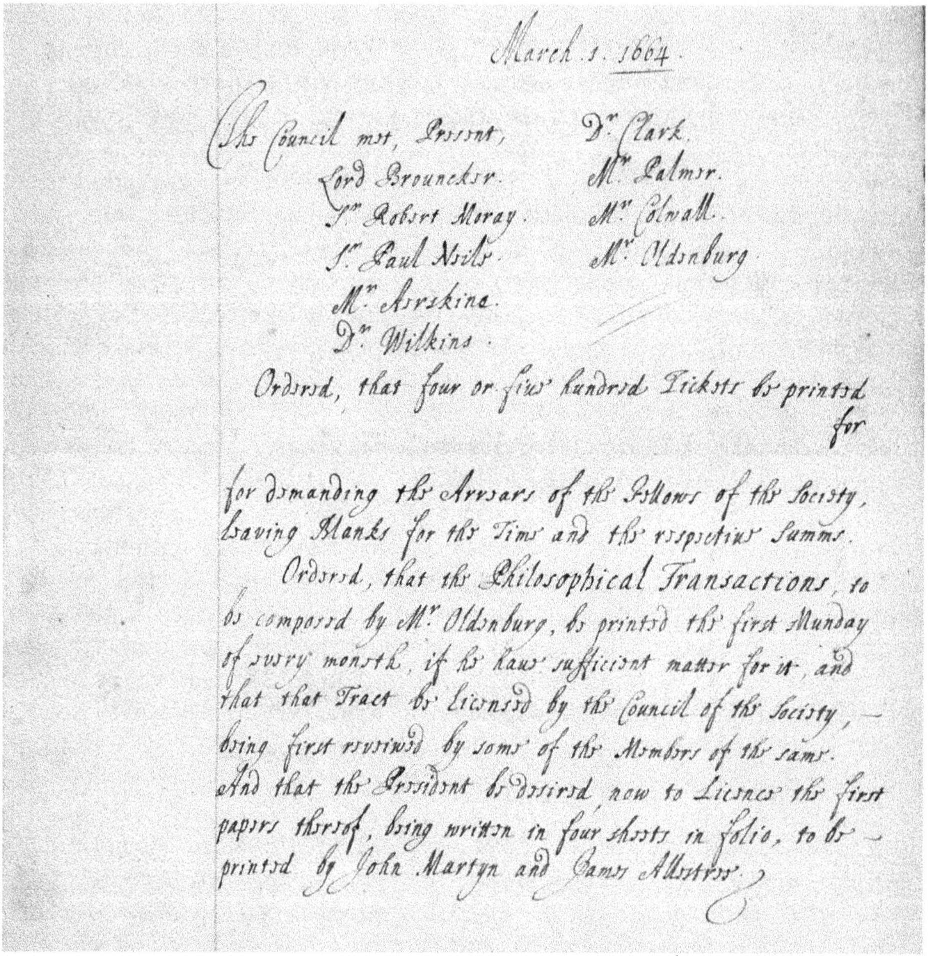

Figure 1.3. The Royal Society's Council minutes of 1 March 1664 (Julian date; excerpt). Note the phrase, '*first revised by some of the Members of the same*,' which represents an important development in the history of peer review. (© The Royal Society).

series of letters from 1664 and 1665 between Oldenburg, editor of the *Philosophical Transactions*, and Boyle, a founding Fellow of the Royal Society: ...

'*[Huygens] hath been written to, to communicate freely to ye Society; what new discoveries he maketh or wt new Expts he tryeth, the Society being very careful of regist'ring the person and time of any new matter, imparted to ym, as the matter itselfe; whereby the honor of ye invention will be inviolably preserved to all posterity.*' [48]

'*[By registering and giving due honour] I thence persuade myselfe, yt all Ingenious men will be thereby incouraged to impart their knowledge and discoveryes, as farre as they may, not doubting of ye Observance of ye Old Law, of Suum cuique tribuere.'* [49] ...

'I might justly be thought too little sensible to my own Interest, if I should ... neglect the opportunity of having some of my Memoirs preserv'd, by being incorporated into a Collection [such as the continuing issues of the *Philosophical Transactions*]*, that is like to be as lasting as usefull.'* [50]

Both men clearly realized the importance to their contributors of establishing priority of discovery and the journal's role in safeguarding this: ...

'But some here [in Oxford] are a little Jealous yt if our Expts be known elsewhere wthout being before hand registred by you together wth ye Time of their having been made or proposd, they may beget such claimes & disputes.' [51]

'I acknowledge, yt yt yealousy, about the first Authors of Experiments, wch you speak off, is not groundlesse. And therefore offer myselfe, to register all those, you or any person shall please to communicate, as now, wth yt fidelity, wch both of ye honor of my relation to the R. Society (wch is highly concerned in such Experiments) and my owne inclinations doe strongly oblige me to.' [52]

At the time of Oldenburg's letter to Boyle of 24 November 1664, the Royal Society's secretary had clearly been made aware of plans to publish an academic journal which were afoot in France, as evidenced by a passage in that same letter: ...

'My New correspondent... hath given me notice... yt they have a dessein in France to publish from time to time a Journall of all which passeth in Europe in matter of knowledge both Philosophicall and Politicall.'

These developments appear to have accelerated the Royal Society Council's plans to publish its own scholarly journal. Following publication of the *Journal des Sçavans'* first issue in early January 1665, Boyle read out an excerpt at the Royal Society's 11 January 1665 meeting. In early February that year, a sample issue of the *Philosophical Transactions* was presented, leading to approval by the Society's Council on 1 March and subsequent publication of the first issue on Monday 6 March 1665.

Of particular note is a paper in that first issue, specifically a reproduction of a set of hand-written 'Directions for Sea-men, bound for far Voyages' (1661; see figure 1.4) by Lawrence Rooke (1622–1662), the Professor of Astronomy and Geometry at Gresham College, which was prefaced by a review of the Royal Society's *raison d'être*: ...

'It being the Design of the R. Society, for the better attaining the End of their Institution, to study Nature *rather than* Books*, and, from the Observations,*

Figure 1.4. '*Directions for Sea-men, bound for far Voyages,*' Lawrence Rook (1661) (© The Royal Society).

made of the Phaenomena *and Effects she presents, to compose such a History of Her, as may hereafter serve to build a Solid and Useful Philosophy upon; They have from time to time given order to several of their Members to draw up both* Inquiries *of things Observable in forrain Countries, and* Directions *for the*

Particulars, they desire chiefly to be informed about. And considering with themselves, how much they may increase their Philosophical *stock by the advantage, which* England *injoyes of making Voyages into all parts of the World, they appointed that Eminent Mathematician and* Philosopher *Master* Rooke, *one of their Fellowes and* Geometry *Professor of* Gresham Colledge *(now deceased to the great detriment of the Commonwealth of Learning) to think upon and set down some* Directions for Sea-men *going into the* East *and* West-Indies, *the better to capacitate them for making such observations abroad, as may be pertinent and suitable for their purpose; of which the said Sea-men should be desired to keep an exact* Diary, *delivering at their return a fair Copy thereof to the* Lord High Admiral of England, *his Royal Highness the* Duke *of* York, *and another to* Trinity-house *to be perused by the* R. Society *Which Catalogue of Directions having been drawn up accordingly by the said Mr.* Rook, *and by him presented to those, who appointed him to expedite such an one, it was thought not to be unseasonable at this time to make it Publique, the more conveniently to furnish Navigators with Copies thereof.'* [53]

Interestingly, despite lacking a reliable method for longitude determination at sea, as we will see in detail in the remainder of this book, the first direction emphasized, ...

'To observe ye declination of ye Compass, or its variation from ye meridian of ye place, frequently marking withall ye latitude and longitude of ye place, where ever such observation is made, as exactly as may be, and setting down ye method, by which they made them." [54]

Until 1752, when the journal became an official publication of the Royal Society [55] and a 'Committee on Papers' was established to handle its editorial aspects, including peer review, the *Philosophical Transactions* was managed by its editors as an unofficial, private undertaking. Its first editor, Oldenburg, bankrolled the publication at his personal expense. As becomes clear following perusal of his extensive correspondence (which includes 3176 published letters dated between 1641 and 1677 [56]), Oldenburg aimed at realizing Hartlib's ideal of a universal 'office of address.' In his roles as editor and mid-century intelligencer, Oldenburg extended his reach beyond Europe and its periphery to the American Colonies.

He ran the journal as a for-profit venture, since the Royal Society did not provide him with a salary. Unfortunately, his profits only just allowed him to cover the rent of his house in Piccadilly [57], leaving room for little else. Nevertheless, his income from the journal provided him with a keen incentive to continue with its publication. Often referring to himself as the journal's 'compiler' or sometimes as its 'author,' he published a total of 136 issues before his death in 1677 [58], only missing issues in July–November 1666 on account of the Great Plague of London (1665–1666) and in June and August 1667. During those latter periods, he was incarcerated in the Tower of London, having been accused of treason and spying for the Dutch during the Second Anglo–Dutch war of 1665–1667 because of his extensive correspondence with foreign scholars and contributors to the *Philosophical Transactions*.

Apart from a monetary incentive, however, Oldenburg's editorial leadership at the *Philosophical Transactions* was driven by his desire to save time and labour. Just like the *Journal des Sçavans* in France, the printed *Transactions* replaced much of the extensive scientific correspondence among leading scholars, with Oldenburg in England and Colbert in France as central characters. However, the new, printed journals did not fully overshadow the need for scientific communication through letters. Letters continued to play important roles in the editorial process and as 'prototype' for final publication. Hand-written letters and formal journal publications adapted to become complementary means of communication among the educated members of society.

Oldenburg clearly intended his journal to become a collection of scientific advances shared among its readership, given his description of the journal as '*one of these [natural] philosophical commonplace books.*' [59] The rather odd arrangement that led Oldenburg to publish the *Philosophical Transactions*, ostensibly on behalf of the Royal Society but at his own expense, was likely beneficial to both parties. The Royal Society's approval provided an air of authority and authenticity to the publication, while the Society benefited from being able to communicate the scientific advances of its Fellows without direct responsibility for the journal's contents. This latter aspect was important during a time—in the wake of the Interregnum of England, Scotland, and Ireland (1649–1660), between the execution of King Charles I and the reinstatement of the monarchy with King Charles II—when the potential for censorship represented a real danger to the Royal Society's operations. At this time, publications were heavily regulated, and the concept of a free press had not yet been enacted in England.

Following the initial successes of the *Journal des Sçavans* and the *Philosophical Transactions*, learned societies in other European countries followed suit with their own scholarly and literary journals, including *Acta eruditorum* (Leipzig, 1682), *Nouvelles de la république des lettres* (Amsterdam, 1684), *Giornale de'letterati* (Ferrara, 1688), *De Boekzaal van Europa* (Rotterdam, 1691–1694), and *Galeria di Minerva* (Venice, 1691–1694). [60] By 1699, thirty scientific journals had been launched; by 1790, there were 1052. [61]

1.5 The 17th Century: early modern pinnacle of human ingenuity

In the preceding sections, I have attempted to sketch a broad overview of the scientific and scholarly environment that had emerged by the 17th Century, a stimulating environment slated to witness major breakthroughs in the sciences. In chapters 2 through 6, our narrative will take us through some of the highlights of human creativity and resilience: the long quest for a reliable means to determine one's longitude at sea on extended voyages away from land. This aspect of the history of science provides, above all, unique insights into the inner workings of the scientific enterprise during the early years of the modern scientific method. Then, as now, scientific research, exchanges, and dissemination transcended national borders; scientists from a wide range of backgrounds were brought together by their curiosity—conditions that have changed little over the past 300 years. Scientific pursuits require access to the brightest minds and the latest developments. The Scientific Revolution and the significantly changed inquisitive attitudes during the Renaissance and the European Enlightenment formed the basis for the establishment of productive and collaborative multinational scientific networks. If anything, those conditions remain in place today, with the scientific enterprise having become even more international than during the period covered by our narrative.

Chapter 2 is, in essence, a brief history of cartography. Its aim is to establish the context of the times. The development of global coordinate grids originated in Greek Antiquity, although the idea took independent root in the Islamic world and East Asia. Establishing an absolute reference grid requires accurate determination of longitudes and latitudes. Latitude determination has been practiced since at least the Phoenicians and the early Polynesian navigators, several centuries BCE. With successive Greek scholars such as Eratosthenes, Hipparchus, Marinus of Tyre, and Ptolemy having laid the foundations of mathematical geography and map projection, European developments stagnated in the Middle Ages, when religious intolerance and wars dominated the public consciousness. In the Islamic world and under the Chinese Imperial Dynasties, major progress continued, however, propelled by al-Khwārizmī, al-Bīrūnī, and al-Idrisi, as well as by Zheng He's navigators, predominantly on the Indian Ocean, where rudimentary longitude determination at sea was practiced routinely. Despite its inherent dangers, European sailors including Columbus continued to navigate by 'dead reckoning,' that is, sailing by knowing one's latitude and compass course while accounting for the effects of the wind and currents, also known as the *three L's*: lead, lookout, and latitude. While sailing by dead reckoning today has become a rather sophisticated undertaking, in its early incarnation the process consisted of throwing a log overboard and trying to adjust one's course for drift and currents by estimating the rate at which the log drifted away, and in which direction. Accurate longitude determination at sea became an urgent issue with a series of early-16th Century papal decrees which divided the newly discovered lands between Spain and Portugal based on poorly specified meridians. Improved mapping approaches and, particularly, Mercator's impact on map projections eventually returned European geographers to the forefront of global cartographic advances. Chapter 2 should be read in tandem with the Epilogue, where we explore the historical and political developments that eventually led to the adoption of the Greenwich meridian as global reference line.

Having established the prevailing boundary conditions, chapters 3 through 6 focus in detail on the technical advances leading to the construction of reliable timepieces for use on board rocking and pitching ships at sea. The turn of the 17th Century saw a step change in scientific thinking. Developments in science and technology, in particular Galileo's invention of the pendulum as a timekeeping device and its application in the pendulum clock, first achieved by Huygens, set the tone for the Scientific Revolution. European governments announced rich rewards to anyone who could solve the intractable 'longitude problem.' As we will see in chapter 3, significant efforts by the educated elite revealed not only new laws of physics, but also deeply entrenched positions and personal struggles, once again showing that scientists and scholars throughout history have always been just as humanly fallible as we are today.

Chapter 4 focuses on the second half of the 17th Century, an era that witnessed sustained progress in the development of a practical timepiece for use at sea. Accuracies reached levels of better than 10 seconds a day by application of the novel, widely introduced anchor escapements. Major breakthroughs were made, not only by Huygens but also by his competitors. We will place particular emphasis on the contributions by Hooke and Alexander Bruce (1629–1681). Increased competition triggered accelerated progress in both practical construction and theoretical understanding of pendulum operation, with developments cycling through the concepts of the conical, parabolical, and compound pendulums, explorations of the compound pendulum's centre of oscillation, *en passant* yielding novel insights into centrifugal motion and the centripetal force, the theory of evolutes, and the realization that 'cycloidal' trajectories approximate isochronicity.

Major sea trials were undertaken by both the French Academy and the Royal Society across the Mediterranean and the Atlantic Ocean on a number of occasions. While English scientists made sustained progress in their scientific understanding of harmonic oscillators, their French counterparts made major strides in cartography of the known 17th Century world, unfazed by the range of international conflicts playing out across their spheres of influence.

In chapter 5, we will see that following his publication of *Horologium Oscillatorium* in 1673, Huygens temporarily refocussed his scientific explorations on a detailed understanding of the tautochrone problem, that is, of finding the curve a pendulum bob needs to trace for the pendulum's operation to become isochronous. This naturally led to a shift in emphasis to spring-driven oscillators, which in turn enormously aided practical efforts to design an accurate marine timepiece. Surprisingly, perhaps, these efforts were eventually abandoned, once again in favour of continued development of pendulum clocks with triangular and tricord pendulum suspensions. Scientists, scholars, and clockmakers on both sides of the English Channel continued to compete, both scientifically and in their pursuits of the commercialization of their designs. Long-range sea trials from the northern Dutch Republic to the Cape of Good Hope offered significant promise of breakthroughs in clock design and performance, but reality overtook wishful thinking.

Huygens remained the leading scholar in the pursuit of a solution to the longitude problem until his death in 1695, although he never saw his final invention, the perfect marine balance, taken to the test. Developments of techniques other than horological advances had continued during the Huygens–Hooke controversies, although they did not generate a large following. Alternative proposals, entertained from the mid-17th Century onwards, included a wide variety of suggestions, from a renewed focus on astronomical position measurements to outright wacky ideas grounded in alchemy.

In chapter 6, we will first consider one of the main scientific disagreements between Huygens and Newton, the controversy about the nature of the gravitational force on Earth, and how the extended voyages undertaken with Huygens' clocks on board ships of the Dutch East India Company contributed to scholarly enlightenment. We will next place Newton's contributions to solving the longitude problem using stellar position determinations in the era's context, efforts in which John Flamsteed (1646–1719), the first 'Astronomer Royal,' played a major role. We will also take a broader look at other attempts that did not involve horological developments. This final chapter focuses specifically on the period between Huygens' death and the passing of the British Longitude Act in 1714, which completes our narrative.

References and notes

[1] Dasgupta P 2007 *Economics: A Very Short Introduction* (Oxford: Oxford University Press)

[2] Sagan C 1980 *Cosmos* (Episode 6), PBS (USA)

[3] Russell B A W 1945 *A History of Western Philosophy* (New York: Simon & Shuster)

[4] Ross S 1962 Scientist: The story of a word *Ann. Sci.* **18** 65–85

[5] Mencken H L 1919 *The American Language: A Preliminary Inquiry into the Development of English in the United States* 2009 edn (New York: Cosimo Classics) pp 36–7

[6] *Ibid.* 36; footnote 70

[7] Galilei G 1632 *Dialogue Concerning the Two Chief World Systems*: dedication to the Grand Duke of Tuscany

[8] The word 'science' at the basis of this suggestion traces back to Latin roots (scientia, sciens, scire); T F Hoad (ed) 1996 *The Concise Oxford Dictionary of English Etymology* 2003 edn. (Oxford: Oxford University Press)

[9] Anonymous 1834 *Quart. Rev.* **51** 58–61; cited by Ross S 1962 *op. cit.* 71–2

[10] Todhunter I 1876 *William Whewell, D. D., Master of Trinity College, Cambridge: An account of his writings, with selections from his literary and scientific correspondence* vol 1 (London: MacMillan & Co.) p 92

[11] *Ibid.* (Ross S 1962 op. cit. 72; footnote *)

[12] Whewell W 1840 *The Philosophy of the Inductive Sciences* vol 1 (Cambridge: John W Parker, J & J Deighton) cxiii

[13] On Leonardo da Vinci and Coreggio *Blackwood's Edinburgh Mag.* **48** 273 (July–December 1840)

[14] Thomson Sir W (Lord Kelvin) 1890 *Mathematical and physical papers* vol **II** (Cambridge: Cambridge University Press) 318

[15] Faraday M quoted in: *Notes and Records of the Royal Society* **16** 216 1961

[16] The 'problem' of information overload appears to go back to Antiquity, given that Seneca the Younger (Lucius Annaeus Seneca; c. 4 BCE–65 CE) is reported to have complained already, 'distringit librorum multitude' (*The abundance of books is distraction*); Blair A M 2011 *Too Much to Know: Managing Scholarly Information before the Modern Age* (New Haven, CT: Yale University Press). Even earlier, King Solomon muses in *Ecclesiastes* 12:12 (c. 450–180 BCE) in the Bible's Old Testament, 'Of making books there is no end.'

[17] Quoted in: De Solla Price D J (ed) 1963 *Little Science, Big Science* (New York: Columbia University Press) p 63

[18] Regazzi J J 2015 *Scholarly Communications: A History from Content as King to Content as Kingmaker* (New York: Rowman & Littlefield) p 34

[19] Nielsen M 2011 *Reinventing Discovery: The New Era of Networked Science* (Princeton, NJ: Princeton University Press) pp 172–5

[20] Meadows A J 1974 *Communication in Science* (London: Butterworths) pp 50–7

[21] 1676-10-24: Newton, Isaac—Leibniz, Gottfried (via Oldenburg, Henry); in: Hall A R 1980 *Philosophers at War: The Quarrel Between Newton and Leibniz* (Cambridge: Cambridge University Press) p 66

[22] *Ibid.*

[23] Dury J (undated) *Description of Inventions: The Hartlib Papers* 18/2/49A-56B; www.hrion-line.ac.uk/hartlib/view?docset=main&docname=18B_02_49 [accessed 26 May 2017]

[24] Anonymous (undated) *Proposals about Double Writing: The Hartlib Papers* 71/7/2A; www.hrionline.ac.uk/hartlib/view?docset=main&docname=707_02 [accessed 26 May 2017]

[25] Dobbie A M O 1986 *Comenius' Pampaedia or universal Education* (Dover, DE: Buckland) 20

[26] Comenius J A 1640s *De rerum humanarum emendatione consultatio catholica. IV. Pampaedia*

[27] Dobbie A M O 1986 *op. cit.* 19

[28] Bots H 2005 Martin Mersenne, 'secrétaire général' de la République des Lettres (1620–1648) Les grands intermédiaires culturel de la République des Lettres. *Études de réseaux de correspondances de XVIème au XVIIIème siècles* ed C Berkvens-Stevelinck, H Bots and J Haseler (Paris: Honoré Champion) p 175

[29] Goodman D 1994 *The Republic of Letters: A Cultural History of the French Enlightenment* (Ithaca, NY: Cornell University Press) p 20

[30] Young J T 1998 *Faith, Alchemy and Natural Philosophy: Johann Moriaen, Reformed Intelligencer, and the Hartlib Circle* (Aldershot, UK and Brookfield, VT: Ashgate) p 248

[31] Viala A 1985 *Naissance de l'écrivain: Sociologie de la littérature à l'âge classique* (Paris: Minuit) Chapter 1 (*L'essor des académies. Le réseau académique—Les nouveaux doctes—La légitimation et la dérive*)

[32] Bergin T G (ed) 1987 *Encyclopedia of the Renaissance* (Oxford and New York: New Market Books)

[33] Udías A 2003 *Searching the Heavens and the Earth: The History of Jesuit Observatories* (New York: Springer) p 5

[34] Middleton W E K 1971 *The Experimenters: A Study of The Accademia del Cimento* (Baltimore, MD: The Johns Hopkins Press) pp 327–8

[35] Galluzzi P (ed) 2001 *Scienziati a Corte. L'arte della sperimentazione nell'Accademia Galileiana del Cimento (1657–1667)* (Livorno: Sillabe)

[36] Magalotti L 1666 *Saggi di natvrali esperienze fatte nell'Accademia del cimento sotto la protezione del serenissimo principe Leopoldo di Toscana e descritte dal segretario di essa academia* (Florence: Giuseppe Cocchini all'Insegna della Stella) see Middleton W E K 1971 *op. cit.* 88

[37] Syfret R H 1948 The Origins of the Royal Society *Notes and Records of the Royal Society of London* **5** 78

[38] *Ibid.* **5** 75

[39] *Ibid.* **5** 79

[40] *Ibid.* **5** 80

[41] Evans R J W 1977 Learned Societies in Germany in the Seventeenth Century *Eur. Stud. Rev.* **7** 129–51

[42] McClellan J E III 1985 *Science Reorganized: Scientific Societies in the Eighteenth Century* (New York: Columbia University Press)

[43] Quoted in: Weld C R (ed) 1848 *A History of the Royal Society, with memoirs of the Presidents* vol **I** (London: The Royal Society) p 148

[44] Boyer J 1894 The Founder of the First Scientific Journal *Popular Science Monthly* (March) 690–2

[45] The Royal Society Publishing *Philosophical Transactions – the world's first science journal*; http://rstl.royalsocietypublishing.org [accessed 25 May 2017]

[46] Vittu J-P 2002 La formation d'une institution scientifique: le Journal des Savants de 1665 à 1714 *Journal des Savants* **1** 179–203

[47] Royal Society Council minutes, 1 March 1665

[48] 1664-11-24: Oldenburg, Henry–Boyle, Robert W; Royal Society Library: MS OB No. 26

[49] 1664-12-03: Oldenburg, Henry–Boyle, Robert W; Royal Society Library: MS OB No. 28

[50] 1665: Boyle, Robert W–Oldenburg, Henry; Royal Society Library, quoted in: Mabe M A 2001 Revolution of Evolution? Digital Myths and Journal Futures: Shifting Fact from Fiction *Charleston Conf. Proc.* 2001 ed K Strauch (London: Taylor & Francis) pp 37–50

[51] 1665-08-27: Boyle, Robert W–Oldenburg, Henry; Royal Society Library: MS B1 No. 89

[52] 1665-08-29: Oldenburg, Henry–Boyle, Robert W; Royal Society Library: MS OB No. 33

[53] Rooke L 1665–1666 Directions for Sea-Men, Bound for Far Voyages *Philos. Trans.* **1** 140–3 (reproduced from hand-written directions, 1661)

[54] For a full transcript, refer to www.nma.gov.au/exhibitions/exploration_and_endeavour/ take_no_ones_word_for_it/transcript_directions_for_sea_men [accessed 27 May 2017]

[55] McDougall-Waters J Moxham N Fyfe A 2014 *Philosophical Transactions: 350 years of publishing at the Royal Society (1665–2015) exhibition catalogue*; https://royalsociety.org/ ~/media/publishing350/publishing350-exhibition-catalogue.pdf [accessed 25 May 2017]

[56] Hall A R and Hall M B (ed) 1965–1986 *The Correspondence of Henry Oldenburg* 13 vols (Madison, WI: University of Wisconsin Press; London: Mansel; London: Taylor & Francis)

[57] Bluhm R K 1960 Henry Oldenburg, F. R. S. (c. 1615–1677) *Notes and Records of the Royal Society* **15** 183

[58] McDougall-Waters et al. 2014 *op. cit.*

[59] 1669-04-02: Oldenburg, Henry–de Sluse, René-François Walter; quoted in: Yeo R R (ed) 2014 *Notebooks, English Virtuosi, and Early Modern Science* (Chicago, IL: University of Chicago Press)

[60] Vittu J-P 2005 Du Journal des Savants aux Mémoires pour l'histoire des sciences et des beaux-arts: l'esquisse d'un système européen des périodiques savants *Dix-septième siècle* **228** 530

[61] Kronick D A 1976 *A History of Scientific & Technical Periodicals: The Origins and Development of the Scientific and Technical Press, 1665–1790* 2nd edn (Metuchen, NJ: Scarecrow Press) p 78

Chapter 2

Global development of mathematical geography

2.1 Coordinate systems

The idea of covering maps with a positional grid is often attributed to the ancient Greek astronomer and geographer Claudius Ptolemy (Κλαύδιος Πτολεμαῖος; c. 100–c. 170 CE). However, the concept of a global coordinate system which allowed travellers to uniquely identify their position goes back as far as the Phoenicians in the eastern Mediterranean around 600 BCE and the Polynesians some 200 years later. Both sea-faring civilizations—the Phoenicians were the first sailors to circumnavigate the African continent—used observations of celestial objects to calculate their latitude. The Phoenicians may, in fact, have inherited this knowledge from their predecessors, the Minoan civilization of Crete. It has been suggested that the Minoans used the star Kochab—which was located at the north celestial pole at the height of the Minoan civilization, around 1450 BCE—as their guide to travel, possibly as far as North America and India [1].

Determination of one's latitude is fairly straightforward, and the Minoans had apparently mastered this skill to within a mile, a conclusion reached on the basis of the positioning of the stone circles found at Stonehenge, Almendres in present-day Portugal, and Callanish in the Outer Hebrides off the coast of Scotland [2]. Latitude determines one's position with respect to the Equator, so that one needs to determine the height of the Sun at local noon or (in the northern hemisphere) that of Kochab or Polaris, presently the North Star, transiting the local meridian—the great circle that passes through the celestial North and South Poles and the observer's zenith (as well as her nadir, directly opposite the local zenith direction)—at night. Using increasingly sophisticated instruments, such as gnomons, quadrants (a quarter circle marked at regular intervals, equipped with a dial), or astrolabes, astronomers and navigators became highly practiced at determining stellar inclinations, i, that is, stellar heights above the horizon. Use of such devices allowed northern observers to easily determine the altitude of Polaris above the local horizon; the latitude followed directly, $lat. = 90° - i$.

Daytime latitude determinations based on the Sun's meridian transit involve a few additional, yet fairly straightforward calculations. At local noon, one measures the Sun's displacement from the local zenith (known as the zenith angle, z_s) or alternatively its maximum altitude or 'elevation,' α_s. On the dates of the spring and autumn equinoxes—when the lengths of day and night are the same—the measured zenith angle corresponds directly to the observer's latitude. On any other day, the tilt of the Earth's inclination with respect to the ecliptic plane—that is, the orbital plane in which the Earth moves around the Sun—must be taken into account. On the day of the summer solstice, one must add 23.45° to the measured zenith angle, since the Earth is tilted away from the Sun; at the winter solstice, one must subtract the Earth's 23.45° tilt from the measured zenith angle. Between these benchmark dates, a simple correction, given by the Sun's declination for the day, δ, which is readily available in tabulated form and approximately corresponds to the Sun's latitude on the sky, provides a direct measure of one's latitude,

$$\cos z_s = \sin \alpha_s = \sin \Phi \sin \delta + \cos \Phi \cos \delta \cos h,$$

where Φ represents the observer's latitude and h is the solar hour angle, the difference between the Sun's actual position and the local meridian; $h = 0$ when the Sun reaches it maximum daily elevation.

2.1.1 Navigation in the Western Pacific

Determining their positions accurately was particularly important for the Polynesians, given the vast distances across open water they had to cover. Polynesian navigators used celestial bodies—the Sun, the Moon, and the brightest stars—as well as ocean swells, currents, and wave patterns for direction. The Polynesians subdivided the sky into sections of a 32-point compass [3, 4]: see figure 2.1 for a modern example. Celestial bodies rise in a given compass section in the East and eventually set in the opposite section in the West. Observations of a wide range of celestial bodies therefore aided the Polynesian sailors in steadying their direction of travel. Although the basic principles underlying this approach are simple, stellar positions on the sky vary with latitude; seasoned navigators were highly capable of recognizing and correcting for latitudinal changes, a technique known as 'navigation azimuth.'

Of course, navigation based on observations of the sky was only possible in clear and relatively stable conditions. But master Polynesian navigators were well prepared for clouds and stormy weather, too. They knew that ocean swells follow seasonal patterns, which allowed them to use the rocking of their boats to determine their heading and direction at any time. The best Polynesian navigators are said to be able to distinguish among five different types of gentle swells; they would hence know from a change in ocean movement if they had drifted off course

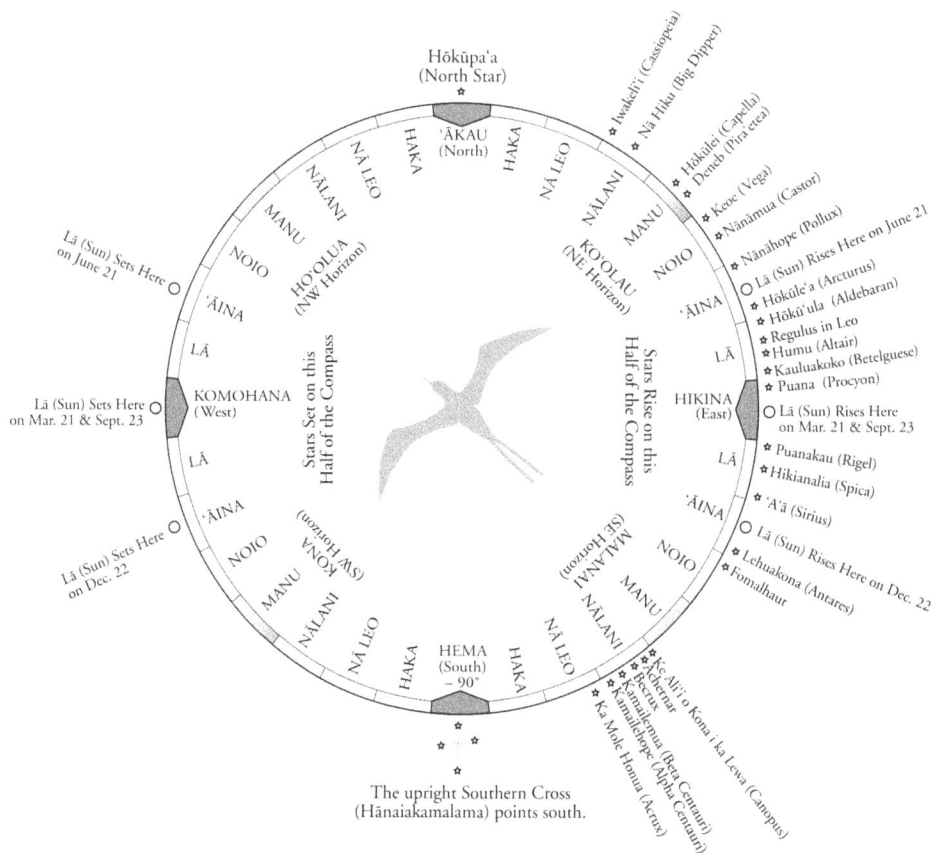

Figure 2.1. Modern Polynesian star compass. (© Charles Nainoa Thompson; reproduced with permission.)

or whether land of any description might be near [5], even with their eyes closed [6, 7].

Almost unique among sea-faring nations [8], Pacific islanders also used 'living seamarks' [9], commonly referred to as 'aimers' in Micronesia. These could be a school of porpoises [10], a row of whales [11], the flight of birds [12], or other natural occurrences, which were associated with specific places between islands and along ocean currents: ...

'These signposts in mid-ocean consist of swarms of fish, flocks of birds, groups of driftwood, or conditions of wave and sky peculiar to certain zones of the sea. Hundreds of such traditional betia [seamarks] were stored up in the race memory as a result of cumulative experience of generations' [13].

Some of these seamarks can be readily understood by realizing that enormous schools of fish follow the heated ocean currents and the algae trails feeding them. In turn, ocean birds would feast on this natural bounty, so their flight paths were also tightly defined. Other trajectories would allow certain specific bird species to make their way from one archipelago to another. Polynesian navigators knew which species were going their way and followed along; the schools would also lead them back home and feed them on the return leg.

In his 1916–1926 ethnographic study of the Gilbert Islanders, the inhabitants of present-day Kiribati in Micronesia, Arthur Grimble elaborated, writing that ...

'[these] seamarks are found along routes between islands and indicate to the navigator that he was at a certain point along his route. For example, the seamark called "the swarming of beasts" consisted of "an extraordinary number of sharks" and indicated the canoe was "a day's sail downwind of land." Other marks include a region where flying fish leaped in pairs, a zone of innumerable jellyfish, an area of numerous terns, an area of sharks and numerous red-tailed tropic[al] birds, a place marked by a school of porpoises, a place where pairs of porpoises point their heads "in the direction of the passage into Tarawa lagoon"' [14].

These skills, passed on orally from generation to generation, continue to be practiced today, in spite of our access to modern navigational aids.

2.1.2 The ancient Greeks

Returning now to the world of Antiquity, European maps based on scientific cartographic principles first appeared in the third Century BCE, a major development driven by the ancient Greek polymath Eratosthenes of Cyrene (Ἐρατοσθένης ὁ Κυρηναῖος; 276–195/194 BCE), in present-day Libya. Now commonly referred to as the 'Father of Geography,' in 245 BCE—at the tender age of 30—Eratosthenes was appointed librarian at the fabled Library of Alexandria, where he created the first world map based on the geographic knowledge of his day, including parallels and meridians on his map for the first time. Figure 2.2 shows an early 20th Century reconstruction of Eratosthenes' first world map by H G Wells in his illustrated *Outline of History* (1919–1920), where the most important parallels and meridians are indicated.

Eratosthenes' use of a grid pattern was meant to provide direct links to every location in the known world—the 'oecumene,' from the Greek οἰκουμένη, literally 'inhabited'—and allow the user to assess the approximate distance to any place of interest. This was greatly facilitated by his other major contribution to early mathematical geography, the first ever measurement of the Earth's circumference, which he made to within approximately 10–30% of the currently accepted value [15]. In fact, Eratosthenes arrived at longitudes by *'transforming distances into their angular values in relation to the circumference of the globe'* [16]. In his three-volume

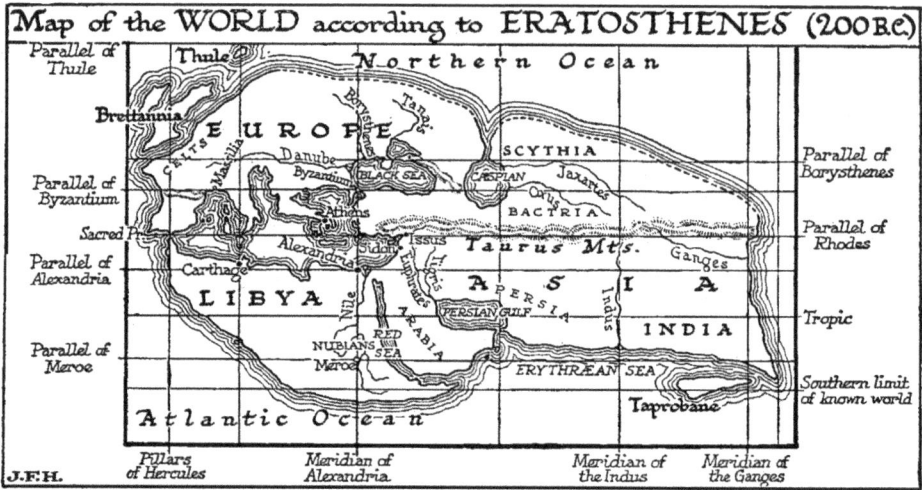

Figure 2.2. Early 20th Century reconstruction of Eratosthenes' map of the known world from ca. 194 BCE. (Wells H G 1920 *The Outline of History: Being a Plain History of Life and Mankind I* (New York: Macmillan).)

work, *Geographika (Γεωγραφικά)*, now lost to history, he provided the first atlas of more than 400 cities and their locations [17].

The ancient Greek astronomer Hipparchus of Nicea (Ἵππαρχος; 190–120 BCE) built on Eratosthenes' pioneering insights. However, the celebrated Greek geographer Strabo (Στράβων; 64/63–c. 24 BCE) often complained that Hipparchus was overly and often unfairly critical of the accuracy and internal consistency of his predecessor's work, particularly in his three-volume treatise *Against the 'Geography' of Eratosthenes (Πρὸς τὴν Ἐρατοσθένους γεωγραφίαν)*, of which only fragments survive [18, 19]. Hipparchus was convinced that accurate and internally consistent geographic maps should be based only on astronomical measurements of latitudes and longitudes, and on triangulation.

He adopted a zero meridian through the Aegean island of Rhodes, in the eastern Mediterranean, proposing that one could determine positions East and West from his reference meridian by comparing the local time to an 'absolute' time, referenced at his zero meridian. He even proposed a method to determine absolute time through observations of lunar eclipses. His method consisted of measuring the start and end times of lunar eclipses, and subsequently finding the difference between this absolute time and the observer's local time. This was the first known realization that one's longitude can potentially be determined by means of accurate timekeeping. Hipparchus was well ahead of his time; the viability of his method was doomed to fail, however, since one needed accurate timepieces for the method to become practicable. As we will see in chapters 3 through 6, it would take until well into the 18th Century before clocks could be manufactured that were sufficiently stable at sea for this method to yield accurate longitude determinations.

Hipparchus, like his predecessor Eratosthenes, was a true pioneer. He pursued geographic latitude determination on the basis of stellar position measurements rather than from the Sun's altitude, a technique that had become standard at the time. He was the first to suggest that geographic longitude could be determined based on simultaneous observations of lunar eclipses from distant places. Putting his theory into practice, he compiled a 'table of *climata*'—latitudes correlated with the length of the longest solstitial day—including numerous improved latitude determinations [20].

It is in this context that we need to consider the commonly held notion that Ptolemy may have been the first to produce maps covered by positional grids in his celebrated eight-volume work, *Geographia* (*Γεωγραφικὴ Ὑφήγησις* or *Geographical Guidance*; sometimes incorrectly referred to as the *Cosmographia*), the most complete compilation of geographic knowledge in the second Century CE. Clearly, in view of the brief history of mapmaking in Antiquity sketched above, this was not an altogether novel idea. In fact, Ptolemy is said to have borrowed it from a contemporary, Marinus of Tyre (Μαρῖνος ὁ Τύριος; c. 70–130 CE). Marinus, the Hellenized geographer, cartographer, and mathematician, hailed from the port of Tyre in the Roman province of Syria—the oldest and largest city in Phoenicia, located in present-day southern Lebanon. Indeed, Ptolemy's references to Marinus' work are the only surviving records of the latter's influence on Ptolemy; none of his maps, treatises, or other texts have survived.

Ptolemy also adopted the idea of subdividing a circle into 360 parts (μοῖραι) or degrees from his illustrious predecessors, an idea which had first been proposed by Hipparchus. Given Ptolemy's casual adoption of this concept, it appears that the idea was already commonly accepted, although Ptolemy did not have any specific or technical terms in his vocabulary that corresponded to a degree; instead, he was frequently forced to refer to '*parts, of which there are 360 to the equator.*' In his positional tables, he further subdivided the degree into 12 parts, equivalent to five minutes of arc each, which was indeed a high measurement accuracy for the times.

Ptolemy was generous in attributing his modern ideas and practices to Marinus. Nevertheless, he questioned some travellers' tales of journeys to Africa and elsewhere [21], based on diary notes of a wide cross section of travellers which Marinus had accepted at face value, because the latter had consulted '*too many conflicting volumes, all disagreeing*' (*Geographia*, Book I): ...

'*However, when it is apparent that he is agreeing with, rather than querying, some untrustworthy matter, either about a matter of draughtsmanship or by relying on preconceived solutions, it is right to bring forward objections to his reasoning and substitute other answers more in keeping with the highest standards of his work*' [22].

Ptolemy also pointed out mistakes in the provincial boundaries recorded by his friend. Even so, Lloyd Brown, curator of maps at the University of Michigan's Clemens Library in the 1940s, was rather forgiving of Ptolemy's attitude: ...

'Ptolemy was both tolerant and gentle in his criticism of Marinus; both men had their own ideas of the incredible and at times there is little to choose between their conceptions of what made sense and what nonsense. ... Marinus was a good man in Ptolemy's estimation, but he had allowed himself to be led astray in his scientific investigations. The world had expanded, to be sure, but not as much as Marinus indicated on his map' [23].

Marinus introduced improvements to the construction of maps, corrected many errors based on travellers' tales and introduced by his predecessors, and developed a system of nautical charts around 114 CE, although the exact date is unclear. Ptolemy, in turn, revised, corrected, supplemented, and generally improved Marinus' maps [24]. Ptolemy reported a number of Marinus' opinions, among others that the World Ocean was separated into an eastern and a western part by Europe, Asia, and Africa. The inhabited world, he thought, encompassed the latitude range from Thule (Θούλη, often identified with Norway) at 63°N latitude, through his reference parallel at Rhodes to the region of the Ethiopians. The latter was referred to as *Agisymba*, an unidentified country in Africa believed to be located north of Lake Chad, at 16°25′ South.

Strabo refers to Thule in relation to Eratosthenes' estimate of the extent of the known world, ...

'Now Pytheas of Massilia tells us that Thule, the most northerly of the Britannic Islands, is farthest north, and that there the circle of the summer tropic is the same as the Arctic Circle. But from the other writers I learn nothing on the subject—neither that there exists a certain island by the name of Thule, nor whether the northern regions are inhabitable up to the point where the summer tropic becomes the Arctic Circle' [25].

In Book IV, chapter 5, of his *Geographica*, Strabo subsequently concludes that, ...

'Concerning Thule, our historical information is still more uncertain, on account of its outside position; for Thule, of all the countries that are named, is set farthest north.'

A few decades afterwards, Pliny the Elder (Gaius Plinius Secundus; 23–79 CE) wrote in Book II, chapter 75, of his *Natural History*, ...

'The farthest of all, which are known and spoken of, is Thule; in which there be no nights at all, as we have declared, about mid-summer, namely when the Sun

passes through the sign Cancer; and contrariwise no days in mid-winter: and each of these times they suppose, do last six months, all day, or all night.'

He next attempted to identify its location more accurately, …

'Last of all is the Scythian parallel, from the Rhiphean hills [an imaginary mountain range in the North] *into Thule: wherein (as we said) it is day and night continually by turns (for six months)'* [26].

The Scythians were Iranian Eurasian nomads who traversed large regions across central Eurasia around the heyday of Greek Antiquity; they are thought to have roamed the steppes from approximately the ninth until the first Century BCE. Ancient Greek historians referred to the Scythians as a people that lived north of the Black Sea and the Caucasus Mountains. As regards the southernmost limit of the known world, Ptolemy wrote that Agisymba was a four-month journey south of the Fezzan—present-day southwestern Libya—and home to large animals, such as rhinoceroses and elephants, as well as many tall mountains. In the *Cambridge History of Africa*, Agisymba is similarly identified: …

'In AD 90 a traveller, probably a trader called Julius Maternus [Matiernus], profiting from the improved relations between the Romans and the Garamantes [inhabitants of the Fezzan] *at this time—no doubt as a result of [Suellius or Septimus] Flaccus's success* [in 86 CE]*—made his way through the land of the Garamantes to the land of Agisymba, where there were rhinoceroses'* [27].

In longitude, Marinus' world stretched from the Isles of the Blessed, also known as the Fortunate Isles, semi-legendary islands in the Atlantic Ocean near the Canary and Cape Verde Islands [28], to Shera (China)—his maps were the first in the Roman Empire to include China. Marinus assigned longitudes and latitudes to all places on his world map, choosing the westernmost land known to him as his zero meridian. Following Eratosthenes but improving his coordinate grid, Marinus' reference latitude (although not latitude zero!), which he called his 'fundamental parallel,' ran from the Sacred Promontory (Cape St Vincent) in present-day southwest Portugal through the Strait of the Columns (the Strait of Gibraltar) and Rhodes to the Gulf of Issus, also known as the Gulf of Alexandretta or the Gulf of İskenderun, in the eastern Mediterranean. Ptolemy followed suit and adopted similar longitude and latitude references; the longitude of the Isles of the Blessed remained the prime meridian until well into medieval times [29] (see also the Epilogue).

Ptolemy adopted the latitude of the Earth's Equator as his zero latitude, following his predecessors. While one can choose one's zero longitude arbitrarily, a property which caused numerous political complications in the centuries to come (for a discussion, see the Epilogue), the position of one's zero-degree latitude is determined by the Earth's rotation and its orbit around the Sun. The celestial bodies in our solar system, particularly the Sun, the Moon, and the planets, pass almost directly overhead at the Equator. The northern- and southernmost parallels of the Sun's apparent motion throughout the year are referred to as the Tropic of Cancer and the Tropic of Capricorn, at northern and southern latitudes determined by the inclination of the Earth's polar axis with respect to the ecliptic plane. Unlike modern conventions, however, in which terrestrial latitudes are expressed in units of degrees North or South of the Equator, Ptolemy's latitudes were given in terms of the length of the longest day (in the summer), in hours and minutes. The Equator was set at 12 hours of midsummer daylight; for greater distances from the Equator, the length of the longest day increases, so that the North Pole was thought to have 24 hours of summer daylight.

On Ptolemy's map in the *Geographia*, the Isles of the Blessed are positioned some seven degrees (560 km) too close to Spain in longitude, while the East–West extent of the Mediterranean region is too great. This led Ptolemy to amicably chastise Marinus, ...

'Marinus says of the merchant class generally that they are only intent on their business, and have little interest in exploration, and that often through their love of boasting they magnify distances' [30].

Marinus estimated a length of 180 000 *stadia* for the Earth's Equator—significantly less than Eratosthenes' 252 000 *stadia*—which corresponds to a circumference of the Earth that is some 17% smaller than the modern value, assuming that he used Italian stadia of approximately 185 metres each [31].

Ptolemy is generally credited with the first use of the terms 'longitude' (μῆκος) and 'latitude' (πλάτος) in their modern sense, but since Marinus' works have not survived the ravages of history, we cannot be certain that Ptolemy did not adopt these terms from Marinus as well. Note that the longitude and latitude in Ptolemy's world specifically referred to the extent of the known world at that time; both terms have long lost all traces of their original constraints.

Despite Ptolemy's liberal borrowing from his predecessors, his compilation of 27 maps in the *Geographia* became a benchmark world atlas in the history of cartography. Ptolemy emphasized the importance of establishing a global coordinate grid or 'graticule,' thus allowing map users to determine locations on Earth exactly and not in relation to other known places. To achieve this, he followed Eratosthenes' lead in proposing to relate the coordinate system on Earth to that in the sky, which turned about Polaris and which was used to measure latitude in the northern hemisphere. Through his two seminal works, an astronomical guide to the motions of the stars and planets across the sky from the perspective of a stationary Earth—now known as the *Almagest* (originally titled *Mathēmatikē Syntaxis*, *Μαθηματικὴ Σύνταξις*)—and the *Geographia*, he contributed significantly to both celestial and terrestrial mapping and thus reinforced the importance of astronomy to geography.

Indeed, in the *Almagest* he announced his intention to use his astronomical insights to determine the positions of the main urban centres on Earth: ...

'Now that the treatment of the angles [between ecliptic and latitudinal circles] has been methodically discussed, the only remaining topic in the foundations [of the treatise] is to determine the coordinates in latitude and longitude of the cities in each province which deserve note, in order to calculate the [astronomical] phenomena for those cities. However, the discussion of this subject belongs to a separate, geographic treatise, so we shall expose it to view by itself [in such a treatise], in which we shall use the accounts of those who have elaborated this field to the extent which is possible. We shall list for each of the cities its distance in degrees of that meridian from the meridian through Alexandria, to the east or west, measured along the Equator (for [Alexandria] is the meridian for which we establish the times of the positions [of the celestial bodies])' [32].

Ptolemy compiled information from many disparate sources. Residing in Alexandria, in the Roman province of Egypt, he was in an excellent position to benefit from many travellers' accounts. During the second Century CE, Alexandria was both commercially and intellectually the wealthiest and most international city in Greek Antiquity. Sailors and caravans from anywhere in the known world would come together there, providing a wealth of information about remote destinations. As part of his *Geographia*, Ptolemy provided an alphabetical index of latitudes and longitudes for some 8000 important locations in Europe, Africa, and Asia, which enabled him to draw the most complete map of the known world at that time. He believed that his maps encompassed roughly a quarter of the Earth's surface, and he was indeed quite close in that assessment. Since he had adopted the Fortunate Isles as his zero longitude rather than Alexandria, the relative coordinates were often incorrect—but Ptolemy knew this and actively worked at correcting these issues.

Ptolemy's maps have familiar orientations, with North at the top and East to the right. This was by design and not by accident, because the known world at his time was concentrated in the northern hemisphere, so that on a flat surface it would be easier to explore the world in this orientation. His meridians are equally spaced by *'the third part of an equinoctial hour, that is, through five of the divisions marked on the Equator.'* In practice, this meant that the 12-hour extent of the known world was covered by 36 meridians at intervals of five degrees each at the Equator and converging at the North Pole.

Ptolemy divided the Earth's habitable surface north of the Equator into 21 parallels, from the Equator to Thule. The parallel representing the southern limit of the habitable world was just as far south of the Equator as the parallel through Meroe (Μερόη)—an ancient city on the eastern bank of the Nile, some 200 km northeast of Khartoum in present-day Sudan—was to the north. As for his meridians, the 21 parallels trace circles parallel to the Equator at equal intervals, marked by both the number of equinoctial hours and fractional hours of daylight on the longest day of the year, and the number of full and fractional degrees north of the Equator.

The first parallel north was located at a distance of '*the fourth part of an hour*' from the Equator and '*distant from it geometrically about 4°15'*.' Only three of the northern parallels were directly associated with existing places; all others were placed on the Earth's surface based on theoretical considerations. *Clima I per Meroe*, at 17° N latitude and also known as the royal seat and principal metropolis of Ethiopia, was defined as 1000 miles south of Alexandria and 300 miles from the 'torrid zone.' *Clima II per Syene* is associated with Syene (modern Aswan, Egypt), considered one of the few scientifically located parallels on the Summer Tropic, while the popular reference parallel *Clima IV per Rodo* traverses at 36° N latitude through Rhodes.

Eratosthenes and Strabo had established the southern limit to the habitable world at the parallel through the easternmost African promontory, Cape Guardafiri in present-day Somalia. The area was known as the 'cinnamon-producing country' and the country of the *Sembritae* [Senaai], 'immigrants' or 'foreigners,' who were first mentioned by Eratosthenes in the context of Egyptian refugees who settled on tracts of land between the Nile and its major tributaries, some 20 days' travel north of Meroe in what is likely present-day Ethiopia. It also passed through Taprobane, usually considered the southernmost part of Asia, located in the far south of Sri Lanka.

Geographia, the most famous collection of classical maps of the world, was lost during most of the Middle Ages, at least to the Western world. Ptolemy's master-piece was, however, preserved in Arab translations since the ninth Century, as well as in Greek in the Islamic states, exerting significant influence on the development of cartography and mathematical geography in the Islamic world through luminaries such Abū al-Ḥasan ʿAlī ibn Mūsā ibn Saʿīd al-Maghrib (علي بن موسى المغربي بن سعيد or Ibn Said; 1213–1286), Abu Ishaq Ibrahim ibn Muhammad al-Farisi al Istakhri (al-Istakhri, استخري; died 956), Muḥammad Abū'l-Qāsim Ibn Ḥawqal (محمد أبو القاسم بن حوقل or Ibn Hawqal; died c. 978), and Mahmud ibn Hussayn ibn Muhammed al-Kashgari (محمود بن الحسين بن محمد الكاشغري or al-Kashgari; 11th Century), among others. We will return to these develop-ments in chapter 2.2.1.

In the West, the *Geographia* resurfaced in the late 13th Century, when it was subsequently further updated and referenced: see figure 2.3. There are no earlier known original Greek copies of the *Geographia* in existence. In the summer of 1295, the Byzantine monk Maximus Planudes (Μάξιμος Πλανούδης; c. 1260–c. 1305) was in search of a copy of the *Geographia* for the monastery associated with the Church of the Holy Saviour (Ἐκκλησία τοῦ Ἁγίου Σωτῆρος Ἐν τῇ Χώρᾳ) in Chora (Asia Minor), in present-day Istanbul. He seems to have succeeded in his quest fairly easily, given that a Greek copy [33] at the monastery from the late 13th Century includes a note identifying Planudes as its former owner [34]. In addition, so-called '*Heroic verses*' added to the manuscript describe the monk's pursuits to locate Ptolemy's largely neglected work. They also reveal his disappointment when he eventually found a copy that lacked maps, but which only included Agathoaemon of Alexandria's (Ἀγαθοδαίμων Ἀλεξανδρεὺς; probably second Century CE) comments on the latter's construction of his world map. Agathoaemon was most likely a contemporary of Ptolemy; his name is referenced in some of the earliest copies of the *Geographia*, e.g., …

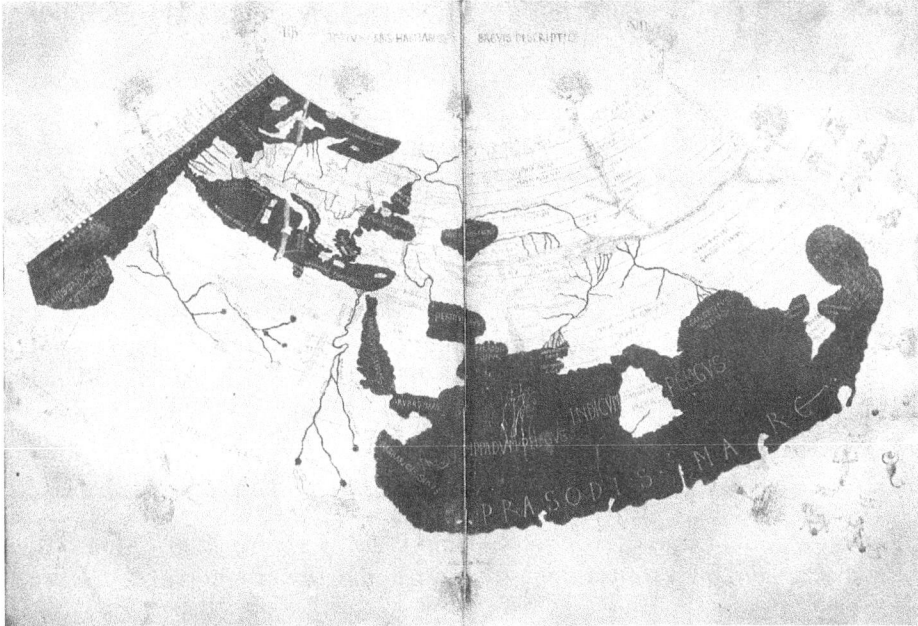

Figure 2.3. Mid-15th Century (1450–1475) Florentine world map based on Jacobus Angelus' Latin translation (1406) of Maximus Planudes' late 13th Century, rediscovered Greek manuscripts of Ptolemy's *Geographia*. Specifically note the parallels and meridians. (Credited to Francesco di Antonio del Chierico; British Library Harley MS 7182, ff 58v–59.)

'From the eight books of Geographia of Claudius Ptolemaeus, the whole habitable world Agathodaemon of Alexandria delineated' [35].

Having found a copy of the *Geographia* without maps, Planudes proceeded to procure and assemble the 27 maps that should have accompanied the text [36].

The few surviving Byzantine copies of the *Geographia*, some predating Planudes' copy by up to two centuries, all consist of eight *Books*. Ptolemy's introduction to these early copies immediately sets the mark; he starts by outlining two very influential definitions, those of *chorography* and *geography*. Chorography, he explains, is selective and regional, *'even dealing with the smallest conceivable localities, such as harbours, farms, villages, river courses, and the like.'* Geography, on the other hand, deals with *'a representation in picture of the whole known world, together with the phenomena which are contained therein.'* Ptolemy understood the primary function of geography as 'making,' so that—to him—geography and cartography were one and the same: …

'It is the prerogative of Geography to show the known habitable Earth as a unit in itself, how it is situated and what is its nature; and it deals with those features likely to be mentioned in a general description of the Earth, such as larger towns and great cities, the mountain ranges, and the principal rivers.'

The task of the cartographer, in Ptolemy's opinion, is to map the world *'in its just proportions,'* to the correct scale.

The first Latin translation after the Byzantine resurfacing of the *Geographia* appeared as the *Geographia Claudii Ptolemaei* in 1406, or possibly in 1407, most likely without the maps [37]; it was compiled by Jacobus Angelus of Scarperia (c. 1360–1411), the Italian scholar and Renaissance humanist, in Florence, Italy. Angelus dedicated this translation first to Pope Gregory IX and then, upon his papal accession in 1409, to Pope Alexander V. Although these copies of the *Geographia* lacked maps, the three oldest surviving copies of the *Geographia* that actually do include maps date from the same era [38]. These manuscripts include the *Urbinas Graecus 82*, currently in the Biblioteca Apostolica Vaticana in Rome, the *Seragliensis 57* in the Sultan's Library in Istanbul, and the *Fragmentum Fabricianum Graecum 23* in the University of Copenhagen's Universitetsbiblioteket. Given the successively decreasing clarity of the special features common to all three manuscripts, it has been suggested that they were copies of each other [39], successively copied in the order introduced above.

The known land masses are all included on Ptolemy's world map (figure 2.3), but some are shaped significantly differently compared with our modern world view. In fact, based on a careful assessment of the positions of the 80 best-known (and, therefore, most reliably located) of the 6345 cities included in Ptolemy's table, it transpires that Ptolemy's map is stretched in longitude by a factor of 1.428 [40, 41]. This is the origin of the apparent misshapen land masses, including the horizontally stretched appearance of Italy. Russo has suggested that if we release the assumption that the Fortunate Isles coincided with either the Canary or the Cape Verde Islands, or with Madeira, but would entertain that they may have been identified with the Lesser Antilles, this might explain Ptolemy's systematic error (stretch) in longitude across the Mediterranean [42]. On the other hand, it would not be surprising if his longitude errors originated from the tall tales of sailors who had misjudged their locations; it is notoriously difficult to determines one's speed across open water without the aid of landmarks to rely on, particularly in the presence of varying weather, wind, and sea conditions.

2.1.3 Ptolemy's map projections

In any case, what really stands out in Ptolemy's world map are the gently curving lines of latitude and, perpendicularly to these parallels, the corresponding meridians. Indeed, Ptolemy's most crucial innovations in and contributions to mapmaking, represented by his beautiful world atlas, consisted of a dedicated effort to find the most suitable projection of the Earth's spherical configuration onto a flat surface. In Book I of his *Geographia*, he pointed out that there are two ways of mapping the world, either by representing it on a sphere or by drawing it on a flat surface. ...

> *'When the Earth is delineated on a sphere, it has a shape like its own, nor is there any need of altering [it] at all.'*

However, the level of detail that could be shown on a sphere is limited by the size of the sphere, which would soon become unmanageable. In his endeavour to represent the world on a flat surface, Ptolemy followed Marinus' lead. Marinus had invented the concept of equirectangular projection [43], a type of projection still used in mapmaking today. Although he borrowed some of the basic ideas from Eratosthenes, he had rejected all previous methods devised to project spherical coordinates onto a flat surface while maintaining congruity—that is, while maintaining the original angles, shapes, and distances. Unfortunately, there are almost no surviving records of what these previous methods entailed. The only known exception is a non-technical, verbal description by Strabo of a *graticule* that could be used for construction of a world map: ...

'But [a world map] requires a large globe, so that the aforesaid segment of it [containing the habitable world], being such a small fraction of it, will be sufficient to hold the suitable parts of the oecumene with clarity and give an appropriate display to the spectators. Now if one can fashion [a globe] this large, it is better to do it in this way; and let it have a diameter not less than ten feet. But if one cannot make [a globe] of this such or not much smaller, one ought to draw [the map] on a planar surface of at least seven feet. For it will make little difference if instead of the circles, i.e., the parallels and meridians with which we show the climata and directions and other variations of the placements of the parts of the Earth relative to each other and to the heavens, we draw straight lines, with parallel lines for the parallels, and perpendicular lines for the [meridians] perpendicular to them. [This is permissible] because the intellect is able easily to transfer the shape and size seen by the sight on a planar surface to the [imagined] curved and spherical [surface]. The same will apply to oblique circles [on the globe] and straight lines [corresponding to them on a map]. And though it is true since the meridians everywhere, since they are all described through the pole, nevertheless it will not matter if on the surface one makes the straight lines for the meridian bend together only a little. For even this is not necessary in many situations when the lines [representing the meridians and parallels on the globe] are transferred to the planar surface and drawn as straight lines, nor is the convergence [of the meridians] as conspicuous as the curvature [of the globe]' [44].

From this passage, it becomes clear that Strabo considered the simple equirectangular projection as well as a second projection where the meridians converge towards the north, although it is not clear whether they would only curve 'inwards' at the highest latitudes or if the meridians would already be inclined with respect to the Equator.

Despite Marinus' criticism of the failure of previously designed methods of projection, Ptolemy complained that his predecessor eventually opted for the least satisfactory method to solve the problem. Marinus' simple projection maps meridians to vertical straight lines at constant intervals and latitudinal circles to horizontal straight lines, also at constant intervals, which thus results in significant

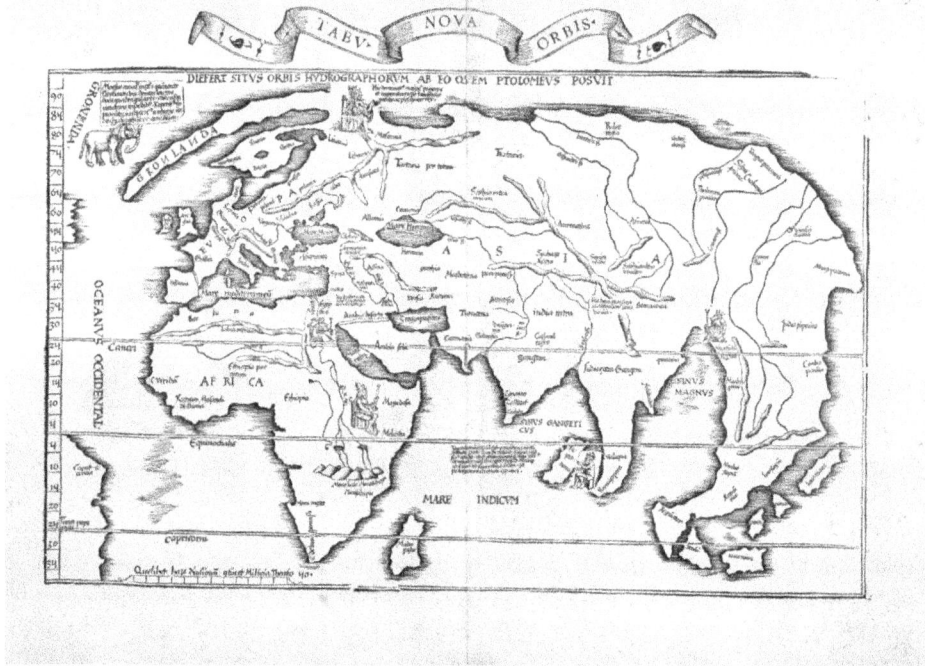

Figure 2.4. *Tabu. Nova Orbis/Diefert Situs Orbis Hydrographorum Ab Eo Quem Ptolomeus Posuit.* Second 'modern' world map in equirectangular projection by the French physician and mathematician Laurent Fries (1485–1532). From the 1535 Lyon edition of Ptolemy's *Geographia*, edited by M Servetus and published by M and G Trechsel, printed from the 1522 woodblocks of the first Fries edition. The map is based on the 1513 Waldseemüller world map. (Courtesy of Götzfried Antique Maps, Tettnang, Germany; https://www.vintage-maps.com/en/antique-maps/world-maps/fries-world-maps-1535::833.)

distortions of the landmasses at increasing latitudes. The resulting map was characterized by an equirectangular projection at a latitude of 36°N: see figure 2.4 for one of the few surviving examples.

The transformation of the spherical coordinates, λ and φ, which correspond, in turn, to the longitude and latitude of the position requiring projection to horizontal and vertical coordinates on a flat map, x and y, follows

$$x = (\lambda - \lambda_0)\cos \varphi_1;$$

$$y = \varphi - \varphi_1,$$

where λ_0 and φ_1 are the map's central meridian and the standard latitudes where the scale of the projection is accurate; the latter apply to both northern and southern parallels.

Projecting spherical maps onto flat sheets is anything but trivial. All projections inherently distort the resulting maps. Marinus' simple projection does not result in unit cells covering equal areas, nor is it angle-preserving ('conformal'). Ptolemy was well aware of this problem, for projections in general, and dedicated much of his working life to finding workable solutions, attempting to retain a semblance of spherical proportions on his flat maps. A significant fraction of Book VII of his *Geographia* is dedicated to detailed analysis of the three projections he recommended to be used for the construction of a map of the world, increasing in both complexity and precision.

To reduce the significant distortions and inaccuracies introduced by Marinus' equirectangular projection, he proposed a more accurate representation by adopting concentric, equidistant parallels on a 'conic-like' surface centred on the Earth's axis of rotation and passing through the Rhodes and Thule *climata*. In this representation, the meridians are straight lines which converge towards the poles. ...

'We shall do well to keep straight lines for our meridians, but to insert our parallels as the arcs of circles, having one and the same centre, which we suppose to be the North Pole, and from which we draw the straight lines of our meridians, keeping above all else similarity to a sphere in the form and appearance of our plane surface.'

Nevertheless, in Ptolemy's maps the apex in the northern hemisphere is not coincident with the North Pole but located some 25 degrees higher, which is caused by his choice of scaling along the parallels with increasing latitude. Ptolemy continued, ...

'The meridians must not bend to the parallels, and they must be drawn from the same common pole. Since it is impossible for all of the parallels to keep the proportion that there is in a sphere, it will be quite sufficient to observe this proportion in the parallel circle running through Thule and the equinoctial, in order that the sides of our map which represent latitude may be proportionate to the true and natural side of the Earth. (...)

Now indeed we are not permitted to carry the lines which are to be drawn as meridians through in one straight course to the parallel [opposite to Meroe] but only to the equator ...; and with the arc [opposite to Meroe] divided in both directions into 90 parts or segments, equal in size and number to those taken on the parallel of Mero[e] we can then draw to these marked points the intervening straight lines from those points marked in the Equator the course of which will seem deflecting towards the south on the other side of the Equator' [45].

Projecting such a conic-like surface onto a plane, one is left with a network consisting of circular parallels and straight, converging meridians. While better than Marinus' simple projection, this conic projection—a prototype of a projection now known as the 'Bonne pseudoconic projection'—is limited in practice to mapping either the northern or the southern hemisphere on any one map. This was only a minor concern for Ptolemy, since he applied this projection technique only rigorously to the known, northern hemisphere.

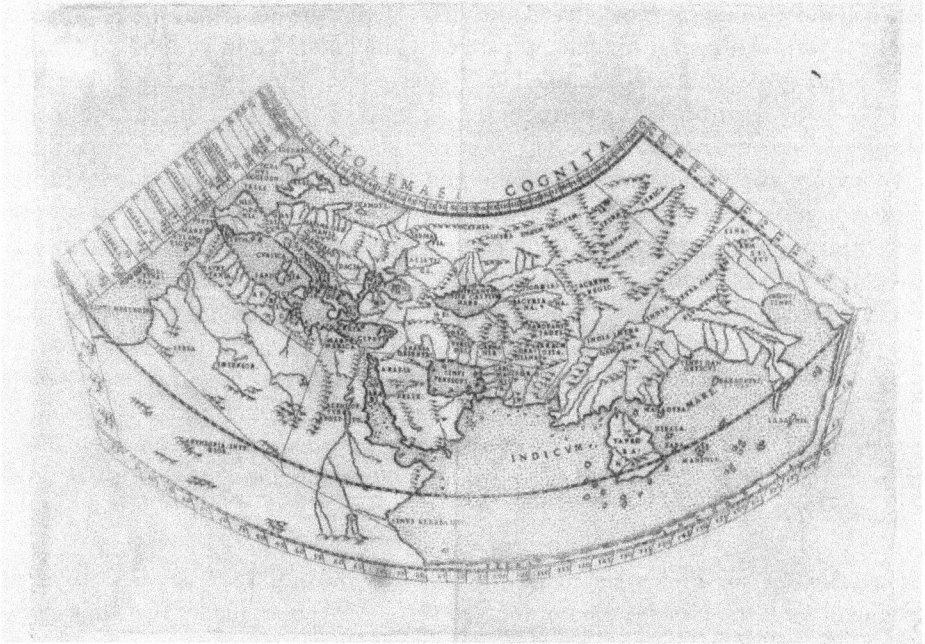

Figure 2.5. *Ptolemaei Cognita*: Ptolemaic world map by the Italian alchemist, physician, and cartographer Girolamo Ruscelli (1504–1566) from *La Geografia di Claudio Tolomeo Alessandrino*, showing the known world described in Ptolemy's *Geographia*. The projection applied is a revised conic projection. This map was printed in Venice in 1574 by G Ziletti; it is based on the 1548 *Geographia* edition of A Mattioli and was newly engraved for this edition. (Courtesy of Götzfried Antique Maps, Tettnang, Germany; https://www.vintage-maps.com/en/antique-maps/world-maps/ruscelli-ptolemy-world-map-1574::223.)

To represent the smaller known parts of the southern hemisphere on the same map, he described a circular arc parallel to the Equator, which was located at the same distance to the south as Meroe is to the north. He divided this southern parallel into parts in a similar way as the Parallel of Meroe and joined the intersections to the corresponding points on the Equator: see figure 2.5 for an excellent 16th Century example.

The equations required for projection of a 'flattened cone' are more complex than those for the simple equirectangular projection. In addition to λ and φ, we also need to know the height, h, of the apex point above the centre of the cone. The transformation equations then become [46]

$$x = \csc(\sec^{-1} h + \varphi)\cos \varphi \sin\left(\frac{\lambda}{\sqrt{h^2 - 1}}\right);$$

$$y = \csc(\sec^{-1} h + \varphi)\cos \varphi \cos\left(\frac{\lambda}{\sqrt{h^2 - 1}}\right).$$

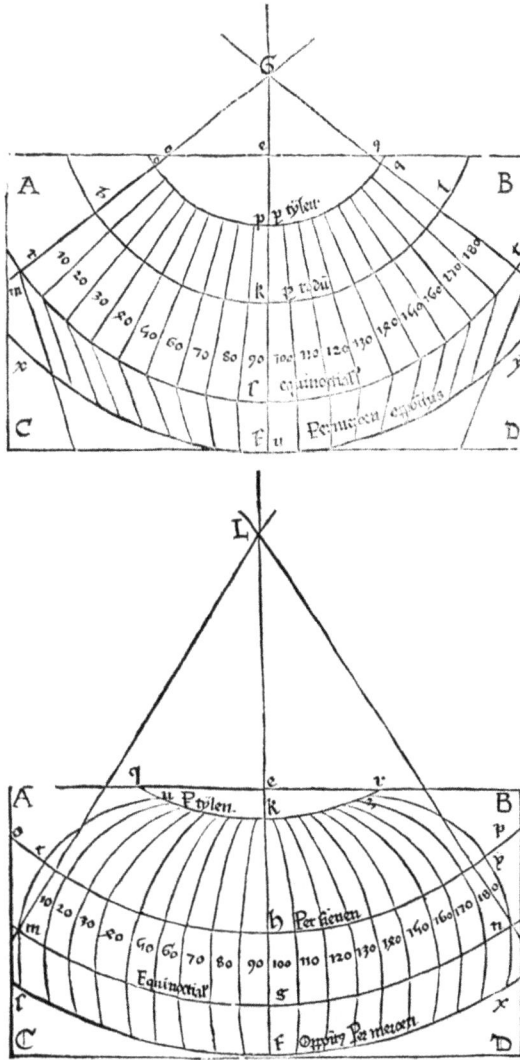

Figure 2.6. Comparison of the simple (*top*) and modified (*bottom*) conic projections proposed by Ptolemy. (*Ptolemy,* Cosmographia, *Ulm: Lienhart Holle,* 1482; digital images courtesy of the Bell Library, University of Minnesota.)

The second type of projection described in Ptolemy's *Geographia* is more complicated. It is generally referred to as a 'cloak' (chlamys; χλαμύς), a modified, or a pseudo-conic projection; the latter designation was introduced by Marie-Armand d'Avezac de Castera-Macaya (1800–1875), the French geographer and archivist, in 1863 [47]. The parallels in this projection follow those in the simple conic projection, but the meridians are curved inwards to the poles, so that the resulting map provides a more realistic rendering of Earth's surface: see figure 2.6 for a comparison with the simple conic projection.

'We shall be able to make a much greater resemblance to the known world in our map if we see the meridian lines, that we have drawn, in that form in which meridian lines appear on a globe, when the axis of the eyes is imagined as directed upon a motionless globe through a point before the eyes in which occurs the intersection of that meridian and that parallel which divides respectively the longitude and the latitude of the known Earth into two equal parts, and also through the centre of the globe, so that the extreme parts which lie opposite each other appear and are perceived by the eye in like condition. (...)

That a greater likeness to a sphere is achieved by this method than by the former will be self-evident. When the sphere stands motionless before the eyes, and is not revolved (which necessarily holds true for a plane map), and the eye rests on the middle of the object, once certain meridian which, because of the globe's position, lies at the middle of the plane passing through the axis of the eye, will exhibit the appearance of a straight line, while those on either side appear inflexed with their concave side towards it, and the more so as they lie farther from it, which is also observed here with exact analogy, just as it is also seen that the symmetry of the parallel arcs keeps the proper ratio of one to another, not only in the equatorial line and the parallel of Thule (as was done in the former case), but in the others also, as closely as they can be made—the difficulty of doing this is evident—and that the conformity of latitude as a whole serves towards a true, general longitude ratio, not only in the parallel drawn through Rhodes but in all of the parallels' [48].

As implied by this passage, the pseudo-conic projection renders areas delineated by successive parallels and meridians of approximately (but not quite) the correct ratio, so that this projection is also known as Ptolemy's *homeotheric* ('equal-area') projection [49]. ...

'Although for these reasons this method of drawing the map is the better one, yet is less satisfactory in this respect, that it is not as simple as the other; ... Since this is so, even though for me both here and everywhere the better and more difficult scheme is preferable to the one which is poorer and easier, yet both methods are to me retained for the sake of those who, through laziness, are drawn to that certain easier method' [50].

In other words, Ptolemy's second projection was rather difficult to implement. The first known published maps using it were created by Donnus Nicolaus Germanus (c. 1420–c. 1490), a German cartographer, when generating copies of Ptolemy's *Geographia* around 1470. Figure 2.7 shows a beautiful surviving example of this modified conic projection in the form of an early map by Sebastian Münster (1489–1552).

Ptolemy also described a third type of projection, which was even more complex than his second. His discussion of this 'modified perspective projection' [51] appears

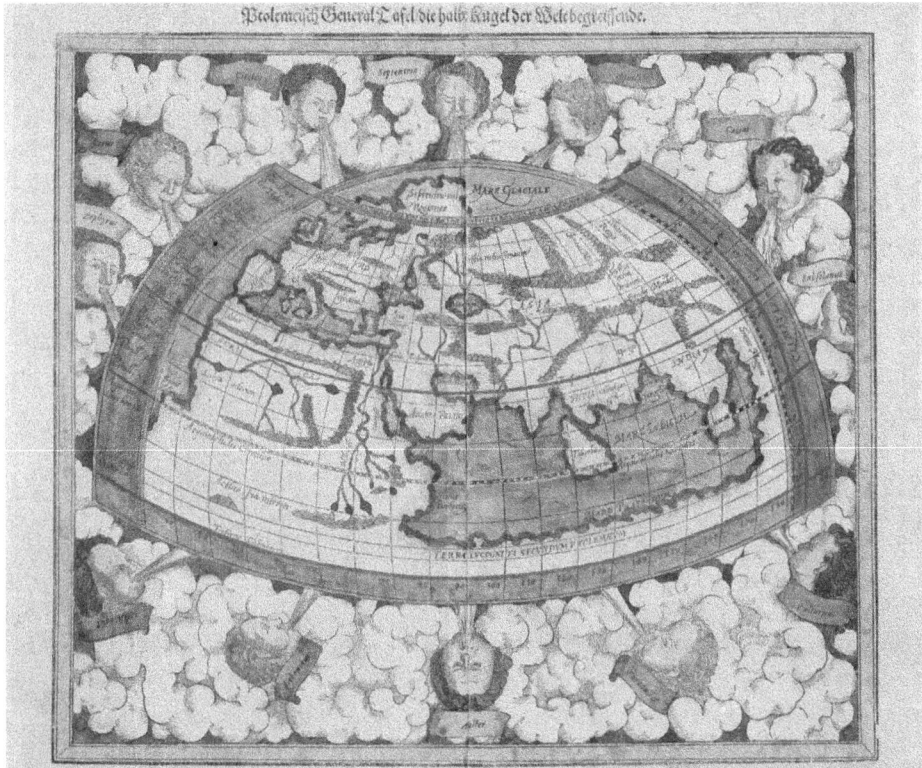

Figure 2.7. *Ptolemeisch General Tafel, die halbe Kugel der Welt begreiffende.* Ptolemaic woodcut world map employing the modified conic projection. This is the second world map created by Sebastian Münster (1489–1552), a former monk and highly influential 16th Century cartographer, which he used in his *Cosmographia* from 1588 onwards. (Courtesy of Götzfried Antique Maps, Tettnang, Germany; https://www.vintage-maps.com/en/antique-maps/world-maps/muenster-ptolemaic-world-map-1588::11588.)

almost as an afterthought, given that it was published in Book VII rather than Book I where the other projections are discussed. It was very rarely, if ever, employed by mapmakers until late in the 16th Century. The resulting world map, designed to show '*the hemisphere of the Earth in which the* oecumene *is located*,' was oval in shape owing to the azimuthal perspective adopted.

Despite Ptolemy's extensive criticism of the imperfect cartographic methods adopted by Marinus, he only used the simple conic projection himself for his large-scale world map in the *Geographia*; for the remaining 26 maps, he adopted the equirectangular approach, using the base of each map as his reference.

2.2 Early cartography and mapping

2.2.1 Advances in the Islamic world

The principles of Hellenist cartography gradually spread beyond their core region. As such, medieval Islamic cartography was heavily influenced by Hellenist ideas from its inception in the eighth Century CE [52]. Islamic scholars had access to

Ptolemy's masterpieces, the *Almagest* and *Geography*, from the ninth Century, but they did not follow Ptolemy's principles of mapmaking unquestioningly [53]. Instead, early Arabian and Persian cartography followed Muḥammad ibn Mūsā al-Khwārizmī (محمد بن موسى خوارزمى or al-Khwārizmī; c. 790–c. 850), the Persian mathematician, astronomer, and geographer in the House of Wisdom in Baghdad, one of the major intellectual centres during the 8th–13th Century Islamic Golden Age.

Al-Khwārizmī's calibration shifted the reference meridian 10 degrees to the East with respect to that adopted by Ptolemy. Around 833 CE he compiled his *Book of the Depiction of the Earth*, which resembled Ptolemy's *Geography* to a large extent [54]. It included the geographic coordinates of 545 cities, as well as regional maps of the Nile, the Island of the Jewel (*Jazīrat al-Jawhar*, also known as the Island of Sapphires; a semi-legendary island thought to be located in the Sea of Darkness, near the Equator, at the eastern end of the inhabited world), the Sea of Darkness, and the Sea of Azov north of the Crimea. The earliest surviving copy of al-Khwārizmī's work dates from 1037 [55]; he clearly states that his maps are based on an earlier map, but this was unlikely a direct copy of Ptolemy's *Geographia*, given the difference in prime meridians adopted between both works.

Al-Khwārizmī modified many of Ptolemy's coordinates beyond this simple transformation, particularly their latitudes. He extended his maps eastwards of Ptolemy's Golden Peninsula (Malaysia) and also eastwards of the phantom peninsula known as the Dragon's Tail, which replaced Ptolemy's unknown eastern shore of the Indian Ocean. His maps employed a rectangular projection of the type subsequently adopted by many Muslim cartographers and advocated by the Persian cartographer Suhrāb (شهراب; Ibn Sarābiyūn, ابن سرابيون; died 930 CE) in his early-10th Century treatise, *Marvels of the Seven Climes to the End of Habitation* ('*Adjā'ib al-aḳālīm al-sab'a ilā nahāyat al-imar'a*; عجائب الأقاليم السبعة إلى نهاية العارة).

Suhrāb suggested to place a lateral scale of 180 degrees at both the top and bottom of the map, and a vertical scale on both sides. The latter would be divided into a northern segment spanning 90 degrees and a segment of 20 degrees south of the Equator. Having outlined the seven *climata* on the map, the locations of the cities could be added based on their known coordinates, using a pair of weighted strings. This resulted in an orthogonal projection which retained the distances along the Equator and the meridians, but which was characterized by a greater East–West stretch than Marinus' equirectangular projection. Although there are no surviving maps using this projection, the scale at the top of the rectangular world map in the early 11th Century *Book of Curiosities* (see figure 2.8) closely resembles Suhrāb's suggested projection.

Around 956 CE, the Persian historian and geographer Abu al-Ḥasan 'Alī ibn al-Ḥusayn ibn 'Alī al-Mas'ūdī (أبو الحسن علي بن الحسين بن علي المسعودي or al-Mas'ūdī; c. 896–956 CE) extolled the virtues of the full-colour map created by a team including al-Khwārizmī at the court of the caliph Abū Ja'far Abdullāh al-Ma'mūn ibn Hārūn al-Rashīd (ابوجعفر عبدالله المأمون or al-Ma'mūm; 786–833 CE, reigned 813–833 CE) [56]. It contained lavish detail, including 4530 cities and more than 200 mountains. His description and favourable comparison with the ancient Greek maps of Marinus and Ptolemy suggest that it may have been constructed according to the principles of mathematical geography.

Figure 2.8. The oldest surviving rectangular world map from the 11th Century *Book of Curiosities*. (Book 2, chapter 2; MS. Arab. c. 90, fols. 23b–24a; © The Bodleian Library, University of Oxford.)

The 10th Century witnessed significant advances in the field of mathematical geography in the Islamic world. Abū al-Wafāʾ, Muḥammad ibn Muḥammad ibn Yaḥyā ibn Ismāʿīl ibn al-ʿAbbās al-Būzjānī (ابوالوفا بوزجانی بوزگانی or Abū'l-Wafāʾ; 940–998 CE) and Abu Nasr Mansur ibn Ali ibn Iraq (Mansur or منصور; c. 960–1036), both Persian mathematicians and geographers, produced trigonometric results which were applied by, in particular, Abū Rayḥān Muḥammad ibn Aḥmad al-Bīrūnī (al-Bīrūnī or ابوریحان بیرونی; 973–1048) to the main problem in the field, the determination of longitude and latitude. Al-Bīrūnī, the 'Father of Geodesy,' considered one of the greatest scholars of the Islamic world, was favourably disposed of the idea that the Earth rotated on its axis, although he could neither prove nor disprove it [57]: ...

'[R]otation of the Earth does in no way impair the value of astronomy, as all appearances of an astronomical character can quite as well be explained according to this theory as to the other. ... This question is most difficult to solve. The most prominent of both modern and ancient astronomers have deeply studied the question of the moving of the Earth, and tried to refute it. We, too, have composed a book on the subject called Miftah-ilm-alhai'a [مفتاح علم الهيئة; Key to Astronomy], in which we think we have surpassed our predecessors, if not in the words, at all events in the matter' [58].

Nevertheless, and crucial in the context of our narrative, allowing for a rotating Earth enabled him to pursue accurate calculations of longitude and latitude, specifically linking time to longitude. Al-Bīrūnī also experimented with projections and published a short treatise on this topic, *Cartography*, by the young age of 22 (c. 995 CE). He invented what we now refer to as the 'azimuthal equidistant projection,' shown in figure 2.9 (top), which represented a marked improvement with respect to Ptolemy's *analemma* or 'azimuthal orthographic projection,' a map representation as the Earth would be seen from a great distance. One of the earliest world maps by al-Bīrūnī using this projection is shown in figure 2.9 (bottom).

The most important properties of this projection are that any given point on the map is at the proportionately correct distance and located in the correct direction (azimuth) from the centre point. Mathematically, for any given point (θ, ρ) on Earth—where θ represents the angle a line from the centre point (φ_1, λ_0) [latitude, longitude] makes with the vertical direction and ρ is the point's distance from the centre—will project to Cartesian coordinates (x, y) as

$$x = \rho \sin \theta, \; y = -\rho \cos \theta,$$

so that

$$\cos \rho = \sin \varphi_1 \sin \varphi + \cos \varphi_1 \cos \varphi \cos(\lambda - \lambda_0).$$

The direction from the centre point to position (θ, ρ) follows,

$$\tan \theta = \frac{\cos \varphi \sin(\lambda - \lambda_0)}{\cos \varphi_1 \sin \varphi - \sin \varphi_1 \cos \varphi \cos(\lambda - \lambda_0)}.$$

For projections centred on the North Pole, we get the following simplification:

$$\rho = \frac{\pi}{2} - \varphi, \; \theta = \lambda.$$

These developments in cartography were of crucial importance for the travellers of the Islamic world. Transport across the largely trackless desert as well as by sea was very important, given that there were few navigable rivers in the region; only the Nile, Tigris, and Euphrates could be used for trade and transportation by ships of any reasonable size. Navigation skills were, therefore, highly developed. Techniques as simple as holding up one's fingers to the horizon or the use of a simple card and string enabled desert travellers to determine their latitude sufficiently well. Early Muslim sailors used quadrants and rudimentary sextants known as 'kamals,' combined with detailed maps based on mathematical geography principles and detailed information on currents and other natural phenomena. This allowed contemporary sailors to leave the relative safety of coastal waters and traverse the open seas. The ancient Greeks had developed the astrolabe, while Muslim scholars perfected its instrumental design from the eighth Century onwards. It found its way into Europe in the 12th Century through Islamic southern Spain. Islamic scholars were pioneers in accurate direction determination on land, given the importance of the direction to Mecca (known as the *qibla*) for religious observances.

Major developments in mathematical geography, including the implementation of grids on navigational maps, occurred early on during the Islamic Golden Age. The earliest surviving world map from the Muslim or Christian world which

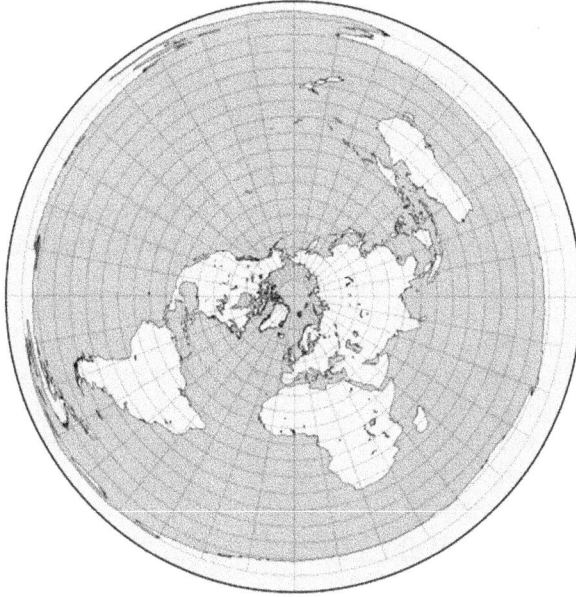

Figure 2.9. Top: Equidistant azimuthal projection. (© Carlos A. Furuti; http://www.progonos.com/furuti/ MapProj/) Bottom: al-Bīrūnī's map of the Distribution of Land and Sea, 1238 CE. South is at the top. (© The British Library Board, Or. 8349 f58r.)

includes a *graticule* dates from the 11th Century's *Book of Curiosities* (see figure 2.8), but it appears that the cartographer was not well versed with its purpose: the grid pattern starts from the left and employs twice the intended scale; moreover, the cartographer seems to have realized his mistake and appears to have given up on this pursuit halfway through [59]. This suggests that the cartographer may have worked on the basis of previous maps that could have included mathematically justified *graticules* [60].

The heyday of Islamic geography was yet to come, however, reaching its apex in the 12th Century with Abu Abdullah Muhammad al-Idrisi al-Qurtubi al-Hasani as-Sabti (أبو عبد الله محمد الإدريسي القرطبي الحسني السبتي) or al-Idrisi; 1100–1165), geographer, cartographer, and Egyptologist at the Sicilian court of the Norman King Roger II (1095–1154; reigned 1130–1154). In 1138, the King commissioned al-Idrisi to compile a book on geography, which had to contain all available data on the locations and climates of the world's main population centres. When completed in 1154, a few weeks before the King's death, the *Nuzhat al-mushtāq fi'khtirāq al-āfāq* (نزهة المشتاق في اختراق الآفاق; known as the *Tabula Rogeriana*)—the centrepiece circular world map—was engraved onto a silver tablet. Al-Idrisi calculated the Earth's circumference at 37 000 km, which is less than 10% different from today's best value. He divided the world into seven *climata*, following his ancient Greek predecessors, supported by 70 longitudinal section maps, which could be combined into a rectangular world map: see figure 2.10.

At the start of the 20th Century, the American scholar Samuel Parsons Scott commented, clearly in awe of al-Idrisi's work, that …

'The compilation of Edrisi marks an era in the history of science. Not only is its historical information most interesting and valuable, but its descriptions of many parts of the Earth are still authoritative. For three centuries geographers copied his maps without alteration. The relative position of the lakes which form the

Figure 2.10. Al-Idrisi's *Tabula Rogeriana* (1154), created from the 70 double-page spreads of the original atlas; North is oriented towards the bottom. (Source: Wikimedia Commons; public domain.)

Nile, as delineated in his work, does not differ greatly from that established by
Baker and Stanley more than seven hundred years afterwards, and their number
is the same. The mechanical genius of the author was not inferior to his erudition.
The celestial and terrestrial planisphere of silver which he constructed for his
royal patron was nearly six feet in diameter, and weighed four hundred and fifty
pounds; upon the one side the zodiac and the constellations, upon the other—
divided for convenience into segments—the bodies of land and water, with the
respective situations of the various countries, were engraved' [61].

Indeed, the *Tabula Rogeriana* is regarded the most accurate map of the known world in pre-modern times. Al-Idrisi had managed to achieve this feat by incorporating all knowledge available at the time of Africa, the Indian Ocean, and the far East, which he had collected by interviewing Islamic merchants and explorers as well as Norman travellers. Although accurate determination of longitude at sea was not achieved until well into the 18th Century, Indian Ocean navigators developed a highly sophisticated method of measuring distances on the open sea parallel or perpendicular to meridians, as well as parallel to the Equator.

Islamic developments in mapmaking naturally found their way to the Indian subcontinent during the heyday of Islamic cartography. In fact, from at least the second century CE, Indian astronomers had already been able to accurately calculate planetary motions across the sky, initially for astrological purposes [62]. Projection of heavenly coordinates on Earth, in particular of parallels, meridians, and the Equator, as well as the use of a prime meridian, based on the longitude of the ancient Indian city of Avanti or Ujjayinī (present-day Ujjain in the state of Madhya Pradesh), was widely employed [63]. Muslim world maps and extensive catalogues containing accurate positions of the most important places started to appear around the start of the 11th Century, heavily influenced by celebrated geographers like al-Bīrūnī.

In fact, in preparation for the construction of his world map, in 1025 al-Bīrūnī completed his masterpiece of Indian mathematical geography, *Tahdīd nihāyāt al-amākin li-tashīh masāfāt al-masākin* (*Determination of the Coordinates of Places for the Correction of Distances Between Cities*), listing the longitudes and latitudes of numerous places of interest. The equivalent list of Indian cities appeared in his astronomical highlight, *Al-Qānūn al-Mas'ūdi* (*The Mas'ūdic Canon*) [64]. Islamic influences continued to play an important role well into 17th Century Indian navigation, including during the Mughal Empire (1526–1540, 1555–1857) when means of determining latitude and longitude—introduced in the 14th Century— were discussed [65], although the pertinent details have been lost in the mists of time... However, as Irfan Habib has highlighted, ...

'[t]he impulse for the compilation of these tables derived only secondarily from
interest in geography. Astronomical observation demanded the determination of
the latitude and longitude of the point from which observations were being made.
The science of astrology, requiring the casting of accurate horoscopes, intensified
the anxiety to have the terrestrial coordinates correctly determined. There was a

religious impulse too: Muslims must pray facing Mecca, and the mosques must be aligned accordingly. The direction in which Mecca lay from any place could, however, be determined only if the latitudes and longitudes of both the places were known. Lists of places with their coordinates were thus compiled independently of the geographers' [66].

A reproduction of what is perhaps the most important of the surviving Indo-Islamic world maps, the *Map of the Inhabited Quarter* (1647), is shown in figure 2.11. Longitude is measured from the island at the top right, which probably represents the Fortunate Isles (or, perhaps, the Canary Islands); note that South is at the top. To its left is the west coast of Africa; the map extends to 'Chin' and 'Mahachin,' which both represent China.

While there is at least some evidence that Indian navigators relied on Muslim-style marine charts and navigation techniques, maps from the Indian subcontinent were not based on mathematical principles until at least the period of British colonial rule. In his 1805 study, *An Essay on the Sacred Isles in the West with Other Essays Connected with that Work*, the British Orientalist and Indologist Francis Wilford (1761–1822) pointed out, …

'Besides geographical tracts, the Hindus have also maps of the world, both according to the system of the Paurán'ics [believers of the Puraanas, the books of ancient Indian history, culture, and civilization]*, and of the astronomers: the latter are very common. They have also maps of India, and of particular districts, in which latitudes and longitudes are entirely out of question, and they never make use of a scale of equal parts. The sea shores, rivers, and ranges of mountains, are represented in general by straight lines. The best map of this sort*

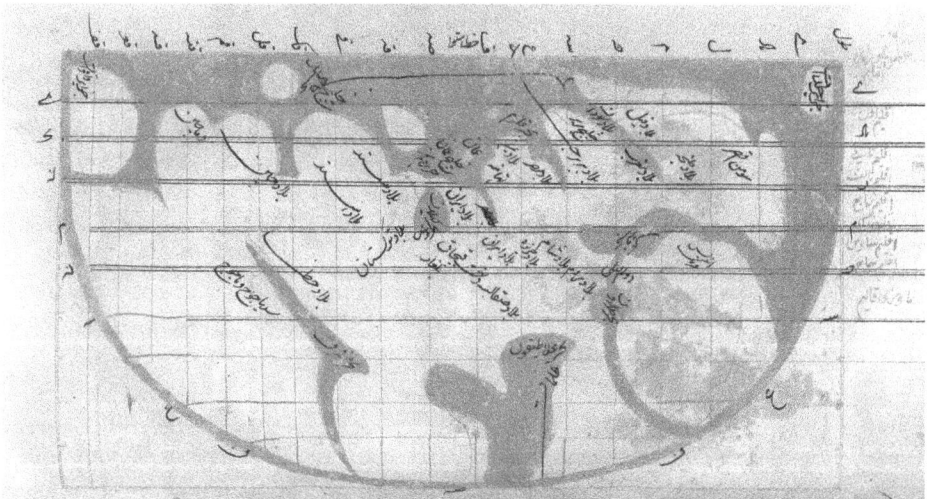

Figure 2.11. Map of the 'Inhabited Quarter,' constructed by Muḥammad Sadiq Ibn Muḥammad Salih Isfahāni (Sadiq Isfahāni) of Janpur (1647). (© The British Library Board, Egerton 1016 f335r.)

I ever saw, was one of the kingdom of Napál, presented to Mr. [Warren]
HASTINGS [first Governor-General of Bengal]. *It was about four feet long,*
and two and a half broad, of paste board, and the mountains raised about an inch
above the surface, with trees painted all round. The roads were represented by a
red line, and the rivers with a blue one. The various ranges were very distinct,
with the narrow passes through them: in short, it wanted but a scale. The valley
of Napál was accurately delineated: but toward the borders of the map, every
thing was crowed [sic], and in confusion.

These works, whether historical or geographical, are most extravagant compo-
sitions, in which little regard indeed is paid to truth…. Geographical truth is
sacrificed to a symmetrical arrangement of countries, mountains, lakes, and
rivers, with which they are highly delighted' [67].

Indeed, most indigenous Indian maps may not have been constructed for
navigation, but they probably served either religious purposes or the needs of the
state.

2.2.2 East Asian navigation and cartography

Although it is often claimed that medieval Islamic developments influenced the
development of Chinese geography under the Mongol Empire, there is mounting
evidence that similar Chinese cartographic advances predate these influences by
more than a millennium.

In fact, the use of maps covered by rectangular grid patterns probably originated
in China [68], most likely with the work of Zhang Heng (張衡; 78–139 CE) [69, 70],
the famous Han Dynasty-era Chinese polymath. Zhang Heng's maps have been lost
in the depths of time, but we can glean some interesting insights from descriptions in
his biography by the scholar Cai Yong (蔡邕; 132–192 CE), who stated that Zhang
Heng *'cast a network about heaven and earth and reckoned on the basis of it'* [71].
Zhang Heng's pioneering work greatly influenced that of his Wei (220–265 CE) and
Jin Dynasty (265–420 CE) successor, the cartographer and Imperial official Pei Xiu
(裴秀; 224–271 CE), who was the first to include a mathematical grid reference and
graduated scale on his maps for improved accuracy in distance estimates [72]. He
adopted the grid as one of his six principles of scientific cartography, on the basis
that …

'[w]hen the principle of the rectangular grid is properly applied, then the straight
and the curved, the near and the far, can conceal nothing of their form from us'
[73].

However, introduction of a grid pattern and establishing the relative positions of
places of interest on land is not the same as determining one's longitude on the open
sea. Ancient Chinese cartographers were well versed in determining their positions
accurately on land, using triangulation as well as observations of the night sky.

Longitude determinations using lunar eclipses were most likely introduced into China by Arab astronomers during the Mongol Empire's Yuan Dynasty (1271–1368). In addition, according to the *Mi Shu Jian Zhi* (秘書監志; *Records of the Palace Library*), the Yuan court ordered officials in Quanzhou (泉州), the official starting point of the Maritime Silk Road during the Yuan Dynasty, to collect Arab compass navigation charts from visiting navigators.

Chinese astronomers were skilled in using lunar eclipses to determine their local longitude, although this was a rather cumbersome process. It required synchronous time measurements across large distances. The lunar eclipse method applied by Chinese astronomers required a base observatory in China. For a given lunar eclipse, observers at both the base observatory and the location of interest identified bright stars which passed through the local meridian at the moment when the Moon started to reappear after the eclipse. Upon their return to the base observatory, both records were compared and additional measurements of stellar meridian passages were obtained to determine the exact time difference of the meridian passages of both stars. The measured time difference corresponded directly to the difference in longitude between both locales.

In addition, Chinese astronomers had long known about the Moon's irregular motion, that is, the Moon's 'equation of time;' these irregularities were first discovered by the astronomer Jia Kui (賈逵; 30–101 CE) during the Eastern Han era (25–200 CE). The Moon's apparent motion takes it 360 degrees around the Earth to its original position among the background stars in approximately 27.3 solar days. This motion corresponds to an average of 13 degrees per day, or slightly more than half a degree per hour. While the 'fixed' stars appear to move westwards because of the Earth's rotation, the Moon appears to describe a retrograde orbit with respect to the stars, covering approximately half a degree per hour in the eastward direction. The Moon appears to speed up when going towards the Sun and seems to slow down when retreating; the change in orbital motion also depends on the Earth's orbital position.

The equation of time is best known in the context of the Sun's motion. It is a correction that needs to be applied to the regular clock time to reflect the 'figure-of-eight' path the Sun traces on the sky throughout the year when we compare its position at local noon based on determining the Sun's highest point on any given day with that of conventional noon (clock time). The equation of time is the result of two competing effects, including the eccentricity of the Earth's orbit—that is, the fact that the Earth's orbit is elliptical rather than circular (which requires an adjustment of the orbit's centre)—and a reduction at the Equator caused by the obliquity of the ecliptic plane, since the Sun's passage is measured on the meridian and not in the Earth's equatorial (rotation) plane which forms the basis of our geographic reference system. The equation of time provides the correction (in minutes) required to convert clock time back to the true local noon: see figure 2.12. The correction required varies throughout the year between −14 minutes and +16 minutes, which—if left uncorrected for—could result in navigational mismatches of almost 900 kilometres at the Equator.

Having noticed the nature of the lunar motion, Chinese astronomers used a variety of interpolation methods to predict the Moon's equation of time, leading to

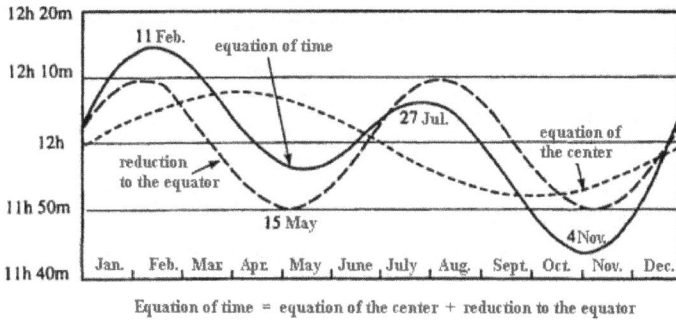

Figure 2.12. The equation of time, showing the competing effects of the 'equation of the (orbital) centre' and the reduction to the Equator required to convert the mean solar time (clock time) to the true solar time. (© Institut de mécanique céleste et de calcul des éphémérides, IMCCE; J. E. Arlot.)

accurate predictions of the Moon's motion across the sky through the piecewise parabolic interpolation proposed by Liu Zhuo (刘焯; 544–610 CE) [74]. This method was further improved upon, using third-degree interpolation (although for the *solar* equation of time), during the Yuan dynasty by Guo Shoujing (郭守敬; 1231–1316), the Chinese astronomer, engineer, and mathematician whom the Jesuit astronomer Johann Adam Schall von Bell (1591–1666) called the 'Tycho Brahe of China.'

By simply measuring the solid angle on the sky between the Moon when it crosses the local meridian and a given star, one can calculate the local longitude. This requires only access to a sextant, without the need for an accurate clock. This thus implies that Chinese astronomers would have been able to accurately calculate lunar positions on the sky since the Yuan Dynasty, which allowed them to compile ephemeris tables listing the positions on the sky of the Moon and a set of reference stars throughout the year. Armed with such tables and with access to a sextant, they could have been able to determine their longitude at sea fairly accurately. This method is, in fact, equivalent to the 'lunar distance method' employed by European astronomers several centuries later (see chapter 6.2.4).

Circumstantial evidence suggests that Chinese navigators may have mastered some degree of longitude determination already in the 14th Century. As part of an expedition undertaken from 1334 to 1339, a traveller from Quanzhou—Wang Dayuan (汪大渊)—successfully made a transoceanic voyage from Mozambique to Sri Lanka using a combination of compass and stellar position measurements called *Guoyang Qianxing Shu* (过洋牵星术; 'star orientation') to determine his ship's latitude and longitude [75, 76]. At that time, East Asian navigators and cartographers already knew East Africa and the Indian Ocean very well, as evidenced by the highly accurate representation of the coasts of East, South, and West Africa on the *Kangnido* (강리도) map (*Yoktae chewang honil kangnido*; *Map of historical emperors and kings and of integrated borders and terrain*, also known as *Honil Gangni Yeokdae Gukdo Ji Do*; *Map of Integrated Lands and Regions of Historical Countries and Capitals*), produced by Gwon Geun (권근) and his team of royal astronomers in Korea: see figure 2.13.

This Korean world map is the oldest surviving global chart among the East Asian cartographic highlights. It dates from 1402, predating any known Chinese or Japanese world maps [77]. The salient feature of this map is that the African continent is represented by a triangular shape. This is markedly different from contemporary Arab or European maps, where the southern half of the African continent would usually be rendered as an extension to the East; this general trend was not remedied until the mid-15th Century.

Nevertheless, Korean cartographers were apparently not concerned with systematic geodetic surveys or coordinate systems until the late 18th Century, when King Chŏngjo (정조: Jeongjo of Joseon; 1752–1800) ordered the director of the astronomical observatory to undertake calendar reform with specific emphasis on the time differences between the peninsula's eight provinces [78]. The earliest reference to both a physical scale on any Korean map and the implied distances between places indicated on the map occur late in Korean history; they are attributed to the cartographer Chŏng (Jeong) Sang-Gi (정상기; 1678–1752). Chinese cartographic influences—particularly of the scale maps of the Taoist monk Zhu Siben (朱思本; 1273–1333), of his Ming-dynasty editor, Luo Hongxian (羅洪先; 1504–1564), or of the Song dynasty's stone-engraved grid map from 1137, the *Yuji Tu* (禹迹图: *Map of the Tracks of Yu*; see figure 2.14)—are non-existent in any surviving records related to Korean cartographic developments. (Similarly,

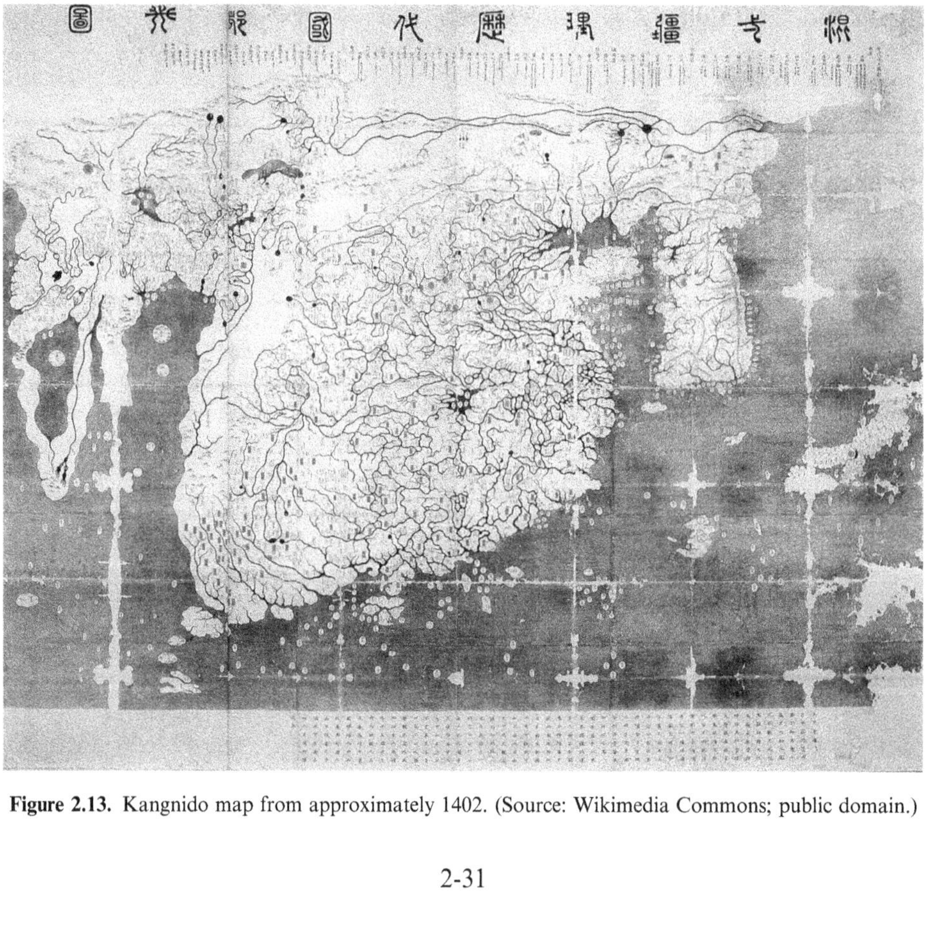

Figure 2.13. Kangnido map from approximately 1402. (Source: Wikimedia Commons; public domain.)

Figure 2.14. Song-dynasty Chinese stele-engraved grid map known as the *Yu Ji Tu* (1137), presently located in Xi'an (西安). The grid consists of 100 *li* squares. (Source: Library of Congress, Geography and Map Division; digital ID: g7821c.ct001493.)

Japanese maps did not include grid lines or lines of longitude and latitude until well into the 18th Century [79].)

Other transoceanic voyages undertaken in the early 15th Century may not have relied on skills to determine longitude at sea. For instance, the fourth expedition (1413–1415) led by Admiral and Chief Envoy Zheng He (鄭和; 1371–1433/5) set off from Sumatra, in present-day Indonesia, sailing 10 days in full wind to reach the Maldives archipelago; from there, they sailed across the Indian Ocean in full wind for 15 days, reaching Mogadishu in East Africa, some 3700 miles from their port of origin [80]: …

'We have traversed more than 100 000 li [50 000 km] of immense water spaces and have beheld in the ocean huge waves like mountains rising in the sky, and we

have set eyes on barbarian regions far away hidden in a blue transparency of light vapours, while our sails, loftily unfurled like clouds day and night, continued their course [as rapidly] as a star, traversing those savage waves as if we were treading a public thoroughfare...' [1]

This voyage appears to have followed well-established Arab and Chinese trade routes rather than breaking new navigational ground.

On the other hand, Gong Zhen writes [81] in *Xiyang Banguo Zhi* (西洋番国志; *Notes on Barbarian countries in the Western seas*) that Zheng He's navigators may have used sightings of the rising or setting Sun and Moon to help determine how far the ship had sailed in an easterly or westerly direction (which is equivalent to determining longitude based on an approach similar to the lunar distance method), while they used stellar elevations to determine both their longitude and latitude.

On 10 October 1432, Zheng He's fleet set sail once again from Pulau Rondo on Sumatra (6° 04′ N, 95° 07′ E) to Sri Lanka [82]:

'Gauging the vertical positions of the given stars above the horizons in the east, west, north, and south, [he] reached Sri Lanka' [83].

The fleet continued to Kuli (Calicut) in India (11° 15′ N, 75° 46′ E), where they arrived on 10 December 1432. Their next leg commenced three days later and involved a 35-day voyage, until 16 January 1433, to Bandar-e Abbas (بندرعباس) in the Strait of Hormuz. They soon proceeded with their next transoceanic leg, crossing the Arabian Sea from Dingde Baxi (present-day Dandi Bandar; 16° 00′ N, 73° 03′ E) to Jabal Khamis in present-day Oman (22° 25′ N 59° 27′ E).

The expedition used the declination of *Zhinü* (織女; the 'weaving girl,' corresponding to the star Vega) and of *Nanmen Shuangxing* (南门双星; two stars in the constellation of Sagittarius). Travelling westwards, they relied on observations of Pollux (in the constellation of Gemini) and Procyon to determine their longitude and latitude [84]. Specifically, Zheng He's navigators based their transoceanic voyages on 10 stars, which enabled them to determine their approximate longitude and latitude at sea. For latitude determination, they relied on the stars closest to the North and South Poles, as well as on stars in Sagittarius; for longitude determination, their guide stars included Pollux in the North West, Procyon in the South West, and Vega and bright stars in Taurus in the East [85]. Since the *Datong Li* (大統曆; the 'great universal system of calculating astronomy')—which was based on Guo Shoujing's pioneering system of calendrical astronomy—had been formally adopted by the Ming Bureau of Astronomy in 1384, it is conceivable that Zheng He and his navigators were familiar with longitude determination at sea on the basis of solar and lunar eclipses as well.

In his capacity of senior administration official, Zheng He led seven expeditions to Southeast, South, and West Asia, and to East Africa, between 1405 and 1433, which were aimed at establishing a Chinese presence and to enforce Imperial control over trade in the Indian Ocean and extend the Yongle (永樂) Emperor's tributary

---

[1] From a tablet erected by Zheng He in Changle, Fujian Province, China (1432)

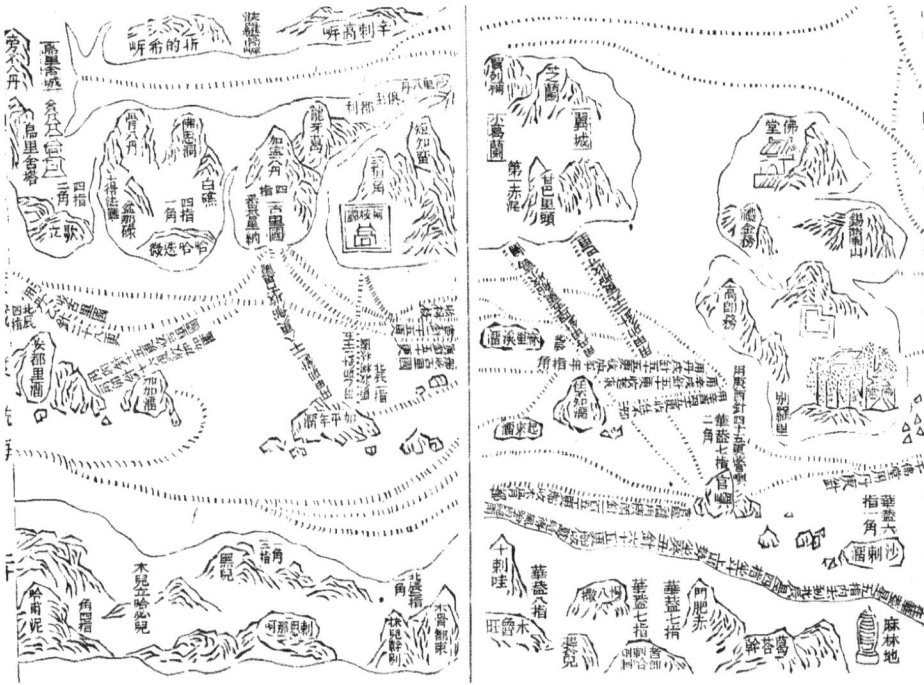

Figure 2.15. Section of the *Wubei Zhi* oriented with East at the top. India is located at the top left, Sri Lanka at the top right, and Africa along the bottom. (Source: Wikimedia Commons; public domain.)

territories. Zheng He's sailing charts, known as the *Mao Kun Map* (鄭和航海圖), consisted, in fact, of four maps, one each centred on Sri Lanka, South India, the Maldives, and a fourth map showing some 400 km of the East African coast, covering the coastal waters to a southern latitude of six degrees. Figure 2.15 shows a section of one of Zheng He's maps, taken from a published collection of his maps from the 17th Century, entitled the *Wubei Zhi* (武備志; *A Treatise on Armament Technology*, 1621; published 1628).

Zheng He's sailing charts were designed—likely with the help of Arabic-speaking pilots with detailed knowledge of the African coast—with the specific purpose in mind that they would be used to sail along certain routes. As such, their positions vary in orientation to align with the ocean currents and winds, as required of a sailing chart, specifically from the user's perspective. The aim was clearly to provide positional information in the shortest time possible, as also evidenced by the orientation of the geographic features: they are seen from the user's orientation. Sailing instructions are offered using a 24-point compass system with a Chinese symbol for each point, combined with a sailing time or distance. Local currents, winds, and depth soundings are taken into account.

The 24-point compass system resembles the system used for navigation in the Western Pacific and which was used already by the ancient Polynesian navigators. In a Chinese cultural context, the compass points are based on the 12 directions of the *Earthly Branches*, a system for timekeeping. For navigation purposes, 12 points were insufficient, prompting the introduction of 24- or even 48-point compasses by inserting intermediate directions.

As early as the Yuan and early Ming Dynasties, the Chinese had already invented an approach to timekeeping that was reasonably accurate and worked on land as well as at sea. They were indeed far ahead of European developments, which would not bear fruit until the 17th and 18th Centuries, as we will see in chapters 3 through 6. A particularly promising Chinese timing device was the *cháng míng dēng* (长明灯; everbright lantern), which operated similarly to the common candle clocks in use at the time. Xie Jie (謝杰), a contemporary Ming author, elaborates on what was known as the compass navigation method of the Zhangzhou (漳州) people: ...

'The compass cabin burns everbright lanterns day and night, five geng [更] each night and five geng each day. So a ship sailing for 12 Chinese hours burns the equivalent of a total of 10 geng.'

The accuracy of this timekeeper was remarkable for the times. Over the course of 24 hours, or 10 *geng*, a lantern would consume one catty (斤) of oil, equivalent to approximately 500–600 grams, so that its burn rate could be determined to better than 0.1斤 by reading off the amount of oil consumed on a graduated glass oil reservoir. Chinese navigators could thus keep track of the local time in their home port while out at sea. Comparing the ship's local times of sunrise or sunset with the home port's time based on the lantern, they could obtain a rough estimate of their longitude difference. Incense coils with known, carefully calibrated combustion timescales had been used for the same purpose since the sixth Century and became commonplace in the Song Dynasty (960–1279) [86]; both timing devices were used for periods of up to weeks or even months. To obtain a more accurate measurement of the time (and longitude) difference with respect to the home port, water clocks (*mashang benchi xinglou*; 马上奔驰行漏) were employed to time the difference between a home port's benchmark time (e.g., the time of sunrise) and that at the ship's location. This was done most accurately by weighing the amount of water consumed between both calibration moments [87].

After the death of the Yongle Emperor in 1424, Zheng He's expeditions were discontinued and a period of decline in cartographic developments commenced. China began to look inwards, adopting a policy of isolationism that lasted a few hundred years. The new Hongxi (洪熙) Emperor (1378–1425) ordered that Zheng's fleet be burnt, along with all records, thus ending the 'Age of the Sea.' In addition, the Hongxi Emperor expressly prohibited overseas travel; anyone who disobeyed the order was killed.

Traditional Chinese cartographic skills became more advanced in the late Ming Dynasty (1368–1644) under the influence of new ideas introduced by the European Jesuit missionaries from the early 1600s onwards. However, major new initiatives were not seen until well into the Qing Dynasty (1644–1912), when the Kangxi (康熙) Emperor (1654–1722) realized that Chinese maps were not sufficiently accurate for navigation and territorial purposes. Therefore, in 1708 he sponsored a geodesy and mapping programme using astronomical observations and triangulation measurements [88]. The final product, the first on-the-spot survey map called the *Huang*

Yu Quan Lan Fen Sheng Tu (皇與全覽分省圖; *Kangxi Provincial Atlas of China*, 1721/2), took well over a decade to complete.

2.3 Towards reliable navigation across the open seas

2.3.1 Portolan charts

While Arab and East Asian navigational prowess saw major new developments throughout the Middle Ages, much of Europe was languishing in an environment dominated by religious intolerance and wars, with progress in science, technology, and innovation largely stifled—a period of stagnation commonly referred to as the 'Dark Ages.' Nevertheless, towards the end of the 13th Century, Western European navigators—particularly in Italy, Spain, and Portugal—went beyond the old Ptolemaic maps that had been in common use until that time and developed a novel type of marine chart based on direct observation using a mariner's compass. Invented in Europe by 1300 [89] and known as 'portolans'—based on the Italian word *portolani*, a 'collection of sailing directions related to ports'—these highly accurate maps, even by modern standards, were used routinely until well into the 16th Century.

The inspiration for this new development in cartography is often attributed to Alfonso X 'the Wise' of Castile (1221–1284), who ordered charts and compasses for practical marine use. However, recent research [90] has questioned that assumption: the oldest extant portolan chart appears to have appeared out of nowhere at the end of the 13th Century, while there is no indication that simpler, earlier versions ever existed. It is therefore now considered plausible that these charts found their origin in a tradition that is now lost [91]. The first portolan charts were made in the Balearic Islands by Jewish cartographers, including Abraham Cresques (1325–1387) on Majorca, author of the *Catalan Atlas* (1375), and in the Italian city states of Genoa and Venice.

Figure 2.16 shows the oldest surviving portolan chart in the collection of the U.S. Library of Congress (c. 1320–1350), drawn on vellum, that is, on sheepskin scrolls; it shows a projection of the Mediterranean and western Black Seas. The oldest known portolan chart is the *Carta Pisana* (c. 1275–1300; probably 1296), which depicts a similar area. Earlier medieval European maps held a religious significance and purpose—for instance, the *mappa mundi*, which had been drawn based on theological directives. Portolan charts, on the other hand, incorporated direct observation and first-hand navigational experience. Prior to the appearance of the first portolan charts, Mediterranean sailors relied on compass measurements and their experience and knowledge of the local sea conditions. Magnetic compasses, invented some time during the 12th Century in both Europe and the Islamic world [92–95], were commonplace on the sea-faring vessels of the time, allowing sailors to maintain their course during overcast conditions; they were usually mounted on gimbals to counteract the ocean swells. Before portolan charts became the norm, sailing records merely included lists of ports in the order they would encounter them, as well as notes on the estimated directions, sailing times between ports, and a few drawings of features such as headlands, capes, harbours, and other recognizable landmarks. This practice continued until well into the 17th Century, as evidenced by the following passage from William Dampier's *A New Voyage Round the World* (1697): …

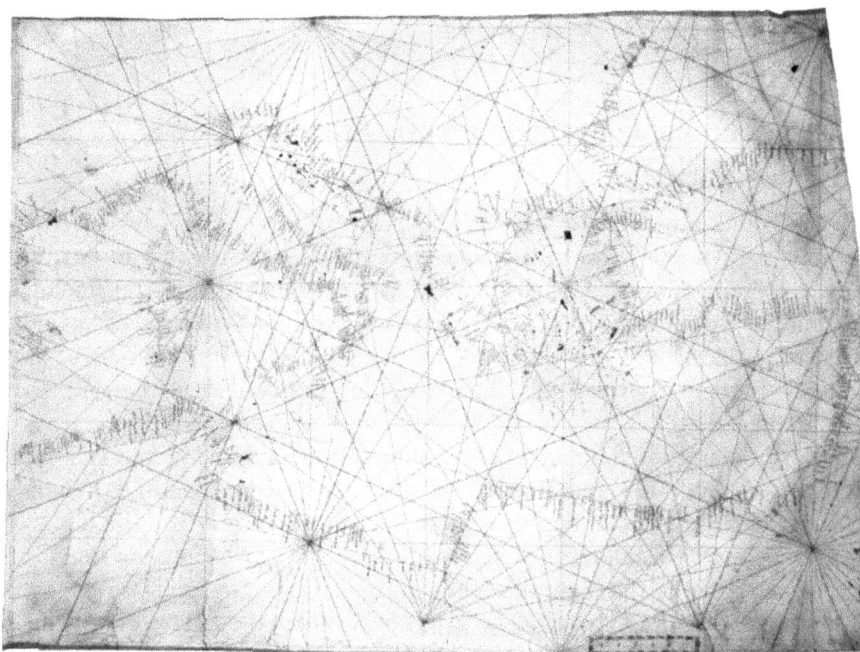

Figure 2.16. Oldest portolan chart in the U.S. Library of Congress collection, showing the Mediterranean and western Black Seas. (Source: Library of Congress, Geography and Map Division.)

'On the back of the town, a pretty way up in the country, there is a very high mountain, towering up like a sugar-loaf, called Monte Christo. It is a very good sea-mark, for there is none like it on all the coast. The body of this mountain bears due South from Manta. About a mile and half from the shore, right against the village, there is a rock, which is very dangerous, because it never appears above water; neither doth the sea break on it, because here is seldom any great sea; yet it is now so well known, that all ships bound to this place do easily avoid it. A mile within this rock there is good anchoring, in 6, 8, or 10 fathom water, good hard sand, and clear ground: and a mile from the road on the west side, there is a shoal running out a mile into the sea' [96].

With the invention of the portolan chart, a new era of mapmaking had arrived in Europe. For the first time, new marine navigation routes could be explored, out of direct sight of the coast. Few portolans remain today; those that do cover the Mediterranean and Black Seas, as well as a section of Europe's Atlantic coastal area, with the highest accuracies achieved in the charts' original heartland of the Mediterranean and Black Sea coasts. The early 20th Century historian of cartography Sir Charles Raymond Beazley (1868–1955) called portolan charts *'the first true maps,'* even going so far as to say that *'in them, true cartography, the map-making of the civilised world, begins.'* [97]—a bold claim indeed...

Most interesting in the context of our discussion of the development of marine navigation, they are covered not by a grid of longitude and latitude lines, but by

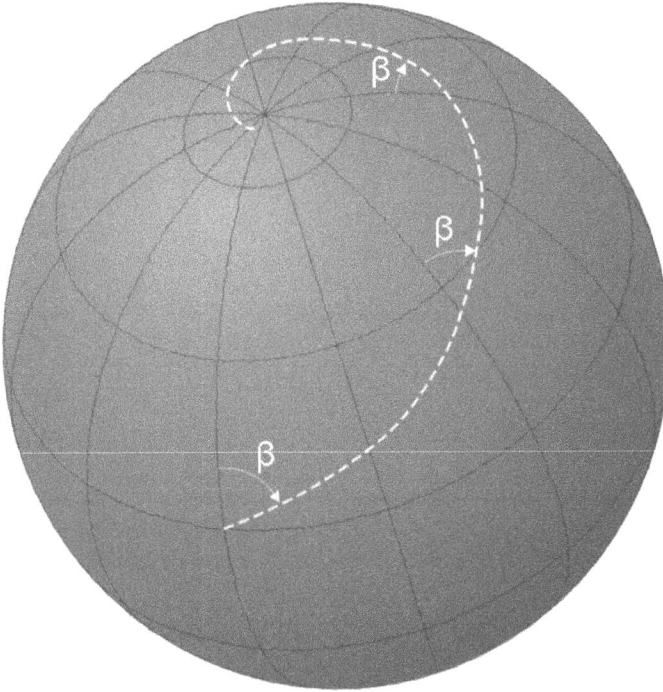

Figure 2.17. Representation of a loxodrome spiraling towards the North Pole (Source: Wikimedia Commons; CC BY-SA 2.5.)

systems of, usually, 16 to 32 directional or 'rhumb' lines, radiating across the charts [98]. These lines are better understood as 'windrose' lines, since portolan charts were not based on any global map projection. In fact, ...

'the word [rhumb line] is wrongly applied to the sea-charts of this period, since a loxodrome [an arc crossing all meridians at the same angle: see figure 2.17] gives an accurate course only when the chart is drawn on a suitable projection. Cartometric investigation has revealed that no projection was used in the early charts, for which we therefore retain the name "portolan" ...' [99]

... and also that a number of errors can be traced to false compass readings caused by variations in the Earth's magnetic field. In addition, most portolan maps included a *physical* scale representation, so that realistic distances could be read off the charts. This represented a first in cartography, although the actual scale, nicknamed the 'portolan mile' [100], varied among maps—it was *approximately* 1.25 km but ranged from 1.16 km for the *Carta Pisana* to 1.31 km for the 1339 *'Dulcert'* map—and required one to know the latitude it applied to [101]. It is thought that portolan charts were mosaics of individually created components, which seem to resemble very early representations equivalent to the Mercator projection, a type of map projection that would only become commonplace almost 300 years later.

Figure 2.18. The Pizzigano brothers' 1367 portolan chart. (Source: Biblioteca Palatina, Parma: Ms. Parm. 1612; public domain.)

The charts' windrose lines allowed sailors to navigate away from coastal view and cross the Mediterranean. They would first identify their destination on the portolan chart and note which line—and, hence, which compass direction—represented the most appropriate angle between the origin and destination ports. Once they had determined the ship's direction, they would sail by 'dead reckoning' towards the coastal waters near their destination. Varying winds and small course deviations would have led the ship to reach a coastal area near but likely not precisely at the destination port; the chart's accompanying description of the coast line and the approach to the destination port would allow the navigator to reach his final destination.

Shown in figure 2.18, the Pizzigano brothers Domenico and Francesco (or, perhaps, a father-and-son team), Venetian cartographers, constructed a portolan chart in 1367 which extended further into the Atlantic Ocean than any previous map had done; their map included Madeira, eight of the Canary Islands, and even the mythical Fortunate Isles, north of the Canary Islands, as well as Saint Brendan's Island. It also includes the islands of 'Brasil' further to the West, shown surrounded by ships and dragons of Arab legend. The map also goes significantly beyond the other boundaries usually adopted for such charts, extending as far North as the mythical Isle of Mam (to the southwest of Ireland) and Scandinavia and as far East as the Baltic and Caspian Seas.

The use of portolan charts and descriptions of coastal features, combined with navigation by dead reckoning based on maintaining a given compass heading, was a proven and tested method of crossing the Mediterranean. Distances across the

Mediterranean from North to South are relatively small, while currents and the effects of tides are minor and magnetic variations are insignificant [102]. However, extending this approach to significantly greater distances across the open ocean would potentially be disastrous [103]. Nevertheless, without a reliable means to determine their location on the open seas, navigators were left to the crude device of dead reckoning for centuries to come.

2.3.2 Dead reckoning and Columbus' legacy

This was precisely the approach taken by Christopher Columbus (c. 1451–1506) when he set off on his fabled voyages across the Atlantic to 'discover' the Americas. Indeed, as an unintended consequence of lacking a reliable means of determining one's position at sea, many lands were 'discovered' by early European mariners, only to be effectively lost again until their rediscovery. The degree of randomness introduced by dead reckoning is reflected very well in the following passage from *The Saga of Erik the Red* by the Icelandic explorer Leif Erikson (c. 970–c. 1020), allegedly the first European to have reached North America: ...

> *'they put out to sea when they were equipped for the voyage, and sailed for three days, until the land was hidden by the water. Then the fair wind died out, and north winds arose, and fogs, and they knew not whither they were drifting and thus it lasted for [a long period of time]. Then they saw the Sun again, and were able to determine the quarters of the heavens; they hoisted sail, and sailed ... through before they saw land. They discussed among themselves what land it could be and [one of them] said that he did not believe that it could be Greenland'* [104, 105].

Upon setting sail to 'discover the East by sailing towards the West,' Columbus would have had access to a variety of contemporary maps, including the (now faded) world map made by Henricus Martellus Germanus (Heinrich Hammer; *fl.* 1480–1496) in 1490/1491, which extended as far as China and, incorrectly positioned, Japan: see figure 2.19. Columbus is said, however, to have based his worldview on the dimensions of the world, or rather those of the *oecumene*, proposed by Ptolemy and Marinus of Tyre. Both men held erroneous views of the Earth's circumference, however, which consequently resulted in Columbus' mistaken idea that the distance from Europe to Asia via a Western course was much shorter than it is in reality.

In his *Geography*, Ptolemy had proposed that the Eurasian continent—specifically the distance from Cape St Vincent on the coast of Portugal to the Cape of Cattigara on the peninsula of India Superior—covered 180 degrees in longitude, while Marinus believed that the continent spanned an even larger area of the Earth's surface, equivalent to 225 degrees [106]. The actual landmass in the northern hemisphere covers 130 degrees to mainland China, while the distance from Western Europe to Japan spans 150 degrees at the latitude of Spain. Most 15th Century European scholars, including the German cartographer and explorer Martin Behaim (Martin of Bohemia; 1459–1507), creator of the *Erdapfel* ('potato'), the

Figure 2.19. Martellus map of the world, 1490. (Source: Beinecke Rare Book and Manuscript Library, Yale University.)

world's oldest surviving globe—created in 1490–1492, not including the Americas, since Columbus did not return to Europe until March 1493—concurred with Ptolemy's estimate of the Earth's circumference, however. Nevertheless, Columbus argued that the ancient cartographers did not know how far the Asian landmass extended. Instead of following the scholarly consensus, he adopted Marinus' estimate of 225 degrees to the East coast of 'Magnus Sinus'—nowadays identified with the Gulf of Thailand and the South China Sea, but which in Columbus' days briefly was equated with the Pacific Ocean—which left only 135 degrees of water to cross. Combined with his adoption of 56⅔ *Roman* nautical miles for the length of a degree of longitude [107] instead of Ptolemy's longer distance of 56⅔ *Arabic* miles, this resulted in an erroneous estimate of the eastward extent of the Eurasian landmass: the Arabic mile—an arcminute of latitude measured along a North–South meridian—is approximately 1.8–2.0 km long, while the Roman mile with which Columbus was more familiar measures 1.480 km.

In addition, Columbus operated on the assumption that Japan ('Cipangu') was much larger, located much farther to the East than its actual position, and much closer to the Equator, as suggested by a passage of hidden text on the Martellus map: ...

'This island is 1000 miles from the continent of the province of Mangi [China]; the people have their own language and the circumference of the island is ... miles'[2].

When he first landed in the New World in 1492, he thought that he had arrived close to Japan. He also believed that further eastwards he would encounter inhabited

---

[2] Recently uncovered by the *Lazarus Project* at the Smithsonian Institution.

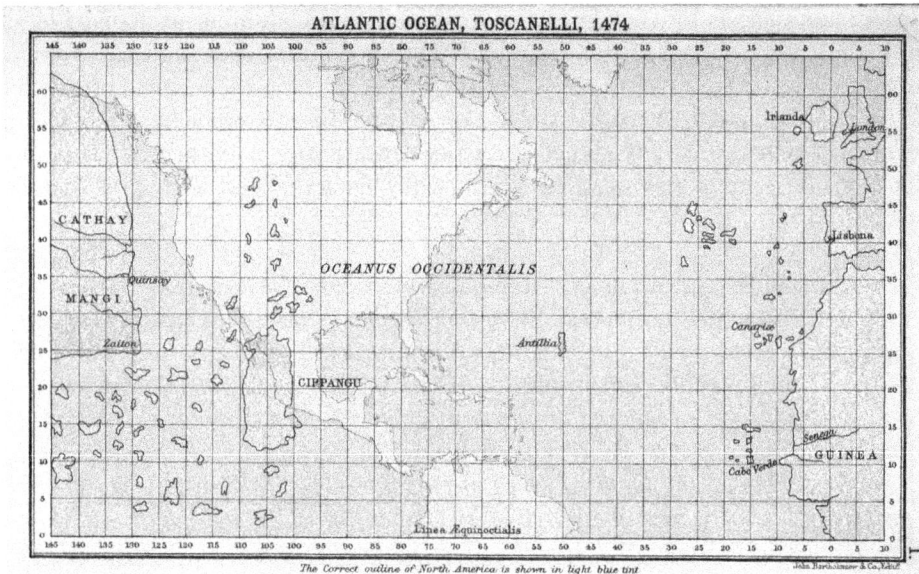

Figure 2.20. Atlantic Ocean according to Toscanelli (1474). The correct outline of North America is shown in light blue. (Source: Batholomew J G and Brooke G C 1911 *A Literary and Historical Atlas of America* (London: Dent) map I; public domain.)

islands. Among the latter, he thought that he might find Antillia, the phantom island said to be located in the Western Atlantic Ocean, also known as the Isle of the Seven Cities, which he expected to lie just to the West of the Azores [108] and which was routinely included on 15th Century nautical maps. Its first occurrence is found on a Pizzigano portolan chart from 1424; see figure 2.20 for a comparison of Columbus' worldview with the actual distances across the Atlantic Ocean.

It thus transpires that the size and circumference of the Earth, as well as the relative proportions of land and water, were still highly controversial and vigorously debated in the 15th Century. In this context, the English philosopher and clergyman Roger Bacon (c. 1219/1220–c. 1292) surmised, for instance, based on Aristotle's manuscripts, that the *oecumene* was necessarily large, but Pablo of Burgos (c. 1351–1435) argued that it could not span more than half the globe. Nonetheless, Columbus persisted in his adoption of Marinus' value for the Earth's circumference and reasoned that East Asia and Cipangu were, respectively, approximately 270 and 300 degrees East, or about 90 and 60 degrees West, of the Canary Islands. He hence estimated a distance from the Canary Islands to Cipangu of approximately 3000 Italian miles (3700 km) [109], which he considered a viable undertaking for his '*Enterprise of the Indies*;' the actual distance is closer to 20 000 km...

Columbus was not the first to moot the suggestion of pursuing a Western expedition to the Indies. The Florentine astronomer Paolo dal Pozzo Toscanelli (1397–1482) is often considered to have conceived the idea; his 1474 world map (for his proposed geography of the Atlantic Ocean, see figure 2.20) was the first to include graduated longitude and latitude lines [110]. Historians' opinions are divided, however, as to whether or not Toscanelli and Columbus exchanged letters about the endeavour prior to Columbus' first expedition. Although Columbus is known to have had access to a copy of a letter dated 25 June 1474 from Toscanelli to his friend Ferdinando Martini, the canon of the Lisbon cathedral, there is no convincing evidence of any direct correspondence between both men. Toscanelli included his famous navigational chart with his letter, as well as a detailed explanation: ...

'The straight lines drawn lengthwise on this map show the distance from east to west; the transverse lines indicate distance from north to south. I have also drawn on the map various places in India to which one could go in case of a storm or contrary winds, or some other mishap... From the city of Lisbon due west, there are twenty-six spaces marked on the map, each of which contains two hundred and fifty miles, as far as the very great and noble city of Quinsay [Hangzhou]. This city is about one hundred miles in circumference ... That city lies in the province of Mangi, near the province of Cathay, in which the king resides the greater part of the time. And from the island of Antillia, which you call the Island of the Seven Cities, to the very noble island of Cipango, there are ten spaces, which make 2500 miles, that is, two hundred and twenty-five leagues. This land is most rich in gold, pearls, and precious stones, and the temples and royal palaces are covered with solid gold. But because the way is not known, all these things are hidden and covered, though one can travel thither with all security' [111].

Toscanelli is said to have received important insights in 1433 from a visiting Chinese ambassador to Florence about the possibility of reaching China via a Western route [112]. Whether or not Columbus had access to these insights remains controversial, however. Toscanelli's 1474 letter was treated as a top-secret document at the Lisbon court, and it was clearly addressed to Martini, not to Columbus: *'Ferdinando Martini canonico ulixiponensi Paulus physicus [dixit]'* [113]. Columbus may have gained access to the letter around 1480 [114], but this suggestion (and any subsequent correspondence resulting from this) is also controversial among historians [115], given that in his *Journal of the Third Voyage*, Columbus maintained, ...

'As your Highnesses know, only a short while ago, no other land was known, except what Ptolemy wrote about, and there was no one in my day who believed it was possible to sail from Spain to the Indies' [116].

If both men had corresponded with each other, one may have expected that Columbus would have referred to that correspondence in his *Journal*. In addition, Toscanelli clearly favoured a westerly course on the Lisbon parallel, but Columbus crossed the Atlantic on that of the Canary Islands. Moreover, upon reaching Hispaniola (nowadays comprising the countries of Haiti and the Dominican Republic), some 60 degrees to the West of Cádiz (southern Spain), he believed that he had arrived at Cipangu; Toscanelli's map implied that he would have travelled only 40 degrees, which casts further doubt on claims that Columbus and Toscanelli were in close contact at any time during the late 15th Century. The jury remains out on this question.

When Columbus first landed in the Lucayan Archipelago (the Bahamas) in 1492 on the island called 'Guanahani' by the locals—either Samana Cays, Plana Cays, or San Salvador Island— …

> '[a]nd he says that by his account that he had gone 1142 leagues from the [Canary] island of [El] Hierro …' [117, 118],

he identified the newly discovered land with 'the Indies' (Asia). The local population —the Arawak or Tainos 'Indians'—told him of a larger island to the south, either Cuba or Hispaniola, which he mistook for Cathay (China) or Cipangu; on his second voyage, he identified Cuba with Southeast Asia. On the basis of the Martellus map, he had expected to encounter Cipangu at approximately 700 leagues West of the Canary Islands. On his fourth voyage to the New World, Columbus identified South America with the Cape of Cattigara in India Superior, as represented on Behaim's globe.

The importance attached by Columbus' financial patrons to finding a westward passage to the legendary riches of the East begs the question as to whether Columbus or any of his navigators independently confirmed the longitudes they had reached on any of his four expeditions to the Americas. Indeed, while determination of longitude at sea was all but impossible in the absence of an accurate timing device, over the course of his four expeditions to the New World, he observed the lunar eclipses of 14 September 1494 (from Saona, Dominican Republic) and 29 February 1504, from his ships' anchorage at St Anne's Bay, Jamaica [119]. A third eclipse occurred while he was on a 29-day overland expedition on Hispaniola, on 21 March 1494. He had scheduled a break in the journey at La Isabella, specifically to observe this eclipse [120]. However, the onset of the eclipse occurred just before 7 o'clock in the evening, local time, as the Moon rose from the sea to the East. It would therefore have been impossible to see the shadow's onset.

Columbus was familiar with Ptolemy's work, particularly with his description of what may be the earliest record of a difference in longitude in history, based on the total lunar eclipse of 20 September 331 BCE. On that date, Alexander the Great's army witnessed a widely reported total lunar eclipse at its encampment near Arbela (Mesopotamia; present-day Erbil in northern Iraq). They recorded the onset of the eclipse two hours after sunset; in Carthage (located 15 km from present-day Tunis in Tunisia), however, the onset of the eclipse occurred at local sunset. This timing difference thus implied a distance between Arbela and Carthage of 30 degrees.

Columbus would thus have been familiar with the use of lunar eclipses as a means of distance and longitude determination. In fact, he is known to have owned a copy of the 1474 astronomical almanac, the *Kalendarium*, of Johannes Müller von Königsberg (better known as Regiomontanus; 1436–1476), which predicted the times of lunar eclipses for the next 30 years. Columbus carried a copy of the almanac with him on at least his first transatlantic voyage, and he is reported to have observed a lunar eclipse while in the Americas:

'He declared also from the observation of his people that when in the year of our Lord 1494 there appeared an eclipse in the month of September, it was seen in Española four hours before that it was visible in Spain [121].' ...

'On September 15th by the mercy of God they sighted an island which lies off the eastern end of Española ... in the middle of a great storm he anchored behind this island ... That night he observed an eclipse of the Moon and was able to determine a difference in time of about five hours and twenty-three minutes between that place and Cádiz' [122]. ...

Columbus himself also reported to have observed this particular eclipse: ...

'In the year 1494, when I was at the island of Saona, which is at the eastern end of the island of Hispaniola, there was an eclipse of the Moon on the 14th of September, and it was found that there was a difference from there to the Cape of St Vincent in Portugal of five hours and more than one half' [123],

a claim he repeated in a letter of 7 July 1503 to King Ferdinand II of Aragon and Queen Isabella I of Castile, ...

'In the year ninety-four I navigated in twenty-four degrees [of latitude] to the westward to the end of nine hours, and I cannot be in error because there was an eclipse' [124].

Indeed, this was confirmed by Bartolomé de las Casas (c. 1484–1566), Bishop of Chiapas, in his *Historia de las Indias*: ...

'From the end of Cuba (that is seen in Hispaniola), which was called the End of the East, and by another name Alpha and Omega, he sailed westwards from the southern part, until he passed the end of ten hours on the sphere, in such a way that when the Sun set to him, it was two hours before rising to those that lived in Cádiz, in Spain; and he says that there couldn't be any error, because there was an eclipse of the Moon on the 14th of September, and he was well prepared with instruments and the sky was very clear that night' [125],

and once again in a later chapter, although there de las Casas reports a longitude difference of 5 hours 23 minutes West of Cádiz [126].

These historical records leave us with an unsatisfactory accuracy of Columbus' longitude determination. Pickering tried to disentangle the underlying causes of the discrepancies. First, the correct longitude difference between Saona Island and Cádiz is 4 hours 10 minutes. If Columbus had based his measurements on Müller's almanac readings, he would have made a small mistake, since Müller's predicted time was 24 minutes late [127]—he would thus have concluded that, at his observation point at Saona, he was 4 hours 34 minutes West of Cádiz.

The six longitude differences reported in the contemporary, primary records cited above include five different measurements which are discrepant with respect to the actual difference in longitude by anywhere from 34 minutes to more than five hours. Local time could be measured to an accuracy of 10–15 minutes, while the eclipse onset might be uncertain by up to 5 minutes; these uncertainties do not add up to even the least discrepant measure of the local longitude.

Historians have long been suspicious of Columbus' reports [128]. One cause for concern is a report that Columbus diverted to Saona to shelter from an approaching storm, a report confirmed by his son Fernando, who added that the storm had already reached them by the time they reached their anchorage. But then, how could they have made accurate observations of the lunar eclipse timings, apparently lost today? And why did Columbus later change his report, suggesting that the night was clear? Pickering has suggested that Columbus' motivation to report clearly fraudulent results was driven by ambition; in his *Book of Privileges* we read that Columbus asked the Spanish crown to confirm his dual status of 'Admiral of the Ocean Sea' and Viceroy of the New World. The Sovereigns' confirmation specifically restricted his Admiral's privilege to only '*the Ocean Sea in the region of the Indies*' [129], which may have motivated Columbus to represent the newly discovered lands as the Asian continent [130].

After all, as Columbus reminded his patrons in his letter of July 1503, the African/ Eurasian landmass—the *oecumene*—was thought to span from 12 to 15 hours of longitude, depending on whether one adopted Ptolemy's or Marinus' calculations [131]. Therefore, Columbus likely anticipated that the remaining section of the globe, nine to 12 hours, was covered by the extent of the 'Ocean Sea' between Europe and Asia, sailing westwards. Although this reasoning may explain why he claimed a longitude difference with respect to Cádiz of nine hours or more, the smaller differences reported by the other primary sources are still too discrepant from the actual value to be attributed to timing uncertainties alone.

These differences are likely owing to the misfortune that bad weather intervened in Columbus' attempt to observe the lunar eclipse of September 1494. In his *Historia de las Indias*, De las Casas points out that Columbus measured the distance between the Canary Island of La Gomera and Dominica at 850 leagues [132]. Adopting Columbus' conversion of 56⅔ leagues to a degree of longitude, these 850 leagues translate into four hours of longitude (although this conversion formally only applies at the Equator). Columbus himself did not report this small longitude himself, but his biographer Antonio Gallo (died c. 1510) did, perhaps based on a personal conversation with Columbus, his son Fernando, or any of the sailors on board of the Admiral's ship. Note that the distance of 1142 leagues quoted in Columbus' *Diario*, when converted to degrees of longitude using the same conversion, translates to 5 hours 23 minutes—as reported by Fernando in his biography and by De las Casas in his *Historia*.

Pickering suggests that Columbus may have been unsatisfied with the 'short' transatlantic distance he had recorded in his *Diario*. At a later stage, he proceeded to add the East–West distance he sailed within the Caribbean islands to the earlier total of 1142 leagues, which would have corresponded to at least 31 leagues from the Bahamas to Cuba [133], resulting in a total longitude difference with respect to

Cádiz of 5 hours 31 minutes—indeed similar to the difference he quoted in his *Libro de Profecías* (*Book of Prophecies*; 1501–1505).

The numbers quoted in the *Profecías* are not all internally consistent, however: the longitude difference Columbus reported for the second lunar eclipse he observed from Jamaica in February 1504 is 7 hours 15 minutes West of Cádiz. The correct longitude difference between both locations is 4 hours 44 minutes. In view of the fraudulent longitudes reported in relation to the September 1494 eclipse, it may well be that Columbus proceeded along similar lines here.

In his July 1503 letter to the Spanish crown, he stated, ...

'When I set out thence to come to Española, the pilots believed that we were going to reach the island of San Juan [Puerto Rico], and it was the land of Mango [western Cuba], four hundred leagues more to the west than they said' [134].

The distance from San Juan Island to Mango is the longest East–West distance Columbus recorded within the Caribbean archipelago. Added to the longest trans-atlantic distance he recorded on his first voyage, 1142¼ leagues (re-determined by Pickering), this yields a distance of 1542¼ leagues from Cádiz, corresponding to a difference of 7 hours 15 minutes in longitude—precisely the difference claimed by Columbus based on his purported observations of the 1504 eclipse [135].

Meanwhile, Amerigo Vespucci (1454–1512) claimed to have made a number of voyages to the New World. Although his claims are disputed by historians, the second and third voyages in 1499–1500 and 1501–1502 most likely happened. Like Columbus, his expedition also practiced 'running down the latitude.' Adopting this approach meant sailing North or South in sight of the coast until the ship reached the desired latitude, then heading East or West.

In a letter written in 1502 (referred to as 'letter IV' by Vespucci scholars) to his patron, Lorenzo di Pierfrancesco de'Medici (1463–1503), Vespucci suggests that, on 23 August 1499, while on his second voyage in the Amazon river delta in Brazil, he determined his longitude based on the lunar distance method: ...

'I maintain that I learnt [my longitude] ... by the eclipses and conjunctions of the Moon with the planets; and I have lost many nights of sleep in reconciling my calculations with the precepts of those sages who have devised the manuals and written of the movements, conjunctions, aspects, and eclipses of the two luminaries and of the wandering stars, such as the wise King Don Alfonso in his Tables, Johannes Regiomontanus in his Almanac, and Blanchinus, and the Rabbi Zacuto in his almanac, which is perpetual; and these were composed in different meridians: King Don Alfonso's book in the meridian of Toledo, and Johannes Regiomontanus' in that of Ferrara, and the other two in that of Salamanca' [136].

Ironically, apparently unaware of Columbus' fraudulently reported longitude measurements, Vespucci most likely plagiarized the latter's five-and-a-half-hour longitude difference, thus perpetuating the longitude fraud [137].

Portuguese and Spanish navigators did not have access to a reliable means of longitude determination at sea in 1495, when Vasco da Gama (c. 1460s–1524), the Portuguese explorer, engaged the celebrated Arab navigator Ahmad ibn Mājid (أحمد بن ماجد; c. 1430–after 1500) at Malindi (in present-day Kenya) to guide him across the Indian Ocean directly to Calicut, in India, the first Indian transoceanic voyage by a European.

Ibn Mājid had developed an approach to navigation on the open seas that could be considered a precursor to longitude determination at sea. In his major navigational treatise, *Kitāb al-fawā'id fī ma'rifat 'ilm al-baḥr wa-al-qawā'id* (كتاب الفوائد في معرفة علم البحر و القواعد.; *Book of Lessons on the Foundation of the Sea and Navigation*), he described three types of sea routes, including mainland routes (always in sight of land), set courses (relying on fixed bearings), and conclusive courses from one specified port to another. He enumerated his 12 practical principles of navigation [138], that is, (1) the lunar mansions, (2) compass rhumb lines, (3) routes, (4) East–West distances along a given parallel, (5) latitudinal variations following the altitude of Polaris, (6) stellar altitude measurements for latitude determination, (7) visible signatures, including water colours and tides, (8) revolutions of the Sun and Moon, (9) winds, (10) seasons, (11) instruments, and (12) captain–crew reactions. In addition, he referred to information a navigator would use for a course set by dead reckoning, including sea snakes, reefs, seabird species, and coastal features, reminiscent of the Polynesian navigation principles we discussed in chapter 2.1.1. Ultimately, ibn Mājid justified his focus on making improvements to marine navigation by pointing out that without directional knowledge, one does not know the direction of the *qibla*, that is, the direction to Mecca, required for Islamic religious practice.

The first generally available European maps that included any part of the American continent appeared from 1506 onwards, when the Venetian cartographer Giovanni Matteo Contarini (d. 1507) joined forces with Francesco Rosselli (1445–before 1513), map engraver and painter of miniatures. Prior to 1506, a few maps had been produced which included recent new discoveries from a number of voyages around the world, including Da Gama's. These included maps by Juan de la Cosa (Juan the Biscayan; c. 1450–1510) from 1500, based on Columbus' second expedition, and the *Cantino* world map from approximately 1502 (see figure 2.21)—but those early maps were considered state secrets.

The year 1507 saw the publication of the limited-edition Waldseemüller map—which was largely based on the Martellus maps and which included the designation 'America' for the first time and showed America separate from the Asian continent—as well as a widely distributed, and therefore highly influential, map by the Dutch cartographer Johannes Ruysch (c. 1460–1533), published as part of the 1507 and 1508 editions of Ptolemy's *Geographia*. In 1513, the Ottoman admiral Ahmed Muhiddin Piri (1465/1470–1553; also known as Pîrî Reis or Hacı Ahmed Muhiddin Pîrî Bey) compiled a highly praised portolan chart of the known world on the basis of eight maps, including '*the map of the western lands drawn by Columbus,*' showing the newly discovered Americas.

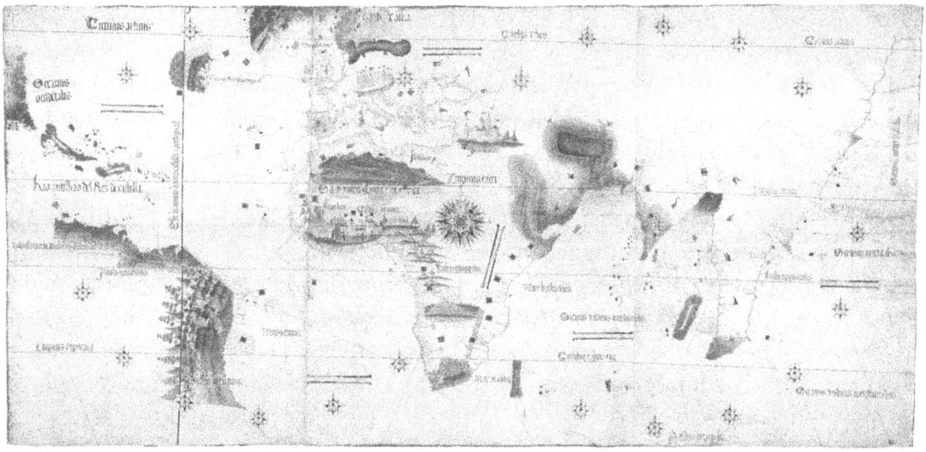

Figure 2.21. The *Cantino* planisphere, c. 1502. (Biblioteca Universitaria Estense, Modena, Italy; public domain.)

2.3.3 Piracy and the Portuguese/Spanish conquest of the New World

As Europe emerged from the Middle Ages and entered the Renaissance, considerable renewed interest in map projections developed, particularly between 1470 and 1669 [139]. Mathematical advances allowed for more innovative and 'realistic' projections. Columbus still largely relied on Ptolemy's maps from Antiquity, which were characterized by erroneous conversions from distances into degrees and vice versa, leading to significantly exaggerated representations of the Mediterranean, in particular. Throughout the transition period and the Renaissance, these problematic renditions were gradually eliminated as improved measurements of terrestrial features and distances became available.

Development of map projections became fashionable, and many schemes were proposed. Many pre-Renaissance maps were based on philosophical rather than mathematical ideas. These included the so-called T–O (*orbis terrarum*) maps (see figure 2.22), which showed the entire known world surrounded by an 'O'-shaped ocean, with the horizontal stroke of the inset 'T' representing the approximate meridian spanning from the Don to the Nile rivers, and its perpendicular stroke the axis of the Mediterranean [140], with the entire 'T' shape separating the known continents of Asia, Europe, and Africa. T–O maps were first described by the seventh Century scholar (Saint) Isidore of Seville (c. 560–636 CE), Archbishop of Seville: ...

> *'The [inhabited] mass of solid land is called round after the roundness of a circle, because it is like a wheel ... Because of this, the Ocean flowing around it is contained in a circular limit, and it is divided in three parts, one part being called Asia, the second Europe, and the third Africa'* [141].

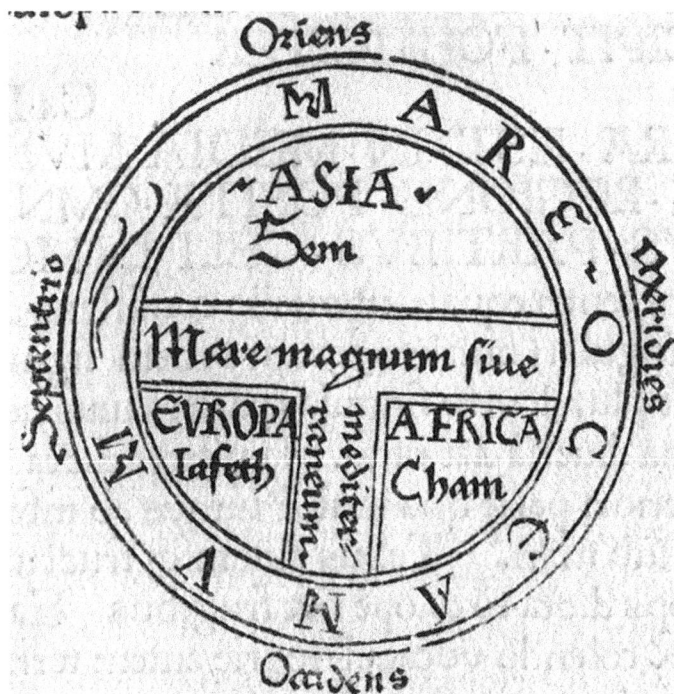

Figure 2.22. T–O style *mappa mundi* from Isidorus' *Etymologiae* (c. 623 CE); the first printed version, by Günther Zainer (d. 1478), appeared in 1472 in Augsburg, Germany. This map identifies the three known continents as populated by descendants of Sem (Shem), Iafeth (Japheth) and Cham (Ham). (Source: Wikimedia Commons; public domain.)

This approach to map making changed dramatically during the transition from the Middle Ages to the Renaissance, particularly after the turn of the 16th Century. The German–Dutch (Flemish) cartographer Gerardus Mercator (1512–1594) became renowned for creating a globe in 1541 and a world map in 1569—see figure 2.23—using a projection which allowed marine navigation along a course of constant bearing by following straight rhumb lines, or loxodromes (see figure 2.17). In order to achieve such a projection on a flat surface, Mercator arranged the meridians and parallels such that a loxodrome intersected the meridians at a constant angle. The main impact of this choice of projection is that East–West distances (and thus also directions and surface areas) are distorted at higher latitudes, because the meridians converge towards the Poles.

Mercator's solution to counteract the effects of this unavoidable distortion was to increase the distances between subsequent parallels in proportion to the increase in the intervals between the meridians with increasing latitude. This adjustment allowed him to retain the correct angular representations and represent land surfaces at the highest achievable accuracy, that is, by employing a *conformal* projection: small areas retain their approximate shapes very well, whereas large areas may be significantly distorted (on current world maps using the Mercator projection, compare the areas covered by Greenland and South America with their actual sizes). In essence,

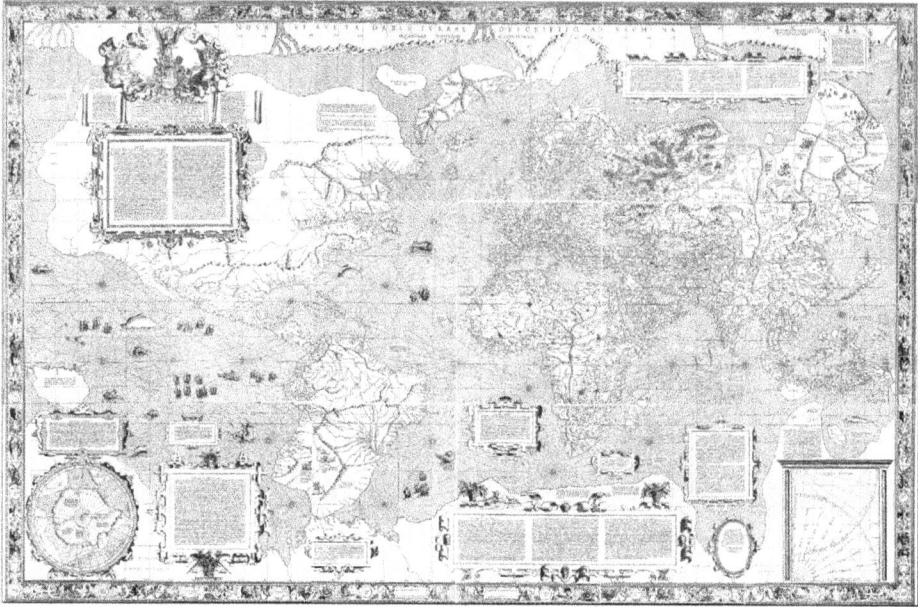

Figure 2.23. Mercator's 1569 world map (*Nova et Aucta Orbis Terrae Descriptio ad Usum Navigantium Emendate Accommodata*) showing latitudes from 66° S to 80° N. The map is composed of 18 separate sheets. (Source: Wikimedia Commons; public domain.)

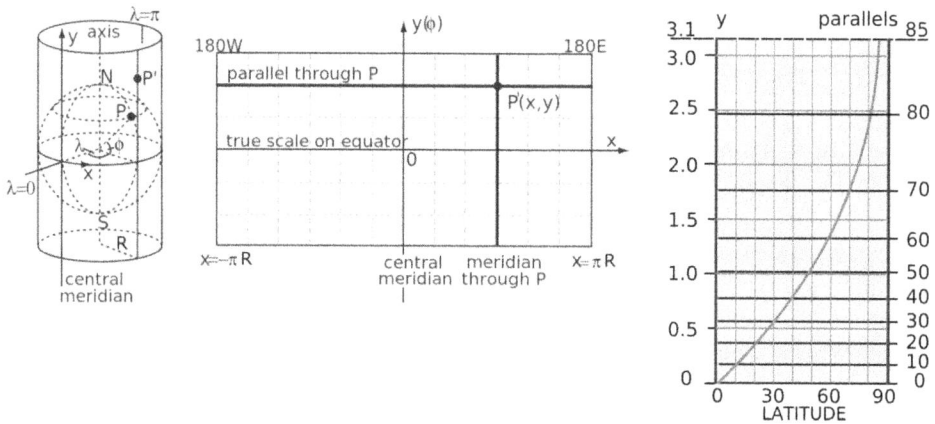

Figure 2.24. (*left and middle*) Geometry of cylindrical map projection. (*right*) Vertical displacement of the latitudinal lines in a Mercator projection. (Credit: Peter Mercator, Wikimedia Commons; reproduction licensed under the Creative Commons Attribution-Share Alike 3.0 Unported license.)

Mercator's projection is what is referred to as a 'regular cylindrical projection,' a projection with equidistant, straight lines of longitude, while the lines of latitude are also straight, parallel to one another, and oriented perpendicularly to the longitudinal lines: see the left and middle panels of figure 2.24 for a graphical impression.

All points on the same meridian, λ, on the globe project one-to-one to the cylindrical surface, x. However, the vertical distance as measured from the Equator depends on the latitude, $y(\varphi)$: see the right-hand panel of figure 2.24. Since the Mercator projection is conformal, by definition, it follows that

$$\frac{\delta y(\varphi)}{\delta \varphi} = \frac{R}{\cos \varphi} = R\sec \varphi,$$

where R is the radius of the cylinder or, equivalently, the radius of the Earth and $\frac{1}{\cos \varphi}$ is the latitudinal scale (stretch) factor. Integration of this differential equation yields

$$y(\varphi) = R \ln\left[\tan\left(\frac{\pi}{4} + \frac{\varphi}{2}\right) \right] = R \ln\left(\frac{1 + \sin \varphi}{\cos \varphi}\right),$$

of which the functional form is shown in the right-hand panel of figure 2.24.

Mercator included a great deal of explanatory text, in Latin, on his 1569 map, particularly in relation to the map's construction: ...

'In making this representation of the world we had three preoccupations.

First, to spread on a plane the surface of the sphere in such a way that the positions of places shall correspond on all sides with each other both in so far as true direction and distance are concerned and as concerns correct longitudes and latitudes; then, that the forms of the parts be retained, so far as is possible, such as they appear on the sphere. With this intention we have had to employ a new proportion and a new arrangement of the meridians with reference to the parallels. Indeed, the forms of the meridians, as used until now by geographers, on account of their curvature and their convergence to each other, are not usable for navigation; besides, at the extremities, they distort the forms and positions of regions so much, on account of the oblique incidence of the meridians to the parallels, that these cannot be recognized nor can the relation of distances be maintained. On the charts of navigators the degrees of longitude, as the various parallels are crossed successively towards the Pole, become gradually greater with reference to their length on the sphere, for they are throughout equal to the degrees on the Equator, whereas the degrees of latitude increase but very little, so that, on these charts also, the shapes of regions are necessarily very seriously stretched and either the longitudes and latitudes or the directions and distances are incorrect; thereby great errors are introduced of which the principal is the following: if three places forming any triangle on the same side of the Equator be entered on the chart and if the central one, for example, be correctly placed with reference to the outer ones as to accurate directions and distances, it is impossible that the outer ones be so with reference to each other. It is for these reasons that we have progressively increased the degrees of latitude towards each Pole in proportion to the lengthening of the parallels with reference to the Equator; thanks to this device we have obtained that, however two, three, or even* *more than three, places be inserted, provided that of these four quantities: difference of longitude, difference of latitude, distance, and direction, any two be observed for each place associated with another, all will be correct in the association of any one place with any other place whatsoever and no trace will anywhere be found of any of those errors which must necessarily be encountered on the ordinary charts of shipmasters, errors of all sorts, particularly in high latitudes. ...'*

Elsewhere, he explains his projection of the polar regions: ...

'As our chart cannot be extended as far as the Pole, for the degrees of latitude would finally attain infinity, and as we yet have a considerable portion at the Pole itself to represent, we have deemed it necessary to repeat here the extremes of our representation and to join thereto the parts remaining to be represented as far as the Pole. We have employed the figure [projection] which is most apt for this part of the world and which would render the positions and aspects of the lands as they are on the sphere.'

In addition, the mapmaker provided detailed explanations as to how to determine distances between two places on the map and how to employ it for marine navigation.

With Mercator's new map projection and the increasing ease of travelling the world, the 'Age of Discovery' had truly arrived. However, sailing the world's oceans was still an adventure which could have fatal consequences in the absence of accurate means to determine one's longitude on the open seas. Development of accurate timing devices that would not be susceptible to a ship's movements became ever more urgent, yet it would take another two centuries before this feat would finally be achieved. That narrative—a tale of human ingenuity, personal sacrifices, clashing personalities, and superb scientific insights—is told in great detail in chapters 3 through 6.

Increases in trade across the world's oceans and the discovery of the New World had unexpectedly created an environment in which accurate longitude determination had become more important than ever before. Trade with the East Indies received a significant boost with the establishment of the British and Dutch East India Companies around the turn of the 17th Century, soon followed by the foundation of the Danish and French East India Companies in 1616 and 1664. (The Portuguese also briefly ran an East India Company, from 1628 to 1633, while the Swedes were late in joining the club, founding their East India Company only in 1731.)

The *East Indiamen*, as their ships were commonly referred to, carried untold riches, e.g., in the form of spices such as nutmeg, mace, pepper, and cloves on their return voyages to European waters. However, without being able to determine their longitude at sea, ships would often sail by seeking the safest course. This implied that they would either be 'running the latitude' or follow predictable, well-established, and predetermined shipping lanes within coastal view to avoid getting lost at sea—thus making them easy prey to either pirates or naval ships sailing under the flags of warring states. Ill-intentioned ships' captains simply had to lie in wait in order to ambush unsuspecting or, at least, poorly armed cargo ships on their return voyages from the East Indies or along the western coast of South America. Equipped with the ability to determine one's longitude at sea, ships would be able to travel more directly and faster to their destinations, which would thus offer both commercial and strategic advantages. Without this ability, however, one's best hope at avoiding being ambushed by pirates or getting lost in unknown waters involved adopting a dead reckoning approach at less popular latitudes.

Navigation in unknown waters was a particularly pressing problem in the Pacific Ocean, a huge body of water that was, in essence, a closed book to European navigators and therefore seen as an obstacle to be crossed as quickly as possible. Spanish activity was greatest in the North Pacific and along the coasts of South and Central America, given that their silver-rich colonies were located there. Maps of these areas were closely guarded state secrets, lest competing nations (or, indeed, pirates!) would get an unfair trading or military advantage: ...

'In this prize I took a Spanish manuscript ... It describes all the ports, roads, harbours, bays, sands, rocks, and rising of the land and instructions how to work the ship into any port or harbour. They were going to throw it overboard but by good luck I saved it—and the Spaniards cried out when I got the book' [142].

In adopting this attitude, the Spanish were not alone. Dutch and Portuguese sailors had spent significant effort to carefully map the western periphery of the Pacific. Their maps, an example of which is shown in figure 2.25, contained valuable commercial information and often also included rudimentary information about the coastlines of Australia and New Zealand, particularly of the territories' western coasts. They were not meant for general circulation; instead, they were for privileged access by East India Company officials only.

In the East Pacific, meanwhile, Spanish sailors learnt that they could follow the tropical easterly winds across the North Pacific to reach the Philippines, an important trading post for the exchange of precious metals from the Americas for Asian products [143]. The return voyage would take them further north, where they could take advantage of the prevailing westerly trade winds.

This mindset of seeing the Pacific as an obstacle that should be dealt with as quickly as possible resulted in continued ignorance of the existence of the Hawaiian island chain; these islands were not 'discovered' by Europeans until January 1778, during Captain James Cook's third and final Pacific voyage (1776–1779); Cook's expeditions were also instrumental in finally proving that an alleged massive 'Southern Continent,' *Terra Australis Incognita*—since Greek Antiquity argued to be required in order to balance the Earth's mass distribution and still included on, e.g., Francesco Rosselli's world map as late as 1508—did not exist at habitable latitudes. Given the predictable courses taken by the richly laden Spanish galleons on their trans-Pacific voyages, they were indeed easy prey to ever-increasing numbers of pirates and buccaneers. This situation, as well as that on the Atlantic Ocean where piracy had turned into a serious problem for the trading nations, soon became untenable.

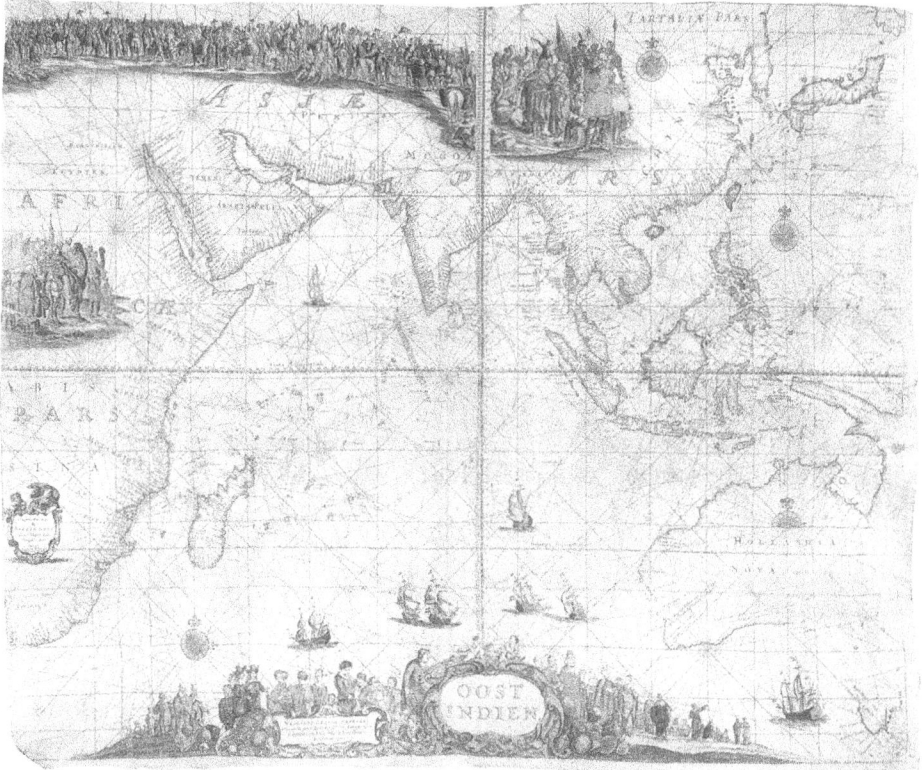

Figure 2.25. Dutch cartographer Pieter Goos' 1660 map of the East Indies, showing the broad outline of Australia ('Hollandia Nova') and the western coast of New Zealand. (Source: Wikimedia Commons; public domain.)

Politically, determination of one's accurate position at sea and in the newly discovered lands became a hot topic in the ongoing rivalry between Spain and Portugal, two major European marine powers in the 15th and 16th Centuries. On 4 September 1479, the Catholic Sovereigns of Castile and Aragon, Ferdinand and Isabella, had signed the *Treaty of Alcáçovas* (also known as the *Peace of Alcáçovas–Toledo*) with King Alfonso V of Portugal and the Algarves (1432–1481), 'the African,' and his son, 'the Perfect' Prince John II of Portugal (1455–1495) to end the War of the Castilian Succession (1475–1479). The Catholic Monarchs had been victorious on land, whereas the Portuguese had established their hegemony at sea [144, 145]. Confirmed on 21 June 1481 with the papal bull *Æterni regis*, issued by Pope Sixtus IV, the treaty granted all lands south of the Canary Islands to Portugal.

However, upon his return from the New World in 1493, Columbus had first arrived in Lisbon, Portugal, where he impressed (by then) King John II with his new discoveries. Mindful of the *Treaty of Alcáçovas*, the Portuguese King announced to the Spanish Sovereigns that all lands discovered by Columbus—who had sailed under Castilian patronage—did, in fact, belong to Portugal. The Catholic Monarchs opted for a diplomatic resolution of this competitive situation, which eventually resulted in the *Treaty of Tordesillas*, signed on 7 June 1494. In the lead-up to the

Treaty's authentication, Pope Alexander VI (Rodrigo Borgia) issued a decree on 4 May 1493 in the form of his *Inter caetera* bull, stating that all newly discovered territories to the West of a Pole-to-Pole meridian 100 leagues West and South of the Azores and Cape Verde Islands would form part of the territory of Castile, with the exception of lands under Catholic rule as of Christmas 1492. Portugal's claims were not included. On 26 September 1493 the pope issued a second bull, *Dudum siquidem*, 'Extension of the Apostolic Grant and Donation of the Indies,' which awarded all mainlands and islands, *'at one time or even yet belonged to India'* to Spain, no matter where they might be located geographically.

In response, and highly disturbed by the explicit reference to India, a major territorial growth area for the Portuguese, the King of Portugal bypassed the pope and opened direct negotiations with representatives of King Ferdinand and Queen Isabella. King John II's aim was that the papal line be moved to the West, and he agreed to adopt the *Inter caetera* bull as the negotiators' starting point. As a result of these negotiations, the meridian line was moved 270 leagues to the West, which meant that eastern Brazil now became part of the Portuguese sphere of influence; ...

> *'both sides must have known that so vague a boundary could not be accurately fixed, and each thought that the other was deceived, [concluding that it was a] diplomatic triumph for Portugal, confirming to the Portuguese not only the true route to India, but most of the South Atlantic'* [146].

Pope Julius II (1443–1523; Giuliano della Rovere), 'the Fearsome Pope,' also known as 'the Warrior Pope,' subsequently sanctioned the updated treaty by means of a bull issued on 24 January 1506, *Ea quae pro bono pacis* [147]. A decade later, and at the insistence of Portugal, which attempted to limit Spain's expansion in Asia, Pope Leo X (1475–1521; Giovanni di Lorenzo de'Medici) pronounced, through his bull *Praecelsae devotionis*, that the line of demarcation applied to the Atlantic Ocean only [148]. (The Protestant marine powers, particularly England and the Dutch Republic, did not recognize the division of the world decreed by a succession of popes, nor did Catholic France [149].)

Note that the original *Treaty of Tordesillas*, as well as the updated version of 1506, identified the demarcation line in terms of leagues West of the Azores and Cape Verde Islands. It remained vague on detail as regards which of the islands should be adopted as point of reference, or exactly which measurement of the league to employ, thus leaving the precise longitude of demarcation, a reference measured in degrees, up in the air. The *Treaty* specified that these issues had to be resolved through a joint voyage, which however never materialized. Instead, at least five different opinions as to the exact location of the line of demarcation surfaced between 1495 and 1524 (see also figure 2.26).

1. At the request of King Ferdinand and Queen Isabella, the cosmographer Jaime Ferrer de Blanes concluded in 1495 that the line was located 18° West of the central island of the Cape Verde archipelago, identified with Fogo. The historian Henry Harrisse (1829–1910) determined that de Blanes' line of demarcation corresponded to a longitude in modern units of 47°37' W, 2276.5 km West of Fogo [150].

2. On the (Portuguese) *Cantino* planisphere of 1502, shown in figure 2.21, the line of demarcation is located between Cape São Roque at the northeastern tip of Brazil and the Amazon river delta. Harrisse located the line at 42°30′ W in modern units [151].

3. In 1518, Martin Fernandez de Enciso (c. 1470–1528), the navigator and geographer from Seville, placed the line of demarcation at a longitude which corresponds to 45°38′ W in modern units, although his descriptions are less than clear. Harrisse concluded that the line identified by De Enciso could also be near the mouth of the Amazon river between 49° and 50° W [152].

4. Finally, two opinions were published in 1524. The Castilian captains Thomas Duran, Sebastian Cabot, and Juan Vespuccius offered their insights to the Badajoz Junta, the authorities charged with settling territorial disputes, although the latter did not manage to reach a clear-cut decision. The ships' captains suggested that the line was located at 22° plus nearly 9 miles West of the centre of the westernmost Cape Verde island, Santo Antão. This corresponds to a modern longitude of 46°36′ W [153]. Independently, Portuguese representatives presented a globe to the Badajoz Junta, showing the line at 21°30′ West of Santo Antão, or 22°6′36″ to the west of the island in modern units [154].

This focus on a demarcation line in the Atlantic Ocean only changed with the Portuguese discovery of the rich Moluccas islands in 1512, which coincide with part of present-day Indonesia. In 1518, Spain insisted that the *Treaty of Tordesillas* divided the Earth into two equal hemispheres, which would favourably place the Moluccas in its own, Western sphere of influence. To resolve this potentially tense conflict situation between both maritime powers, the *Treaty of Vitoria*—agreed

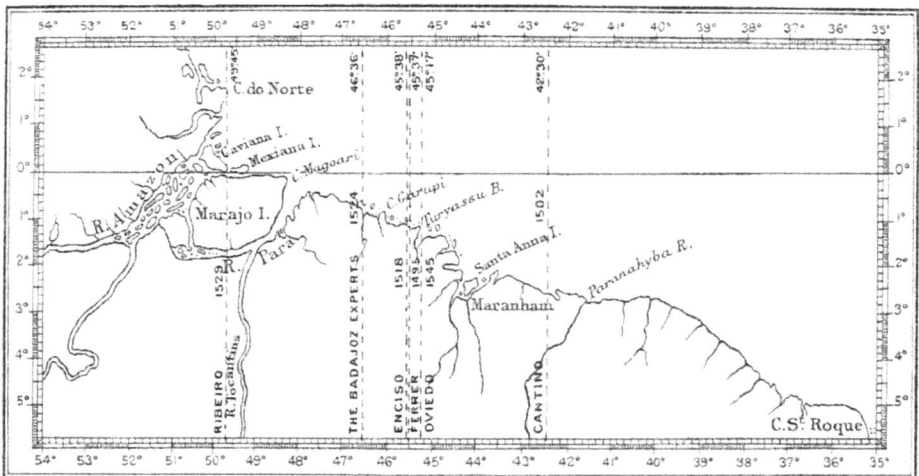

Figure 2.26. Early demarcation lines (1495–1545) associated with the 1494 *Treaty of Tordesillas*. (Source: Harrisse H 1897 *The Diplomatic History of America: Its First chapter* 1452–1493–1494, (London: Stevens), frontispiece.)

upon on 19 February 1524—charged the Badajoz Junta to meet that year to seek a solution to the disagreement on the 'anti-meridian.' That approach to reconciliation failed [155].

Both parties would eventually reconcile, reaching an agreement which was enshrined in the *Treaty of Zaragoza* (also known as the 'Capitulation of Zaragoza') of 22 April 1529. Specifically, Spain would forego its claim to the Moluccas archipelago in return for a payment by Portugal of 350 000 ducats, or approximately 100 kg of gold. The anti-meridian was set at 297.5 leagues or 17° to the East of the Moluccas, with the *Treaty* specifying that the demarcation line passed through the islands of Las Velas (the 'Sails') and Santo Thome [156]. The identity of the latter of these islands is unknown. The Islas de las Velas, on the other hand, can be found in a Spanish history of China from 1585 [157], on the world maps of the Dutch cartographer Petrus Plancius (1594) and of Petro Kærio (1607), and in the 1598 London edition of Jan Huygen van Linschoten's collection of nautical maps. They correspond to a North–South island chain known in the 16th Century as the Islas de los Ladrones ('Islands of the Thieves') [158, 159], later renamed the Mariana Islands. Guam, the southernmost and largest of the Mariana Islands, is located 17°21' to the East of the Moluccas, thus confirming the islands' 17th Century identification.

The *Treaty of Zaragoza* thus implicitly divided the world into unequal hemispheres, awarding territorial claims to Portugal and Spain to sections spanning 191° and 169°. It thus transpired that Portugal got the better deal in terms of the area covered by its sphere of influence, despite the significant remaining uncertainties owing to the wide range of opinions in relation to the location of the *Treaty of Tordesillas* demarcation line. In fact, in the late 19th Century, a dispute about the exact location of the *Tordesillas* line nearly led to a war between the USA and the UK [160]. The disagreement, between Venezuela and the UK, related to the precise boundary of British Guyana in northern South America. Although the issue was eventually settled through diplomacy, it resulted in a major Western review of all methods and measurements adopted by the maritime powers at the time of the *Treaty of Tordesillas*. Accurate longitude determination thus continued to facilitate settlements of major territorial claims.

References and notes

[1] Tsikritsis M 2010 quoted in: Menzies G 2011 *The Lost Empire of Atlantis: History's Greatest Mystery Revealed* (New York: William Morrow) chapter 32; Tsikritsis M 2015 *Astronomy of the Cretomycenaean Culture* (self-published)

[2] *Ibid.*

[3] George M 2012 Polynesian Navigation and Te Lapa—'The Flashing' Time and Mind, *The Journal of Archaeology, Consciousness and Culture* **5** 135–74

[4] *Hawaiian Voyaging Traditions, Star Compasses* http://archive.hokulea.com/ike/hookele/star_compasses.html [accessed 28 April 2017]

[5] George M 2012 *op. cit.*

[6] Gatty H 1958 *Finding Your Way Without Map or Compass* (New York: Dover)

[7] Lewis D 1972 2nd edn 1994 *We, the Navigators: The Ancient Art of Landfinding in the Pacific* (Honolulu, HI: University of Hawaii Press)

[8] In antiquity, the Polynesians were not the only navigators who could read natural signs to find their way across the open ocean. Navigators in the Indian Ocean and the South China Sea also took advantage of the fairly constant monsoon winds to judge their direction, making long voyages viable; Chisholm H (ed) 1911 Navigation *Encyclopædia Britannica* 11th edn p 19

[9] *Ibid.*

[10] *Hawaiian Voyaging Traditions, Holding a Course* http://archive.hokulea.com/ike/hookele/holding_a_course.html [accessed 27 April 2017]

[11] *Traditional Navigation in the Western Pacific: Living Seamarks* (Philadelphia, PA: Penn Museum) www.penn.museum/sites/Navigation/living/living.html [accessed 27 April 2017]

[12] Gatty H 1958 *op. cit.*

[13] Grimble A F and Maude H E (ed) 1989 *Tungaru Traditions: Writings on the Atoll Culture of the Gilbert Islands* (Honolulu, HI: University of Hawaii Press) p 48

[14] *Ibid.* 49–50

[15] Walkup N 2010 Eratosthenes and the Mystery of the Stades *Convergence: Where Mathematics, History, and Teaching Interact* (Math. Assoc. Am.) August; www.maa.org/press/periodicals/convergence/eratosthenes-and-the-mystery-of-the-stades-introduction [accessed 1 May 2017]

[16] Crone G R 1968 *Maps and their Makers* (London: Hutchinson)

[17] Roller D W 2010 *Eratosthenes' Geography* (Princeton, NJ: Princeton University Press)

[18] Berger H 1869 *Die geographischen Fragmente des Hipparch* (Leipzig: B. G. Teubner)

[19] Dicks D R 1960 *The Geographical Fragments of Hipparchus* (London: Athlone Press)

[20] Shcheglov D 2003–2007 Hipparchus' Table of Climata and Ptolemy's Geography *Orbis Terrarum* **9** 159–91

[21] e.g Fraser P M (ed) 1996 *Cities of Alexander the Great* (Oxford: Clarendon Press) pp 97–8 footnotes 42, 43

[22] Ptolemy C 2nd Cent. CE Geographia Book I, Sect. 6; transl. Francis L 1994; http://ota.ox.ac.uk/desc/2422 [accessed 1 May 2017]

[23] Brown L A 1979 *The Story of Maps* (Honolulu, HI: Courier Corporation) p 65

[24] Berggren J L and Jones A 2000 Ptolemy's Geography *An Annotated Translation of the Theoretical Chapters* (Princeton, NJ: Princeton University Press) pp 63–4

[25] Strabo Geographica Book II, chapter 5, §8; Loeb Classical Library edn 1917 **I** 440; https://archive.org/stream/geographyofstrab01strauoft/geographyofstrab01strauoft_djvu.txt [accessed 4 May 2017]

[26] Gaius Plinius Secundus (Pliny the Elder) 77 CE *Naturalis Historia (Natural History)* Book IV, chapter 16

[27] Fage J D 1978 Trans-Saharan contacts and West Africa *The Cambridge History of Africa* vol **2** (from c. 500 BC to AD 1050) p 286

[28] In recent years, statistical analysis has led to the suggestion that these islands may have coincided with the Lesser Antilles, the westernmost islands of the Caribbean; Russo L 2013 *L'America dimenticata. I rapporti tra le civiltà e un errore di Tolomeo* (Milano: Mondadori Università); more conventional scholarship places the zero meridian near El Hierro, one of the Canary Islands.

[29] Wright J K 1923 Notes on the knowledge of latitudes and longitudes in the Middle Ages *Isis* **5** 75–98

[30] Stevenson E L (transl/ed) 1932 (reprint: 1991) *Claudius Ptolemy: The Geography* (New York: New York Public Library) p 33. Unfortunately, this translation (the only complete version in English) contains numerous mistakes: Diller A 1935 Review of Stevenson's translation *Isis* **22** 533–9

[31] Walkup N 2010 *op. cit.*

[32] Dilke O A W (transl) 1987 Cartography in the Byzantine Empire *History of Cartography*, vol **I**, chapter 11 p 183

[33] Vat. Gr. 177 (Rome: Biblioteca Apostolica Vaticana)

[34] Dilke O A W 1987 *op. cit.* chapter 15 pp 258–75

[35] Long G 1842 Agathodæmon *The Biographical Dictionary of the Society for the Diffusion of Useful Knowledge* vol **I** part II (London: Longman, Brown, Green, & Longmans) p 443

[36] *Ibid.*

[37] Clemens R 2008 Medieval Maps in a Renaissance Context: Gregorio Dati *Cartography in Antiquity and the Middle Ages: Fresh Perspectives, New Methods* ed R J A Talbert and R W Unger (Leiden: Koninklijke Brill NV) pp 237–56

[38] Dilke O A W 1987 *op. cit.* chapter 15 258–75

[39] Diller A 1940 The oldest manuscripts of ptolemaic maps *Trans. Am. Philolog. Assoc* **71** 6267

[40] Russo L 2012 Ptolemy's longitudes and Eratosthenes' measurement of the Earth's circumference *Math. Mech. Complex Syst* **1** 67–79

[41] Russo L 2013 *op. cit.*

[42] Russo L 2012 *op. cit.*

[43] Snyder J P 1993 *Flattening the Earth: Two Thousand Years of Map Projections* (Chicago, IL: University of Chicago Press) pp 5–8

[44] *Geographica* Book II, chapter 5 §10 Loeb Classical Library edn 1917 vol. **I**, pp 449–51

[45] Stevenson E L 1932 *op. cit.*, 40, 42–3 (transl)

[46] Weisstein E W 2017 *Conic Projection* Mathworld (Champaign IL: Wolfram) http://mathworld.wolfram.com/ConicProjection.html [accessed 4 May 2017]

[47] d'Avezac-Macaya M-A 1863 Coup d'oeil historique sur la projection des cartes de géographie *Bull. Soc. Géogr.* **V** 483

[48] Stevenson E L 1932 *op. cit.* 43

[49] d'Avezac-Macaya M-A 1863 *op. cit.* 282–3

[50] Stevenson E L 1932 *op. cit.* 45

[51] Snyder J P 1993 *op. cit.* 10–4

[52] Tibbetts G R 1992 The Beginnings of a Cartographic *Tradition, Cartography in the Traditional Islamic and South Asian Societies* ed J B Harley and D Woodward (Chicago, IL: University of Chicago Press) pp 90–107

[53] Edson E and Savage-Smith E 2004 *Medieval Views of the Cosmos: Picturing the Universe in the Christian and Islamic Middle Ages* (Oxford: Bodleian Library, University of Oxford) pp 61–3

[54] Rapoport Y and Savage-Smith E 2008 The Book of Curiosities and a Unique Map of the World *Cartography in Antiquity and the Middle Ages: Fresh Perspectives, New Methods* ed R J A Talbert and R W Unger (Leiden: Koninklijke Brill NV) pp 121–38

[55] *Ibid.* 127

[56] al-Masʿūdī 1894 Kitāb al-Tanbīh wa-al-ishrāf *Bibliotheca Geographorum Arabicorum* 8 (Leiden: Koninklijke Brill NV)

[57] Douglas A V 1973 Al-Biruni, Persian Scholar, 973–1048 *J. R. Astron. Soc. Canada* **67** 209–11

[58] Sachau E C 1888 *Alberuni's India: An Account of the Religion, Philosophy, Literature, Geography, Chronology, Astronomy, Customs, Laws and Astrology of India about A.D. 1030* (London: Kegan Paul, Trench, Trubner & Co. Ltd) p 277

[59] Rapoport Y and Savage-Smith E 2008 *op. cit.* 126–7

[60] *Ibid.* 129

[61] Scott S P 1904 *History of the Moorish Empire in Europe* vol **I** (Philadelphia & London: J. B. Lippincott Co) pp 461–2

[62] Schwartzberg J E 1987 Introduction to South Asian Cartography *History of Cartography*, **II**, Book 1, chapter 15 p 314

[63] *Ibid.* 315

[64] *Ibid.* 323–4

[65] Habib I 1977 Cartography in Mughal India *Medieval India, a Miscellany* 4 (Bombay: Asia Publ. House) 122–34 (also published in Habib I 1979 *Indian Archives* **28** 88–105)

[66] *Ibid.* 128–9

[67] Wilford F 1805 *An Essay on the Sacred Isles in the West with Other Essays Connected with that Work*, 271–2; Note 7

[68] Needham J 1986 *Science and Civilization in China* vol **3** Mathematics and the Sciences of the Heavens and the Earth (Taipei: Caves Books Ltd) pp 106–7

[69] Nelson H 1974 Chinese maps: an exhibition at the British Library *China Q* **No 58** 357–62

[70] de Crespigny R 2007 *Zhang Heng* 張衡 *A Biographical Dictionary of Later Han to the Three Kingdoms (AD 23–220)* (Leiden: Brill) pp 1049–51

[71] Temple R 1989 *The genius of China: 3,000 Years of Science, Discovery, and Invention* (New York: Inner Traditions) p 30

[72] Needham J 1986 *op. cit.* 538–40

[73] Harvey P D A 1980 *The History of Topographical Maps* (London: Thames & Hudson) 133–4

[74] Qu A-J 2000 Why Interpolation? *Historical Perspectives on East Asian Science, Technology and Medicine* ed A K L Chan, G K Clancey and H-C Loy (Singapore: Singapore University Press) pp 336–44

[75] Wang T P 2004 Zheng He, Wang Dayuan and Zheng Yijun: Some New Insights *Asian Culture* **28** 54–62

[76] Morton W S and Lewis C M 2005 *China, Its History and Culture* (New York: McGraw Hill) p 128

[77] *Yoktae chewang honil kangnid, the 'Kangnido'* www.myoldmaps.com/late-medieval-maps-1300/236-yoktae-chewang-honil/236-kandingo.pdf [accessed 10 May 2017]

[78] Ledyard G 1994 Cartography in Korea *History of Cartography* **II** Book 2 chapter 10 p 310

[79] Unno K 1994 Cartography in Japan *History of Cartography* **II** Book 2, chapter 11 pp 414–5

[80] Wang T P 2004 *op. cit.*; Morton W S and Lewis C M 2005 *op. cit.*

[81] Gong Z 2005 *Xiyang Banguo Zhi (Notes on Barbarian countries in the Western seas)* (Beijing: Zhonghua)

[82] *Ibid.*

[83] *Charts of Zheng He's voyages*, also known as the *Mao Kun Map* (鄭和航海圖)

[84] Wang T P 2004 The Charts of Zheng He's Voyages, *Zhongguo Kexue Jishi Shi, Jia Tong Quan (The History of Chinese Science and Technology, Transportation)* ed Xi Fei Long Yangxi, Tang Xiren (Beijing: Science Publisher Beijing) pp 395–7

[85] *Ibid.* 396

[86] Richards E G 1998 *Mapping Time: The Calendar and its History* (Oxford: Oxford University Press) p 52

[87] Han Z H 2002 *Hanhai Jiaotong Maoyi Yanjiu (A Study of Contacts by Navigation and Trade)* (Hong Kong: University of Hong Kong Press) pp 248–9

[88] Hostetler L 2013 Early modern mapping at the Qing court: survey maps from the Kangxi, Yongzheng, and Qianlong reign periods, *Chinese History in Geographical Perspective* ed J Kyong-McClain and Y-T Du (Lanham, MA: Lexington Books) pp 15–32

[89] Lane F C 1963 The economic meaning of the invention of the compass *Am. Hist. Rev.* **68** 615ff

[90] Nicolai R 2016 A critical review of the hypothesis of a medieval origin for portolan charts *PhD Thesis* Utrecht University

[91] *Ibid.*

[92] Taylor E G R 1951 The south-pointing needle *Imago Mundi* **8** 1–7

[93] Schmidl P G 1996–1997 Two early Arabic sources on the magnetic compass *J. Arabic Islamic Stud* **1** 81–132

[94] Guarnieri M 2014 Once upon a time, the compass *IEEE Industr. Electron. Mag* **8** 60–3

[95] Kreutz B M 1973 Mediterranean contributions to the medieval mariner's compass *Technol. Cult* **14** 367–83

[96] Dampier W 1927 *A New Voyage Round the World* 2007 edn (New York: 1500 Books) pp 123–4

[97] Beazley C 1904 The first true maps *Nature* **71** 159–61

[98] Crone G R 1968 *op. cit.* 30

[99] Bagrow L 2010 *History of Cartography* 2nd edn (Piscataway, NJ: Transaction) p 65ff

[100] Nordenskiöld A E 1897 *Periplus: An Essay on the Early History of Charts and Sailing-Directions* (Stockholm: P. A. Norstedt)

[101] Nicolai R 2016 *The Enigma of the Origin of Portolan Charts: a Geodetic Analysis of the Hypothesis of a Medieval Origin* (Leiden: Brill) chapters 5.6.7–5.6.8

[102] Steinberg P E 2010 Portolan Charts *Encyclopedia of Geography* ed B Warf (Thousand Oaks, CA: Sage) 911 https://philsteinberg.files.wordpress.com/2012/05/portolan-charts-encyclopedia-of-geography.pdf [accessed 11 May 2017]

[103] The frequent use of windrose lines on 17th Century world maps is, therefore, most likely purely decorative and not meant for navigational purposes; Steinberg P E 2010 *op. cit.*

[104] Erikson L and Sephton J (transl), before 1265 (transl 1880) *The Saga of Erik the Red (Eiríks saga rauða)*

[105] Wigal D 2007 *Historic Maritime Maps* (New York: Parkstone Press) p 22

[106] Crino S 1930 Schizzi cartografici inediti dei primi anni della scoperta dell'America *Rivista Marittima* **LXIV** (9) Supplemento 48 (figure 18)

[107] Jane C (ed) 1988 *The Four Voyages of Columbus* vol **II** (New York: Dover) p 84

[108] Morison S E 1942 *Admiral of the Ocean Sea: The Life of Christopher Columbus* (Boston: Little, Brown & Co.)

[109] Phillips W D Jr and Phillips C R 1992 *The worlds of Christopher Columbus* (Cambridge: Cambridge University Press)

[110] Vignaud H 1920 *The Columbian tradition on the discovery of America and the part played therein by the astronomer Toscanelli* (Oxford: Oxford University Press) p 41

[111] Brinkbäumer K and Höges C 2006 *The Voyage of the Vizcaína: The Mystery of Christopher Columbus's Last Ship* (San Diego, CA: Harcourt) p 74

[112] Wang T P 2006 *The papacy and ancient China*; www.gavinmenzies.net/wp-content/uploads/ 2011/08/thepapacyandancientchina.pdf [accessed 16 May 2017]

[113] Vignaud H 1902 *Toscanelli and Columbus: The letter and chart of Toscanelli* (London: Sands & Co) p 294

[114] Brinkbäumer K and Höges C 2006 *op. cit.* 74–5

[115] Wiener L 2012 *Africa and the Discovery of America* vol **1** (Outlook Verlagsgesellschaft mbH) 2

[116] de Lollis C 1892–1896 *Raccolta de documenti e studi pubblicati dell Real Commissone Columbina* part I vol **II** p 22

[117] Dunn O and Kelley J E Jr 1989 *The Diario of Christopher Columbus's First Voyage to America, 1492–1493* (Norman, OK: University Oklahoma Press) pp 130–1. *Diario* entry of 2 November 1492, when Columbus' fleet was based off the coast of Cuba. Note that the actual distance sailed was approximately 1230 leagues and the distance as measured on a chart would have been only 1100 leagues (one Italian league is equivalent to 2.67 nautical miles and one Portuguese league covers 3.2 nautical miles). See also Pickering K A 1997 *The Navigational Mysteries and Fraudulent Longitudes of Christopher Columbus*; http:// columbuslandfall.com/ccnav/shd973.shtml [accessed 15 May 2017]

[118] Murphy P J and Coye R W 2013 *Mutiny and Its Bounty: Leadership Lessons from the Age of Discovery* (New Haven, CT: Yale University Press)

[119] Keith Pickering explored Columbus' longitude 'fraud' in detail (Pickering K A 1997 *op. cit.*). Here, I follow his reasoning and I intend to give full credit to his insights.

[120] Hunter D 2011 *The Race to the New World: Christopher Columbus, John Cabot, and a Lost History of Discovery* (New York: St Martin's Press) p 13

[121] Antonio Gallo's (d. c. 1510), chronicler from Genoa, in his 1497 biography of Columbus, translated by Thacher J B 1903 *Christopher Columbus: His Life, His Work, His Remains* (New York and London: G. P. Putnam & Sons) pp 192, 195

[122] Colon F and Keen B 1992 *The Life of the Admiral Christopher Columbus by His Son Ferdinand* (New Brunswick, NJ: Rutgers University Press) p 48

[123] West D C and Kling A 1991 *The Libro de las profecías of Christopher Columbus* (Gainesville, FL: University of Florida Press) pp 226–7

[124] Jane C 1988 *op. cit.* vol II p 82

[125] de las Casas B 1951 *Historia de las Indias* vol **I** (Mexico: Fondo de Cultura Economica) p 390 (transl. Pickering K A)

[126] *Ibid.* 395–396

[127] Müller J 1489 *Kalendar Maister Johannes Kunisperger* (Augsperg: Erhart Radolt) see Pickering K A 1997 *op. cit.*

[128] Morison S E 1942 *op. cit.* vol II p 147

[129] Nader H and Formisano L 1996 *The Book of Privileges Issued to Christopher Columbus by King Fernando and Queen Isabel* vol **2** (Repertorium Columbianum) (Berkeley, CA: University of California Press) pp 73–4 87, 151, 153

[130] Morison S E 1942 *op. cit.* vol **II** pp 140–1

[131] Jane C 1988 *op. cit.* vol **II** p 84

[132] de las Casas B 1951 *op. cit.* vol I p 497

[133] Pickering K A 1997 *op. cit.*

[134] Jane C 1988 *op. cit.* vol **II** p 98

[135] For a different explanation, involving Columbus having made a series of mistakes, see Olson D W October 1992 Columbus and an eclipse of the Moon *Sky & Telescope* **82** 437–40

[136] Formisano L (ed) 1992 *Letters From a New World* (New York: Marsilio Publ) 38–9

[137] Randles W G L 1995 Portuguese and Spanish attempts to measure longitude in the sixteenth century *The Mariner's Mirror* **81** 403

[138] Glick T F 2005 Ibn Majid, Ahmad *Medieval Science, Technology, and Medicine: An Encyclopedia* ed T Glick, S J Livesey and F Wallis (New York and London: Routledge) p 252

[139] Snyder J P 1993 *op. cit.* 43

[140] Crone G R 1968 *op. cit.* 26

[141] Isidore of Seville c. 630 De terra et partibus *Etymologiae* chapter 14

[142] Captain Bartholomew Sharp, reporting on having captured the *Santa Rosario* off the South American west coast, July 1681; Rutherford-Moore R 2007 *The Pirate Round: Early Eighteenth Century Maritime Navigation during the Golden Age of Piracy* (Westminster, MD: Heritage Books) p 18

[143] Baker K 2017 Going global—16th-Century-style China and the Manila galleons trade *J. R. Asiat. Soc. China* **77** 56–67

[144] Bailey W D and Winius G D 1985 In a war in which the Castilians were victorious on land and the Portuguese at sea *Foundations of the Portuguese Empire 1415–1580* vol **I** (Minneapolis MN: University of Minnesota Press) p 152

[145] Newitt M 2005 *A History of Portuguese Overseas Expansion, 1400–1668* (New York: Routledge) pp 39–40

[146] Parry J H 1973 *The Age of Reconnaissance: Discovery, Exploration, and Settlement, 1450–1650* (London: Cardinal) p 194

[147] Davenport F G (ed) 1917/1967, *European Treaties bearing on the History of the United States and its Dependencies to 1648* (Washington DC: Carnegie Institute of Washington) pp 107–11

[148] Parry J H 1973 *op. cit.* 202

[149] *Ibid.* 205

[150] Harrisse H 1897 *The Diplomatic History of America: Its first chapter 1452–1493–1494* (London: Stevens) pp 91–7, 178–90

[151] *Ibid.* 100–2, 190–2

[152] *Ibid.* 103–8, 122, 192–200

[153] *Ibid.* 138–9, 207–8

[154] *Ibid.* 207–8

[155] Blair E H (ed) 1903 *The Philippine Islands, 1493–1803* vol **II** (Cleveland OH: The A. H. Clark Co.)

[156] *Ibid.* vol **III**

[157] González de Mendoza J. (Fr.) 1585 *Historia de las cosas más notables, ritos y costumbres del gran Reino de la China (History of the most remarkable things, rites and customs of the great Kingdom of China)* (Rome: Grassi)

[158] Cortesao A 1939 Antonio Pereira and his map of circa 1545 *Geogr. Rev.* **29** 205–25

[159] Clark J O E (ed) 2005 *100 Maps* (New York: Sterling) p 115

[160] Davidson M H 1997 *Columbus Then and Now: A Life Reexamined* (Norman, OK): University of Oklahoma Press) p 69

Chapter 3

Early insights inspired by Galileo Galilei

3.1 Galileo's influence

3.1.1 Aristotelian physics gives way to a quantitative, mathematical approach

By the late-16th Century, our understanding of the natural world had undergone a paradigm shift. Until that time, the leading European scholars and thought leaders had closely adhered to the centuries-old model of Aristotelian physics. In his treatise *Physics* (Φυσικὴ ἀκρόασις), one of the foundations of Western science and philosophy [1], the Greek philosopher Aristotle (384–322 BCE) had attempted to establish general principles of change (including motion) which apply to all natural bodies—bodies composed of a mixture of the 'natural elements,' earth, water, air, and fire. Terrestrial bodies would rise or fall, depending on the ratio of the four natural elements they were composed of. The heaviest bodies, those containing earth or water, would fall to their 'natural' place, the centre of the Universe.

Although Aristotle's ideas were consistent with human experience in everyday life, they were verbal–logical and qualitative in nature and not based on carefully controlled experiments or measurements. While attending the University of Pisa from the young age of 17, the Italian mathematician and astronomer Galileo di Vincenzo Bonaiuti de'Galilei (1564–1642) started to openly question and criticize Aristotelian theory. For instance, he challenged the Aristotelian idea of the incorruptibly perfect Moon, both from an observational perspective—once he had developed his first telescope, he could clearly see craters and mountains on the Moon's surface—and prior to that also theoretically. If the Moon were perfectly smooth, the edges of its disc would appear fainter than its projected centre owing to the geometry of reflected sunlight. On the other hand, if the Moon's surface were uneven, light would be reflected approximately equally in any given direction, leading to the observed disc of approximately equal brightness everywhere [2].

Galileo, who is usually simply referred to by his first name, had inherited his critical thinking abilities from his father, Vincenzo Galilei (c. 1520–1591), a talented musician, composer, and music theorist with a skeptical view of authority. As a case in point, Galileo's father is said to have exclaimed,

'It appears to me that those who rely simply on the weight of authority to prove any assertion, without searching out the arguments to support it, act absurdly. I wish to question freely and to answer freely without any sort of adulation. That well becomes any who are sincere in the search for truth' [3].

Following in his father's footsteps, Galileo developed a critical mind, basing his understanding of the natural world on an approach that has since become known as the 'scientific method.' As present-day scientists, we interpret this concept as a common approach involving the postulation of hypotheses, subsequent experimental measurements aimed at validating or rejecting these hypotheses, mathematical assessment, and reproducibility. Most historians of science credit Galileo as one of the main drivers of this scientific paradigm shift, wholeheartedly agreeing with the British theoretical physicist Stephen Hawking's assessment that …

'Galileo, perhaps more than any other single person, was responsible for the birth of modern science' [4].

Many readers will be familiar with a famous experiment, attributed to Galileo, in which he is said to have dropped two spheres of different masses—a cannon ball and a musket ball—from the top of the leaning Tower of Pisa to demonstrate that they would hit the ground at the same time. The experiment is supposed to have taken place in 1589, according to a biography of Galileo compiled and published in 1659 by Vincenzo Viviani (1622–1703) [5], his assistant and disciple. However, historical discrepancies call into question whether it actually happened as described. Most importantly, in 1589 Galileo had not yet fully developed the final version of his *law of free fall*, which would only see the light shortly after he had first formulated the *law of the pendulum*. The law of free fall is a direct consequence of the basic physical principle —unknown at the time—that a system's total energy is always conserved, which here refers to the combination of gravitational (or 'potential') and kinetic energy:

In the absence of air resistance and irrespective of their direction of motion (that is, in all non-moving, 'inertial' frames), all bodies fall with the same constant, downward-pointing gravitational acceleration, $g = 9.8 \ m \ s^{-2}$.

Nevertheless, Galileo had already formulated an earlier version of his law of free fall, which stated that bodies composed *of the same material* would fall at the same speed through the same medium [6].

In addition, there is no surviving evidence provided by Galileo himself of him having performed the leaning Tower of Pisa experiment; most historians consider it a so-called 'thought experiment.' In particular, the leading philosopher of science, James Robert Brown, has called it *'the most beautiful thought experiment ever devised'* [7], because it offers multiple novel insights [8]: the experiment is surprising, its conclusion cannot be derived from first principles, and it provides a new experience (a 'mental perception'), which demonstrates a new relation between beliefs. In fact, Galileo's conclusion was at odds with the accepted norms based on Aristotelian physics, which taught that objects fall in direct proportion to their masses, so that heavier objects would attain greater speeds than lower-mass bodies by the time they hit the ground—and they consequently would get there earlier. Galileo's postulated new relation [9] was a new law of nature, a novel insight that we now *believe* (emphasis Stuart's), and we have a new justification ('rational intuition') for that belief.

3.1.2 Galileo's pendulum

This challenge to the commonly accepted Aristotelian theory sets Galileo apart as an innovative thinker, who did not accept the contemporary worldview at face value. Instead, Galileo was a keen observer of the natural world. His first biographer, Viviani, stated that Galileo became interested in the periodic motion of a swinging chandelier in the Cathedral of Pisa—presumably when he was distracted during Mass—around 1581 or perhaps in 1582, still during his student days. (Other narratives suggest [10] that Galileo's observations related to the swinging motions of an oil lamp suspended from the Cathedral's nave ceiling, pushed back and forth by drafts.) He is said to have timed the *period* of the chandelier's swing—the time for one complete arc from the point of origin through its maximum swing on the opposite side and back again—with his pulse beats. This enabled him to discover the crucial aspect of pendulums[1] that make them viable time pieces, a property now commonly referred to as 'isochronism:' a pendulum's period is, at least for small angles of its swing, approximately independent of its *amplitude*, that is, the maximum extent or angle of the swing. From at least 1602, he referred to this isochronism (which is also known as *tautochronism*) as an admirable property of pendulums, although he never managed to demonstrate it conclusively. Galileo did not use the word isochronism himself; instead, he generally wrote about pendulums *'going under the same time'* [11]. In fact, while looking back at his lifetime of achievements in physics, in 1632 Galileo wrote a letter in which he paid tribute to the simple pendulum for allowing him to measure and formulate his law of free fall, stating that …

> *'we shall obtain from the marvellous property of the pendulum, which is that it makes all its vibrations, large or small, in equal times'* [12].

Galileo also deduced that a pendulum's period, *P*, does not depend on the mass of the 'bob' at its end (for his experiments, he used balls of cork and lead, most likely 1–2 ounce musket balls [13]), but (for small swing angles) only on the square root of its length, *L*, that is,

---

[1] The word *pendulum* was first used in the 17th Century. It has its etymological origin in the Modern Latin adjective *pendulus* (c. 1660), 'hanging down,' used as a noun, from *pendēre*, 'to hang.'

$$P = 2\pi\sqrt{\frac{L}{g}}\,,$$

where g is the gravitational acceleration. The key advantage of this latter property is that a pendulum's driving force, which is difficult to regulate precisely, could vary somewhat without seriously affecting its period. Galileo's first notes on the subject of pendulums as potential timekeeping devices date back to 1588. In those notes—for a treatise on motion—he refers in passing to the periodic motion of a pendulum, which was considered a form of 'natural motion' (a commonly used description of spontaneous descent or free fall) and which had been overlooked by contemporary natural philosophers.

At the time, Galileo's father was engaged in studying tensions in strings used for musical instruments, combining his prowess in music theory with practical experiments. Music theory, laid down in antiquity, was in essence a mathematical discussion of harmony, that is, an exploration of the ratios of the lengths of strings that produce consonances. Galilei Senior's most important contribution to the field, based on experimentation rather than pure theory, consisted of demonstrating that the ratio of the tensions of strings tuned one octave apart was 4 : 1 instead of the 2 : 1 ratio music theorists had always assumed. He used strings for his experiments that were either suspended freely or hung over a monochord's base. This required him to attach weights to their ends. In turn, this would result in swinging motions. At the time the elder Galilei carried out his experiments, around 1588 [14], Galileo lived with his parents while providing for his family by engaging in private mathematics tuition. It is therefore indeed possible that Galileo was inspired by his father's experimental work in pursuing his own explorations of pendular motions [15]. In any case, during Galileo's formative years, his father was instrumental in instilling in the young scientist an appreciation of how experimentation may inform theoretical developments.

Galileo's interest in the use of pendulums as timekeepers became more academic around the beginning of the first decade of the 17th Century. On 29 November 1602, he described the concept of isochronism in a letter to his early patron, the Marquis Guido Ubaldo dal Monte (1545–1607), from Padua [16]. Although this represents Galileo's earliest surviving letter on the subject of isochronism, in *The Life of Galileo Galilei* [17], John Elliot Drinkwater Bethune (1801–1851) points out that Galileo apologizes for calling his patron's attention *again* (emphasis mine) to the isochronism of the pendulum, which Dal Monte had apparently rejected as false and impossible in their earlier correspondence:

'You must excuse my importunity if I persist in trying to persuade you of the truth of the proposition that motions within the same quarter-circle are made in equal times. For this having always appeared to me remarkable, it now seems even more remarkable that you have come to regard it as false. Hence I should deem it a great error and fault in myself if I should permit this to be repudiated by your theory as something false; for it does not deserve this censure, nor yet to be banished from your mind—which better than any other will be able to keep it

more readily from exile by the minds of others. And since the experience by which the truth has been made clear to me is so certain, however confusedly it may have been explained in my other letter, I shall repeat this more clearly so that you, too, by making this experiment, may be assured of this truth.'

This debate is more important than it may seem at first glance. As fate would have it, Dal Monte became one of Galileo's most significant opponents, despite his influential early role as the young scientist's patron. Dal Monte was a firm believer in the prevailing Aristotelian worldview, which Galileo of course challenged through his precise physical measurements. On a small scale, this debate reflects the larger struggle between Aristotelianism and the new scientific method or, in other words, the legitimacy of idealization in science versus the use of mathematical constructs to understand the natural world.

In his letter of 1602, Galileo tried to convince his patron by providing a carefully constructed mathematical proof:

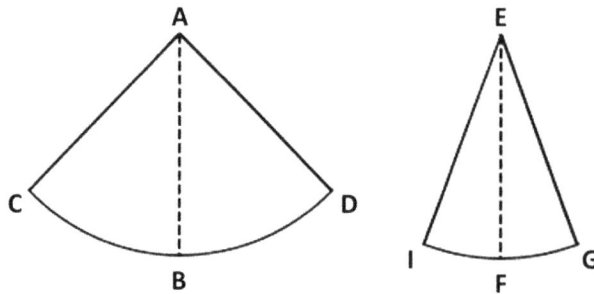

*'Therefore, take two slender threads of equal length, each being two or three braccia long; let these be **AB** and **EF**. Hang **A** and **E** from two nails, and at the other ends tie two equal balls (though it makes no difference if they are unequal). Then, moving both threads from the vertical, one of them very much as through the arc **CB**, and the other very little as through the arc **IF**, set them free at the same moment of time. One will begin to describe large arcs like **BCD**, while the other describes small ones like **FIG**. I am made quite certain of this as follows. The moveable [body] **B** passes through the large arc **BCD** and returns by the same **DCB** and then goes back towards **D**, and it goes 500 or 1000 times repeating its oscillations. The other goes likewise from **F** to **G** and then returns to **F**, and will similarly make many oscillations; and in the time that I count, say, the first 100 large oscillations **BCD**, **DCB**, and so on, another observer counts 100 of the other oscillations through **FIG**, very small, and he does not count even one more—a most evident sign that one of these large arcs **BCD** consumes as much time as each of the small ones **FIG**. Now, if all **BCD** is passed in as much time [as that] in which **FIG** [is passed], though [FIG is] but one-half thereof, these being descents through unequal arcs of the same quadrant, they will be made in equal times. But even without troubling to count many, you will see that*

*moveable **F** will not make its small oscillations more frequently than **B** makes its larger ones; they will always go together.'*

In hindsight, and armed with our present knowledge of the physics of simple 'harmonic oscillators'—systems that, when disturbed from their lowest-energy equilibrium position, are subject to a restoring force, $\vec{F}$, which is directly proportional to the initial displacement, $\vec{x}$, from equilibrium, that is, $\vec{F} = -k\vec{x}$, where k is a positive constant; for springs, this expression is equivalent to *Hooke's law* [2], where k is the spring's stiffness—it is clear that Dal Monte was justified in at least some of his skepticism, but for other reasons than he thought at the time. Dal Monte's objections centred on his assertion that Galileo was an excellent mathematician but a very poor physicist. He contended that mathematics, as an abstract science, did not have any bearing on the real world, unlike physics ('natural philosophy'), and so he was unshaken in his belief that real pendulums do not behave as Galileo had worked out mathematically:

> *'Thus, there are found some keen mathematicians of our time who assert that mechanics may be considered either mathematically, removed [from physical considerations] or else physically. As if, at any time, mechanics could be considered apart from either geometrical demonstrations or actual motion! Surely when that distinction is made, it seems to me (to deal gently with them) that all they accomplish by putting themselves forth alternately as physicists and as mathematicians is simply that they fall between stools, as the saying goes. For mechanics can no longer be called mechanics when it is abstracted and separated from machines'* [18].

The real reason for some of the tension between Galileo's mathematical proof and the behaviour of pendulums in the real world is that pendulums go out of sync rather quickly. Paolo Palmieri has attempted to reproduce Galileo's experiments with pendulum lengths of four to five 'braccia'—corresponding to two to three metres, depending on one's choice of the equivalent of Galileo's braccio—and found that he could barely observe more than one hundred good-quality oscillations; he defines an oscillation as a full swing through the maximum amplitude on the other side and back to its origin. On the other hand, Galileo himself apparently claimed to have counted hundreds of oscillations or 'vibrations,' words he used interchangeably. Significant fine-tuning would have been required for Galileo to achieve the results he describes, even if he meant half-oscillations instead of full swings back and forth, which is not quite clear from his letters. Indeed, Dal Monte had also noticed in his own experiments that real pairs of pendulums go out of sync within a few dozen oscillations, irrespective even of the notion that isochronicity would, in essence, imply that pendulums were *perpetual mobiles*, which they are clearly not.

---

[2] *Hooke's law* is named after Robert Hooke, who first stated it in 1660 as a Latin anagram, rearranged alphabetically, *ceiiinosssttuv*, to which he published the solution in 1678, i.e., *Ut tensio, sic vis*: 'As the extension, so the force.'

Palmieri carefully assesses the historical evidence, as well as his predecessors' interpretations of Galileo's surviving manuscripts, and proceeds to cast doubt on whether Galileo actually performed the experiments he described instead of having imagined them as thought experiments [20,21]. He points out that Galileo seems to suggest that direct observation of the isochronism of pendulums is cumbersome, but numbering the oscillations is more secure. Indeed, Galileo clearly understood that counting was the most robust approach to establishing whether or not his pendulums were isochronous [22]—although Palmieri, in turn, demonstrates that counting oscillations has its own, sometimes serious, problems [23]. However, in his letter of 1602 to Dal Monte, Galileo's attitude is surprisingly careless in suggesting that even without numbering the vibrations, one could easily confirm a pendulum's isochronism by simply observing two different pendulums swing simultaneously.

This prompts us to explore the type of experimental set-up that may have been used by Galileo and his students. Palmieri complains that what little we know about Galileo's set-up is unfortunately unconvincing. The only suggestion Galileo provides about his pendulum arrangement is shown in a drawing that accompanied his 1602 letter to Dal Monte. The sketch suggests that Galileo initially suspended 2–3 braccia-long pendulums adjacent to each other from the ceiling or on a wall of his workshop or bedroom. However, Palmieri questions this drawing, which is preserved only in a third-hand copy from the 19th Century (see the right-hand panel of figure 3.1):

'The diagram of the two pendulums looks suspicious since, in the text, Galileo says that one of the pendulums is removed from the perpendicular 'a lot' [assai], while the other is removed 'very little' [pochissimo], yet the two arcs drawn in the diagram are almost of the same amplitude. Moreover, it is possible that, even if accurate, the diagram only had an illustrative function, and was not intended to describe a real set-up' [24].

The basis of his discomfort with the scene depicted is that Galileo later referred to having used 4–5 braccia-long pendulums, which would imply that he needed a wall of at least 20 braccia (7 metres) wide to allow the pendulums to reach their full extents. This arrangement, in turn, would have required Galileo to stand at a considerable distance from the experiment, making the observations cumbersome and potentially inaccurate. In addition, having to work with such long pendulums would imply that space requirements imposed that he could only have used small-amplitude oscillations of up to 15°, for which deviations from isochronism are small [25]. Galileo's long pendulums with wooden bobs would have decayed to negligible amplitudes within about seven minutes [26]. As a result, the effects of anisochronism would have become undetectable rather quickly. For these reasons, Palmieri concludes that an arrangement such as that shown in the left-hand panel of figure 3.1 would be more plausible, since its gallows-like structure has the advantage of allowing for easier observation of the oscillations of two pendulums relative to each other.

Drinkwater Bethune also questioned Galileo's confidence in having observed clear isochronism, since the periods of oscillations describing larger arcs tend to

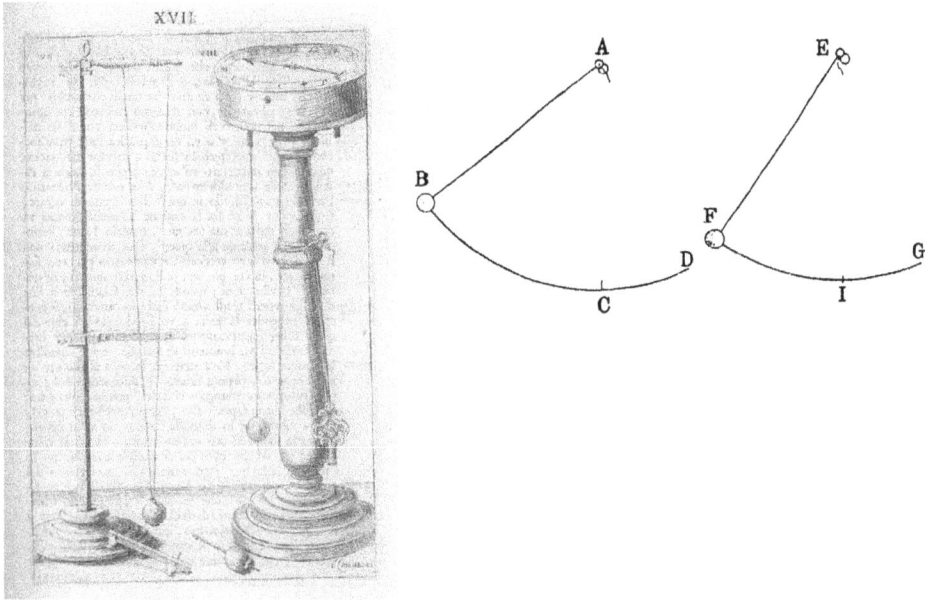

Figure 3.1. (*Left*) pendulum used by the *Accademici del Cimento* (Academy of Experiment), a scientific society briefly established in 1657, for approximately a decade. (Magalotti 1666, *Saggi di naturali esperienze fatte nell'Accademia del cimento sotto la protezione del serenissimo principe Leopoldo di Toscana e descritte dal segretario di essa accademia,* Florence: Giuseppe Cocchini, p. 21.) (*Right*) Galileo's sketch of his double-pendulum set-up in his letter of 29 November 1602 to Dal Monte. (*Le opere di Galileo Galilei: Edizione Nazionale* ed A Favaro, Florence, 1890–1909, **x**, pp. 97–100)

increase rapidly, that is, they tend to slow down relatively quickly. He asserted that it is probable that Galileo's confidence of the perfect equality of these oscillations was based in his belief that the increase of time—which he could not avoid noticing, particularly when considering hundreds of oscillations—would be caused by the increased resistance of the air during the larger vibrations [27]. Indeed, on a number of occasions Galileo candidly concedes that his measurements do not always correspond to his theoretical predictions, but he dismisses such effects as '*accidents*' (although '*occurrences*' may be a more appropriate translation),

> '*Of these accidents [accidenti] of weight, of velocity, and also of form [figura] infinite in number, it is not possible to give any exact description; hence, in order to handle this matter in a scientific way, it is necessary to cut loose from these difficulties; and having discovered and demonstrated the theorems, in the case of no resistance, to use them and apply them with such limitations as experience will teach*' [28].

As a case in point, Galileo seems to have dismissed the effects of friction, given that he commented that the resistance of the air will not affect the period of the oscillation. However, that conclusion was a direct consequence of his incorrect belief that the period of oscillation along all arcs is the same. In his *Dialogo Sopra i Due*

Massimi Sistemi del Mondo (*Dialogue on the Two Chief World Systems*, 1632), he goes to great lengths to explain the intrinsic property of pendulums causing them to slow down and eventually come to a complete stop, irrespective of external factors. One should place these insights into their proper historical context. At the time of Galileo's ground-breaking work on the pendulum, many of his contemporaries still supported the Aristotelian worldview: they would have assumed that the medium (the air) was responsible for keeping the pendulum going, not for slowing it down—since a lighter element, the air would then support the heavier elements of which the bob was composed. The basis of Galileo's theory is represented by a drawing of the shape of an oscillating pendulum composed of a thick rope and a bob, which is reproduced in figure 3.2.

Galileo argued that if the pendulum's suspension was made of a thick *corda* (rope), or even a chain, instead of the lighter and thinner *spago* (string) or *spaghetto* he usually employed [29], the rope's components would behave as if it were composed of many infinitesimally small pendulums distributed along the rope, each with their own, well-defined frequency—in essence, he had stumbled upon the presence of so-called 'latent modes' of oscillation, which describe the shape of an oscillating heavy rope or chain, that is, of a compound pendulum. These frequencies would become increasingly higher as their distance from the centre of oscillation becomes ever smaller, and Galileo argued that this arrangement would lead to the oscillating bob slowing down. Palmieri points out [30] that the bob will be 'restrained' by the many pendulums making up the rope, each of which will naturally want to oscillate increasingly faster with decreasing distance to the centre

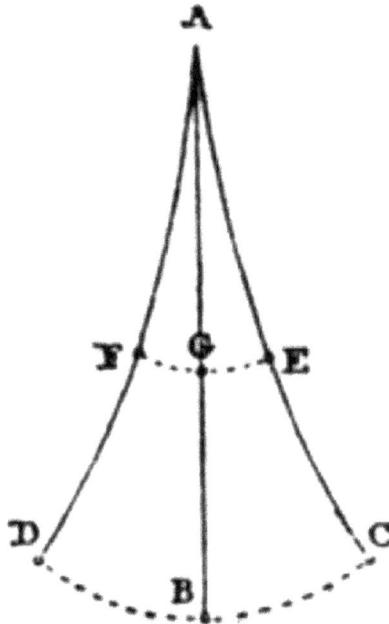

Figure 3.2. Galileo's drawing of the shape of an oscillating pendulum composed of a thick rope and a bob. (Galilei G 1632 *Dialogue on the Two Chief World Systems*)

of oscillation than the bob. In turn, all pendulums will inevitably stop, even in the absence of any external factor such as air resistance or mechanical friction.

Just like Drinkwater Bethune did in 1830, in more recent times a number of scholars have also wondered about the burden of evidence that Galileo had access to. For instance, Naylor writes,

'One of Galileo's most renowned discoveries was the isochronism of the simple pendulum. In the Discorsi, *Galileo used this discovery to good effect—though his claim that the pendulum was isochronous for all arcs less than 180° has created something of a puzzle for the history of science. The question arises as to how far the evidence available to Galileo supported his claims for isochronism. ... Galileo was almost certainly familiar with a much wider range of evidence than he indicated in the* Discorsi. *The examination of the evidence available to Galileo indicates that, though it provided ample support for his thesis, it was certainly not as conclusive as he implies in the* Discorsi. *It also seems clear that Galileo was bound to be aware of this'* [31].

In his letter of 1602 Galileo mentioned, for the first time, the theorem that the times of free fall down all chords drawn from the lowest point of a circle are equal. At this point, he was clearly still working on the validation of his theory, stating confidently that ...

'[u]p to this point I can go without exceeding the limits of mechanics, but I have not yet been able to demonstrate that all arcs are passed in the same time, which is what I am seeking'.

Meanwhile, on 16 October 1604 he suggested in a letter to his friend and patron Paolo Sarpi (1552–1623), that he had achieved a significantly improved understanding of the problem. Alas, the theory he proposed—sometimes referred to as Baliani's theory of motion, first published in 1638 in *De Motu Naturali Gravium Solidorum et Liquidorum* [32]—proved to be false:

A —⌐

'Returning to the subject of motion, in which I was entirely without a fixed principle, from which to deduce the phenomena I have observed, I have hit upon a proposition, which seems natural and likely enough; and if I take it for granted, I can show that the

B —

spaces passed in natural motion are in the double proportion of the times, and consequently that the spaces passed in equal times are as the odd numbers beginning from unity, and the rest. The principle is this, that the swiftness of the moveable increases in

C —

the proportion of its distance from the point whence it began to move; as for instance, if a heavy body drops from A towards D by the line ABCD, I suppose the degree of velocity which it has at B to bear to the velocity at C the ratio of AB to AC. I shall be very glad if your Reverence will consider this, and tell me your opinion

D —⌐

of it. If we admit this principle, not only, as I have said, shall we *demonstrate the other conclusions, but we have it in our power to show that a body falling naturally, and another projected upwards, pass through the same degrees of velocity. For if the projectile be cast up from **D** to **A**, it is clear that at **D** it has force enough to reach **A**, and no farther; and when it has reached **C** and **B**, it is equally clear that it is still joined to a degree of force capable of carrying it to **A**: thus it is manifest that the forces at **D**, **C**, and **B** decrease in the proportion of **AB**, **AC**, and **AD**; so that if, in falling, the degrees of velocity observe the same proportion, that is true which I have hitherto maintained and believed'* [33].

Once we have translated this passage into modern language, it becomes obvious that Galileo made a mistake in his reasoning. Consider the following argument to clarify the fallacy of the proposed theorem. Galileo suggests, in essence, that the velocity at any point is the distance that would be travelled during the next moment of time, assuming that the motion continues in an identical manner as at that particular point. At the point of origin, when the body is at rest, there is no motion. Therefore, Galileo's theory implies that the distance travelled over the next time interval is none. It is thus clear that the body cannot begin to move according to this theorem. Galileo realized his mistake and included a corrected version in his unpublished *Dialogues on Motion* (*De Motu*; started around 1590, with improvements added in 1602–1604).

3.1.3 Practical applications

Throughout Galileo's productive experimental life, the pendulum was often at the forefront of his mind. His inquisitive nature was occupied with solving questions such as what the fastest motion was from a higher to a lower point, and in particular whether a path along a circular arc—such as that traced by a simple pendulum bob—or in a straight line on an inclined plane would be traversed faster, or what the relationship was between the length of a pendulum and its period.

This led, eventually, to his *law of inclined planes*, which he published in his *Discorsi e Dimostrazioni Matematiche Intorno a Due Nuove Scienze* (*Discourses and Mathematical Demonstrations Relating to Two New Sciences*) in 1638. Since bodies in free fall move too fast to measure their motions, Galileo came up with the innovative insight that rolling down an inclined plane is analogous to falling, with as main advantage that the body would be slowed just enough that its motion could be measured more easily. Specifically, he found that the distance travelled, d, is proportional to the square of the time, t: $d = \frac{1}{2}at^2$, where a is the system's acceleration. Compare this latter concept with the gravitational acceleration, g, resulting from the law of free fall. Galileo used geometry to show that the acceleration a of a plane inclined by an angle θ would be $a = g\sin\theta$. This implies that if the inclination angle θ tends to 90°, the motion approaches free fall. Galileo did not consider one important complication, however. Rolling bodies accelerate more slowly than bodies that *slide* down an inclined plane, in the absence of friction. The reduction in potential (gravitational) energy owing to the body travelling down the plane is converted into kinetic energy. For a sliding object, all of the energy is converted into linear motion, but for a rolling body the energy is partially also converted into rotation.

Galileo was often preoccupied with thoughts about the practical use of the pendulum. Although he initially employed pendulums to study the laws of motion, he soon proceeded by designing simple timing devices using free-swinging pendulums, including metronomes for students of music. Shortly after Galileo's letter to Dal Monte, his physician friend Santorio Santorio (1561–1636) invented the *pulsilogium* [34], a stop clock that measured a patient's pulse rate. The device used a pendulum with a leaden bullet at its end, which may well have been the first precision instrument in the history of medicine.

Santorio applied Galileo's insights that the frequency of the pendulum's oscillation was inversely proportional to the square root of its length. The *pulsilogium* was composed of a heavy leaden bob and a silk cord. It worked by adjusting the oscillation frequency by changing the pendulum's length—remember that, under ideal circumstances, the frequency only depends on the length of the cord—until the periodic oscillation was synchronized with the pulse beat [35]. Since the patient's pulse rate corresponded to either the position of a knot in the cord alongside a horizontal ruler (see figure 3.3) or to the location of the hand on a dial, the pulse rate was measured in units of length. Santorio's original version of the *pulsilogium* had its cord wound around a drum; as the drum rotated and the cord length changed, the hand moved around the dial.

Extensive experimentation with his new tool allowed Santorio to derive the circadian rhythm (the 24-hour cycle) of the cardiac frequency. He was already a well-known physician in the Republic of Venice when, in 1603, he published a series of books, entitled the *Methodi Vitandorum Errorum Omnium qui in Arte Medica Contigunt Libri Quindecim* (*Methods to Avoid All Errors Occurring in Medical Art, 15 Books*), in which he introduced his invention:

'*In order to commemorate quickly and exactly my knowledge on the pulse of a patient, I have invented the pulsilogium [pulse meter], which makes it possible to measure exactly the beats of the arteries... and to compare them with the beats of earlier days. ... With the help of the pulse meter, we can monitor at what day and at which hour the pulse deviated in intensity and frequency from its natural state*' [36].

The pendulum law at the basis of these developments, that is, the notion that a pendulum's oscillation frequency is inversely proportional to the square root of its length, was independently derived by Marin Mersenne, Parisian philosopher, mathematician, music theorist, and ordained Jesuit priest. Although Mersenne's conclusions were not substantially different from Galileo's earlier assertions, Mersenne used acoustic experiments to provide physical proof of the pendulum law [37], a feat originally considered impossible by Galileo. The pendulum law is, hence, known as 'Mersenne's law.' Following the publication of Mersenne's treatise, *Les Méchaniques de Galilée*, on 30 June 1634, Élie Diodati (1576–1661)—a Swiss lawyer from Geneva (Swiss Confederacy)—sent a copy to Galileo on 10 April 1635

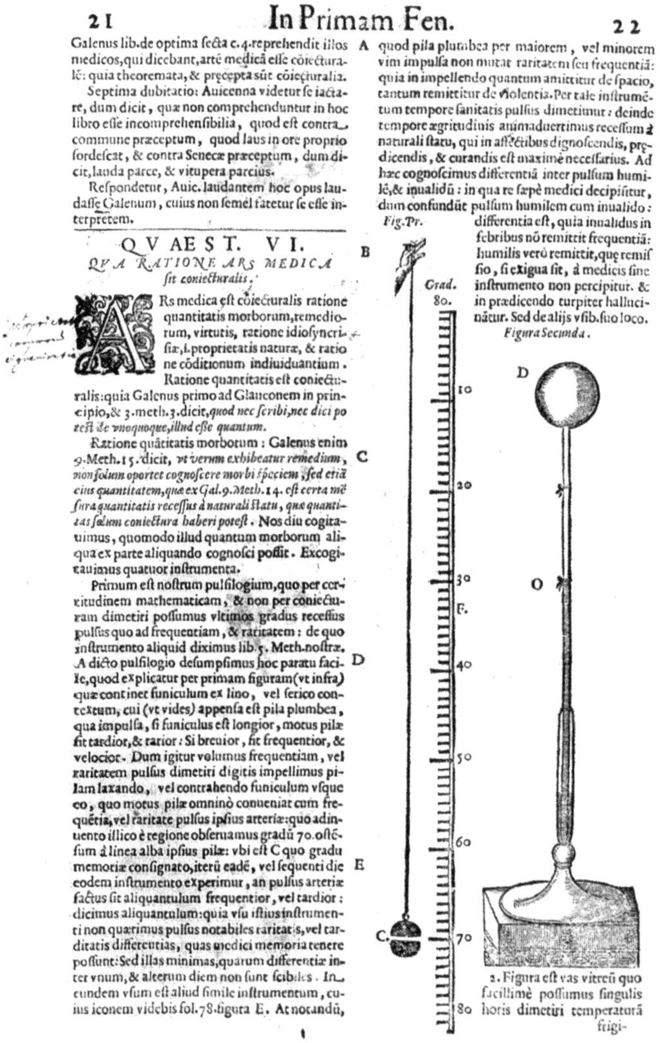

21   In Primam Fen.   22

Galenus lib. de optima secta c. 4. reprehendit illos medicos, qui dicebant, arte medica esse coiecturalé: quia theoremata, & precepta sūt coiecturalia.

Septima dubitatio: Auicenna videtur se iactare, dum dicit, quae non comprehenduntur in hoc libro esse incomprehensibilia, quod est contra commune praeceptum, quod laus in ore proprio sordescat, & contra Senecae praeceptum, dum dicit, lauda parce, & vitupera parcius.

Respondetur, Auic. laudantem hoc opus laudasse Galenum, cuius non semel fatetur se esse interpretem.

QVAEST. VI.
QVA RATIONE ARS MEDICA sit coiecturalis.

Ars medica est coiecturalis ratione quantitatis morborum, remediorum, virtutis, ratione idiosyncrasiae, & proprietatis naturae, & ratione coditionum indiuiduantium.

Ratione quantitatis est coiecturalis: quia Galenus primo ad Glauconem in principio, & 3. meth. 3. dicit, quod nec scribi, nec dici potest de vnoquoque, illud esse quantum.

Ratione quátitatis morborum: Galenus enim 9. Meth. 15. dicit, vt verum exhibeatur remedium, non solum oportet cognoscere morbi speciem, sed etiã eius quantitatem, quae ex Gal. 9. Meth. 14. est certa mē sura quantitatis recessus à naturali statu, quae quantitas solum coiectura haberi potest. Nos diu cogitauimus, quomodo illud quantum morborum aliqua ex parte aliquando cognosci possit. Excogitauimus quatuor instrumenta.

Primum est nostrum pulsilogium, quo per certitudinem mathematicam, & non per coiecturam dimetiri possumus vltimos gradus recessus pulsus quo ad frequentiam, & raritatem: de quo instrumento aliquid diximus lib. 5. Meth. nostrae. A dicto pulsilogio desumpsimus hoc paratu facile, quod explicatur per primam figuram (vt infra) quae continet funiculum ex lino, vel serico contextum, cui (vt vides) appensa est pila plumbea, qua impulsa, si funiculus est longior, motus pilae fit tardior, & rarior: Si breuior, fit frequentior, & velocior. Dum igitur volumus frequentiam, vel raritatem pulsus dimetiri digitis impellimus pilam laxando, vel contrahendo funiculum vsque eo, quo motus pilae omnino conueniat cum frequétia, vel raritate pulsus ipsius arteriae: quo ad inuento illico è regione obseruamus gradu 70. ostēsum à linea alba ipsius pilae: vbi est C quo gradu memoriae consignato, iterù eadē, vel sequenti die eodem instrumento experimur, an pulsus arteriae factus sit aliquantulum frequentior, vel tardior : dicimus aliquantulum: quia visu istius instrumenti non quaerimus pulsus notabiles raritatis, vel tarditatis differentias, quas medici memoria tenere possunt: Sed illas minimas, quarum differentiae inter vnum, & alterum diem non sunt scibiles. In eundem vsum est aliud simile instrumentum, cuius iconem videbis fol. 78. figura E. At notandù,

quod pila plumbea per maiorem, vel minorem vim impulsa non mutat raritatem seu frequentiã: quia in impellendo quantum amittitur de spacio, tantum remittitur de violentia. Per tale instrumentum tempore sanitatis pulsus dimetimur: deinde tempore aegritudinis aniimaduertimus recessum à naturali statu, qui in affectibus dignoscendis, praedicendis, & curandis est maxime necessarius. Ad haec cognoscimus differentiã inter pulsum humilé, & inualidū : in qua se saepe medici decipiūtur, dum confundūt pulsum humilem cum inualido : differentia est, quia inualidus in febribus nõ remittit frequentiã: humilis verò remittit, quae remissio, si exigua sit, à medicis sine instrumento non percipitur. &. in praedicendo turpiter halluci-nātur. Sed de alijs vsib. suo loco.

Fig. Pr.

Grad. 80.

Figura Secunda.

D

O

E

C

2. Figura est vas vitreū quo facillimè possumus singulis horis dimetiri temperaturã frigi-

Figure 3.3. *Pulsilogium* (centre), line with a weight tied to a finger. (Sanctorius, S., 1626, *Commentaria in primam Fen primi libri Canonis Avicennae*, Venice: Sarcina, p 22. Woodcut and text; Wellcome Library, London; Cat. No. L0008488.)

[38]. Diodati was based in Paris, where he helped Galileo publish his manuscripts, given that the latter was unable to do so in his native Italy because of censorship imposed by the Catholic Church.

Perhaps surprisingly given that he first considered the timing implications of pendulums in his youth, it appears that Galileo did not consider their potential applications to mechanical clocks until later in life. Mechanical clocks had been in use for several hundred years by the time Galileo was born (the oldest surviving mechanical clock in Europe dates from 1386), having displaced the much older water clocks. However, they were large, equipped with heavy weights to provide power, and the accuracies of even the best clocks remained inadequate for astronomical applications. The mechanical clocks Galileo must have been familiar with were subject to gaining or losing time in an irregular and largely unpredictable manner. Nevertheless, a number of early adopters had actually suggested their use as a means to determine one's position at sea, particularly when crossing the world's oceans, despite their inherently inaccurate clock mechanisms; in 1594, the English navigator, humanist writer, and mathematician Thomas Blundeville (c. 1522–c. 1606) proposed using '*some true Horologie or Watch, apt to be carried in journeying which by an Astrolabe is to be rectified ...*' [39], but he was indeed far ahead of his time... Based on his experiments with the simple pendulum, Galileo proposed that one could count the oscillations of a carefully calibrated pendulum to accurately measure time intervals in the context of astronomical observations; he subsequently expanded on these thoughts, and in 1636 he proposed that the same principle could potentially be used to find one's longitude.

3.1.4 Galileo's pendulum clock design

It took until 1641, when Galileo had reached the respectable age of 77, before he considered seriously whether a pendulum could be attached to the 'escape' mechanism of a mechanical clock (see below for a technical description) in order to regulate it accurately. By that time, he was completely blind, so instead he described his design to his son, Vincenz(i)o Gamba (later Vincenzo Galilei; 1606–1649), who made a sketch (see figure 3.4) and turned his attention to addressing the problem. Viviani discusses the events leading up to this new focus in Galileo's biography of 1659, 17 years after the Italian scientist's death in 1642,

> '*One day in 1641, while I was living with him at his villa in Arcetri, I remember that the idea occurred to him that the pendulum could be adapted to clocks with weights or springs, serving in place of the usual tempo, he hoping that the very even and natural motions of the pendulum would correct all the defects in the art of clocks. But because his being deprived of sight prevented his making drawings and models to the desired effect, and his son Vincenzio coming one day from Florence to Arcetri, Galileo told him his idea and several discussions followed. Finally they decided on the scheme shown in the accompanying drawing, to be put in practice to learn the fact of those difficulties in machines which are usually not foreseen in simple theorizing*' [40].

By this time, Galileo's ideas about pendulums as potential timekeepers had reached far beyond his native Italy. As we will see, leading thinkers from the Low Countries—encompassing the present-day territory of the Netherlands—had become aware of the potential use of pendulum-based clocks for marine navigation purposes, that is, to determine longitude at sea. The race was on to construct a stable and precise pendulum-based timepiece that would work both on land and at sea.

Figure 3.4. Galileo's pendulum clock design, drawn by Vincenzo Viviani in 1659. Part of the front supporting plate has been removed by the artist to show the wheelwork. (*Opere di Galileo*, **19**, p. 656; Photo: University of Toronto Photographic Services.)

Vincenzo Galilei began construction of the pendulum clock conceived by his father, but he encountered a number of challenges which prevented him from completing his timepiece to the necessary accuracy before his own death in 1649. In addition to the fundamental problem of the pendulum's movement stopping as a result of friction, he encountered difficulties in transferring the energy from the pendulum's oscillatory motion to a cogwheel, which would in turn be employed to adjust the clock's dials.

The basic principle of Galileo's clock design relies on transmission in regular pulses of either the energy in a coiled spring or the potential (gravitational) energy of a descending weight. This is usually achieved—for any mechanical clock—by employing an 'escapement,' a device which allows the energy to 'escape' from the system. Galileo's pendulum clock design included a cogwheel and a pair of curved levers or 'pawls' connected to a fixed rod pendulum. During a full oscillation of the pendulum, one pawl first lifts clear of the cogs, allowing the wheel to rotate until it meets the rigid second pawl. As the latter lever is caught, a small impulse is imparted to the pendulum to keep its oscillation going. The pendulum thus regulates the energy release, while it is also itself recharged periodically. It is crucially important that the pulse is imparted at precisely the right part of the pendulum's swing, at the top of the arc.

Galileo's clock design encountered two main challenges, including the need to reduce the inertia of and the friction between the clock's moving parts, so that only small pulses of energy are required for continuous operation. The second challenge was that the pendulum's motion had to be as smooth and unperturbed as possible, so that its movement would resemble that of a *free* pendulum. Although Galileo's escapement design was not very good in either respect—in practice, the pawls only managed to partially transmit the pendulum's impulse to the cogwheel, which worsened the friction problem—we should keep in mind that neither was the combination of a *verge* escapement[3] and a crude balance wheel in common use in contemporary clocks.

A verge escapement consists of a crown-like wheel with sawtooth-shaped teeth pointing axially forward: see figure 3.5. A vertical rod, the verge, is equipped with two metal plates, the pallets, which drive the wheel by engaging teeth on opposite sides of the wheel. Only one pallet catches the teeth at any given time, rotating the verge by a small angle, meanwhile rotating the second pallet into the path of the teeth on the wheel's opposite side. The second pallet is engaged to rotate the verge back to its original position, thus resulting in oscillation of the verge. Note that development and use of verge escapements predates the pendulum: they were used with hanging weights and possibly in water clocks from at least the 14th Century, perhaps significantly earlier. Development of the pendulum provided a means to precisely regulate such escapements.

In fact, it was the isochronicity of the pendulum which enabled a step change in the development of timepieces. Prior to their use, time was regulated with a verge escapement and a 'foliot,' a periodically oscillating, weight-driven horizontal rod which was the precursor of the balance wheels now commonly used in watches. Such a configuration is intrinsically non-isochronous, and therefore timekeeping could only be done approximately. In this set-up, the escapewheel wheel is driven by a weight, giving it a constant moment of force, M; the foliot has a given moment of inertia, I, in essence a measure of its resistance to acceleration, which is determined by the weights suspended from its extremeties and their relative positions with respect to its axis of rotation. The foliot is driven back and forth, alternately engaging the verge pallets at its times of maximum rotation out of equilibrium. This gives rise to the regular ticks (pulses) of a ticking clock, which correspond to the

---

[3] The word *verge* originates from the Latin word *virga*, 'stick' or 'rod.'

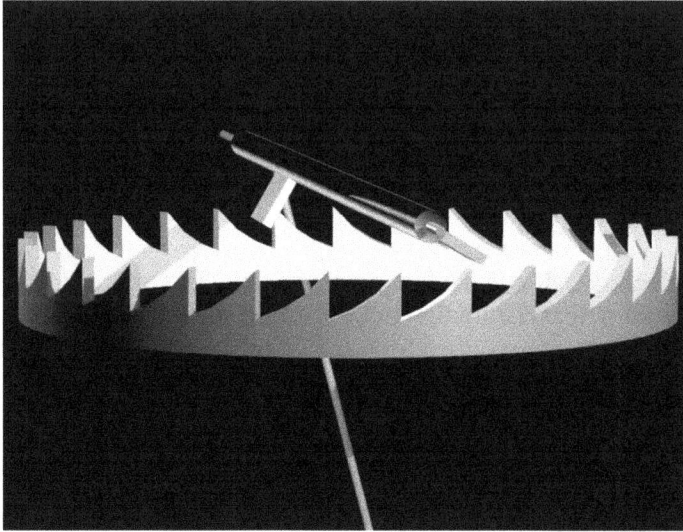

Figure 3.5. Verge escapement fitted with a pendulum bar. (Courtesy Simeon Lapinbleu; GNU Free Documentation License v. 1.2.)

times when the weight-driven motion transfers energy into the clock's operation. In the absence of friction, the period of the foliot's oscillatory motion is given by

$$P = 2\sqrt{2\varphi_{max}\frac{I}{M}},$$

where φ_{max} is the maximum swing angle. The period hence depends on the swing angle, and therefore such clocks are intrinsically non-isochronous. In practice, losses owing to friction and temperature differences wreak further havoc on efforts to use this type of clock for accurate timekeeping.

3.1.5 Galilei's insights reach the Low Countries

Even before Vincenzo Galilei's struggle to make his father's vision of a working pendulum clock a reality, news of the proposed device had reached the Low Countries. Given the Dutch focus on trade as a sea-faring nation, the country's leading scholars immediately appreciated the potential navigational benefits they would be able to gain from developing such an instrument.

In a letter [41] from the Dutch poet and composer Constantijn Huygens (1596–1687)—secretary to two Dutch Royal Princes and father of Christiaan Huygens, the scientist—to Diodati, dated 13 April 1637, Huygens Sr refers to a proposal by Galileo to develop a device to determine longitude at sea, which would require stable operation of time pieces, '*against the agitations of the sea*,' in order to accurately determine the ephemerides of the 'four satellites'—by which he meant the Galilean moons of Jupiter, *viz.* Io, Europa, Ganymede and Callisto—'*as on solid ground*'. Galileo had discovered the moons that now carry his name in 1610 (see figure 3.6), and he believed that their regular motions might enable an absolute calibration of

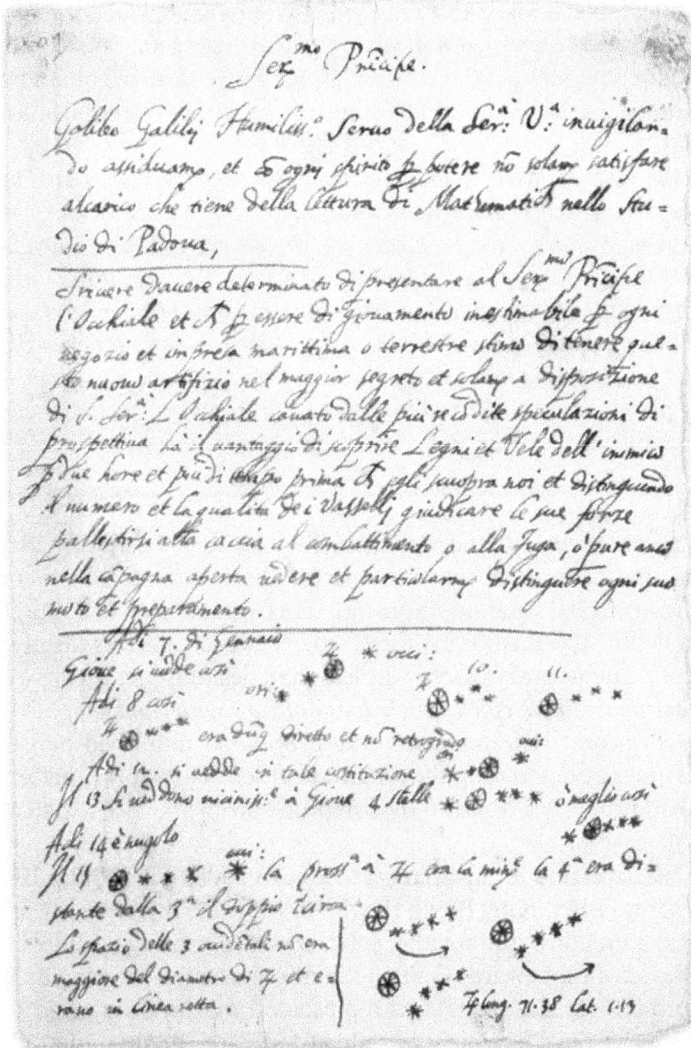

Figure 3.6. Draft of a letter from Galileo to Leonardo Donato, Doge of Venice (August 1609) and Notes on the Moons of Jupiter (January 1610). (University of Michigan Library.)

any clock on Earth. In principle, he was correct. He set out to tabulate their positions with respect to their parent planet, Jupiter, with the intention to give his tables to sailors setting out on long voyages out of sight of any shoreline.

Timekeeping on its own as the basis of longitude determination at sea was not a novel concept; using the newly discovered satellites of Jupiter as an absolute calibration benchmark was. The first suggestion that accurate timekeeping may form the basis of a successful solution to the longitude problem goes back as far as 1530. Regnier Gemma Frisius (born Jemme Reinerszoon; 1508–1555)— Dutch astronomer, professor of medicine and mathematics at the University of Louvain in the Southern Netherlands, and teacher of Gerardus Mercator, the map maker—published his book *De Principiis Astronomiae Cosmographicae* (*On the Principles of Astronomy and Cosmography*) [42] to supplement a combined terrestrial/celestial globe he had produced. In chapter 19 he describes how one may find one's longitude on Earth using the difference between local and absolute time: …

> '*it is with the help of these clocks and the following methods that longitude is found…. observe exactly the time at the place from which we are making our journey…. while we are on our journey we should see to it that our clock never stops. When we have completed a journey of 15 or 20 miles, it may please us to learn the difference of longitude between where we have reached and our place of departure. We must wait until the hand of our clock exactly touches the point of an hour and, at the same moment by means of an astrolabe … we must find out the time of the place we now find ourselves…. In this way I would be able to find the longitude of places, even if I was dragged off unawares across a thousand miles*'.

In essence, he proposed that the timepiece had to be set carefully at the time of departure and that it must be kept running an absolute time, which could then be compared with the local time upon arrival. This would allow one to calculate the distance travelled in the East–West direction. However, he was keenly aware that the accuracy of contemporary clocks was less than desired, because he wrote that '*… it must be a very finely made clock which does not vary with change of air.*' The second edition of *De Principiis Astronomiae Cosmographicae*, published in 1533, includes more detailed notes about finding the longitude at sea. This truly represented the first time anyone had properly addressed the longitude problem, and he had done so with great foresight.

Following his accession to the Spanish throne in 1598, King Philip III followed in the footsteps of his father, King Philip II, who in 1567 had offered a reward to anyone able to propose a practical and reliable solution to determining a ship's longitude at sea. Having heard of the reward, Galileo initially presented his new discovery to the King of Spain in 1612. The King was represented by the *Casa de la Contratación* (*House of Trade*) in Seville, which was entrusted with judging and validating any proposals put forward. The reward on offer, 6000 ducats, a life annuity of 2000 ducats, and 1000 ducats of expenses, testifies to the importance that solving this outstanding navigational problem had been afforded. Surprisingly, therefore, Galileo did not manage to enthuse the King of Spain, and this endeavour came to nothing.

Galileo's proposal was one of two proposals submitted by foreign applicants. The other foreign proposal had been prepared by Michael Florent van Langren (*Lat.*: Langrenus; 1598–1675), a cartographer, astronomer, and mathematician from Flanders in the Southern Netherlands. Van Langren was aware of the severe difficulties in the accurate determination of longitude, both on land and at sea. He created the first known graph of statistical data that showed a large range of previously measured longitude differences between Toledo (Spain) and Rome: see figure 3.7 [44]. He suggested that these estimates could be improved, most notably so at sea, by observing the rising and setting of peaks and craters on the Moon, at any

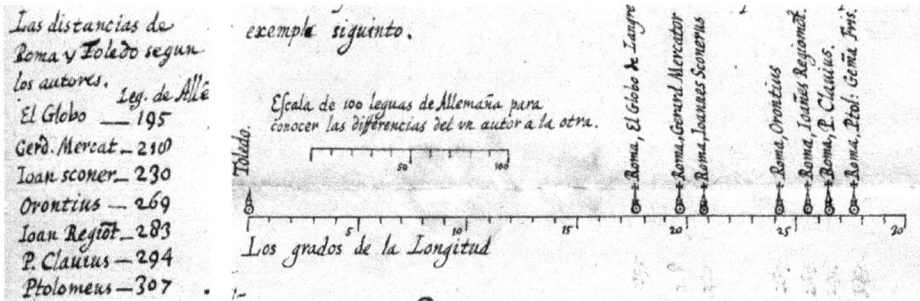

Figure 3.7. Graph and table of longitude determinations by different authors, taken from a letter of Van Langren to Isabella Clara Eugenia of Spain, wife of Archduke Albert of Austria and daughter of Philip II (King of Spain), of early 1628. (Archive of the Counts of Castilfalé, Burgos, Spain; image courtesy of Joaquín Ulargui and Michael Friendly.)

time, that is, not only during lunar eclipses as had been suggested previously [45]. The eight other proposals submitted to the *Casa de la Contratación* by 1634 had all been prepared by Spanish hopefuls, yet no prize was ever awarded beyond the expenses component. This led to a declining interest in the Spanish offer by the mid-1630s.

Meanwhile, Galileo had learnt that the States General of the United Provinces of the Dutch Republic, the 'Staten Generaal,' had offered a competing reward to anyone who could solve the longitude problem. On 1 April 1600, the Staten Generaal had announced the availability of a reward of 5000 carolus guilders, as well as a life annuity of 1000 carolus guilders, for anyone who could provide an adequate solution to what was referred to as the problem of '*finding East and West.*' A year later, the (provincial) States of Holland similarly offered their own longitude prize. The latter prize could be earned in stages: applicants would receive 150 guilders upon submission of their proposals, provided that they were prepared to have their method tested at sea. If six to eight navigators confirmed the practical reliability of the proposed method, the applicant would receive a reward of 3000 guilders as well as a 1000-guilder life annuity. The Staten Generaal increased the monetary value of their prize to 15 000 guilders in 1611 and to 25 000 guilders in 1660; the States of Holland raised their prize money to as much as 50 000 guilders by 1738 [46].

Although Galileo had initially ignored these Dutch opportunities, his attitude changed after the Catholic Church censured him in 1633 for his outspoken pro-Copernican views. He needed a safer, less controversial subject to pursue, which made him return to addressing the longitude problem, although he still pursued this through ephemeris observations of Jupiter's moons—potentially contentious in the eyes of the Inquisition. Around the same time, Dutch friends were hoping to extract him from Italy, to provide him with a safe haven in the more tolerant social strata of the Low Countries. Interestingly, these pursuits were instigated by another persecuted scholar, Hugo de Groot (*Lat.*: Hugo Grotius; 1583–1645), former leader of the Dutch Remonstrants and a diplomat in his own right, who lived in exile in Paris.

De Groot managed to attract the interest of a number of prominent citizens of Amsterdam for his project to provide refuge to Galileo, including of the Dutch astronomer and mathematician Martin van den Hove (*Lat.*: Martinus Hortensius; 1605–1639), full professor 'in the Copernican theory' at the Amsterdam Atheneum (*Athenaeum Illustre*)—the predecessor of the University of Amsterdam—and Willem Jansz Blaeu (1571–1638), Dutch cartographer, atlas maker, and publisher. The group of interested parties was completed with Laurens Reael (1583–1637), a former governor-general of the Dutch East India Company and a former Dutch navy admiral.

In July 1635, De Groot announced to his illustrious friends that Galileo was reasonably confident to have found a way to determine longitude at sea, which formed part of an effort by De Groot to provide the Italian scholar with further credentials in support of his eligibility as a refugee from persecution in the Dutch Republic [47]. Galileo's idea of using a pendulum clock in combination with the eclipses of Jupiter's Galilean satellites was first proposed in a letter he wrote to the leading French lineographer, mathematician, and geostatistician Jean de Beaugrand (1584–1640) on 11 November 1635. Although Galileo's advanced age prevented him from travelling to the Dutch Republic, De Groot had already enlisted Van den Hove and Diodati in persuading Galileo to formally present his discovery to the Staten Generaal [48]. It is important to realize that Galileo's proposal involved the use of a pendulum clock to accurately time tabulated eclipses of Jupiter's moons. He did not consider the use of a pendulum-based timepiece to accurately maintain a ship's home-port time. As we will see later in this chapter, that latter innovation was pursued by Christiaan Huygens a few decades later.

On 15 August 1636, the Italian scholar sent an elaborate and exquisite description of his method of longitude determination at sea to the Dutch Republic's government [49] by means of a lengthy letter passed on by Reael [50], who was seen as Galileo's most important backer in the Dutch Republic. At the same time, he made it clear that he had very little notion of the practical realities of navigating a ship. As designed, he pointed out, his clock could only record the time from midday onwards, and that the proposed observations of Jupiter's moons would require a stable observation platform. In fact, one could not risk losing sight of Jupiter for any length of time during the observations, since the eclipses of its moons would only last of order a minute. He felt that he was now well placed to calculate and tabulate these eclipses with reasonable accuracy, thus making them useful for navigation purposes.

Simultaneously, Galileo wrote to De Groot, pressing upon him the need to make progress towards practical implementation of his proposal given his own deteriorating health, and in particular that he was not driven by monetary concerns, but ...

'my contribution is offered openly, not motivated by greed, but only in pursuit of the admirable joy of the wondrous art of navigation, [to which I hope to offer] something so desired and so useful' [51].

De Groot responded in a letter dated 20 September 1636 [52], saying that he fully supported the *'most exquisite'* discovery Galileo proposed, which he hoped to be successful so that it would be useful '*for all mankind*' and in particular to a country like the Dutch Republic with its sea-faring traditions.

3.1.6 Misfortunes

Once Galileo's letter had arrived at the Dutch Republic's seat of government in The Hague, on 11 November 1636, an *ad hoc* committee chaired by Reael and including Blaeu and Van den Hove was appointed immediately [53]. Initially, they had also intended to appoint Jacobus van Gool (*Lat.*: Jacob Golius; 1596–1667), an Orientalist and mathematician at Leiden University [54], but instead the committee agreed to invite the philosopher and scientist Isaac Beeckman (1588–1637) to join their ranks. Beeckman himself had also proposed the possible use of Jupiter's satellites as a celestial clock, independently of Galileo, in 1631 [55]. The procedure adopted by both the Staten Generaal and the States of Holland in assessing claims of eligibility for their longitude prizes consisted of first appointing an *ad hoc* committee of experts in the theory of navigation (*theoristen*), usually scholars, teachers, and/or surveyors, if warranted followed by a period of scrutiny by experienced sailors (*practisijns*) to ascertain whether the proposed method would be a viable practical solution.

In his April 1637 letter to Diodati, Constantijn Huygens emphasized that stable operation at sea was essential if they were to develop a timepiece that could be employed as a navigation aid, and he expressed his wish that the 'great man' (Galileo) would live to see their successful achievement. In fact, in a letter from March 1637, Galileo pleaded with Diodati to convince Huygens that he had come up with a solution to the stability problem, which Galileo himself forwarded to Reael in June 1637 [56]. This development had been triggered by Reael's probing questions about the potential for stable operation of Galileo's pendulum clock on choppy seas [57]. Galileo's solution involved a marine chair composed of a large universal joint, with one hemispherical component moving inside a second, which in turn was to be fixed to the ship: for an example, see figure 3.8. Both components had

Figure 3.8. Possibly the first illustration of a gimballed (marine) chair on a ship, designed to allow astronomical observations from a stable position. (Besson J 1567–1569 *Le Cosmolabe, ou instrument universel concernant toutes observations qui se peuvent faire,* Paris: Philippe Gaultier de Roville p 244.)

to be separated by water or oil and it was crucial to retain a gap between them, which could be achieved with eight to ten springs, all in all a rather cumbersome contraption [58].

This was not Galileo's first attempt at solving the longitude problem at sea, however. In 1611/1612 he had first realized that the orbits of Jupiter's moons could potentially be used as a highly accurate reference, an idea we already saw he had unsuccessfully tried to monetize by applying to the Spanish court for its reward. Within five years, he had designed a contraption which he hoped would allow seafarers observe the ephemerides of Jupiter's satellites sufficiently accurately. In September 1617, therefore, he put his idea to the test, sitting in a chair in a small boat that was floating in a pool of water on the deck of a ship in the Tuscan port of Livorno. On his head, he was wearing a helmet with a telescope mounted to its eye slit, a *celatone* (from the Italian word 'celata,' which refers to a so-called 'sallet' helmet). Its visor could be adjusted to align the axis of the telescope with the eye of the observer, allowing him to follow Jupiter's moons, while the other (naked) eye could locate Jupiter itself. While Galileo managed to maintain a stable position, he conceded that even on land one's heart rate could cause Jupiter to rhythmically jump out of the telescope's field of view. Although he even sent one of his students out to sea to test the contraption under realistic conditions, the method never became popular and, as we saw already, the officers of the Spanish court-appointed *Casa de la Contratación* could not be convinced of the method's practical viability and accuracy. And, again, they rejected the astronomer's proposal. In the nearly 20 years since the announcement of the Spanish longitude prize in 1598, too many hopeful but woefully inadequate proposals had been submitted to the Spanish Court. The officers at the *Casa de la Contratación* had become experts in finding reasons to reject most of the lunatic ideas they received, and Galileo's proposal did not receive any better treatment. The officers judged that Galileo's proposed method was mostly impractical, because Jupiter's moons could not be traced sufficiently reliably, frequently, or accurately from moving ships to be relied upon for navigation, and certainly not during the daytime or in overcast conditions.

Huygens continued to impress on Diodati that their expedient development of an accurate timepiece could potentially result in significant geographic revisions. Diodati, meanwhile, had become impatient by the perceived slow progress of the committee of *theoristen*, while he was also less than pleased that Galileo's insights had been leaked, through correspondence between Beeckman and Mersenne, to one of the Italian scholar's main competitors, the French astronomer, astrologer, and mathematician Jean-Baptiste Morin de Villefranche (*Lat.*: Morinus; 1583–1656) [59]. However, Huygens—in his role as secretary to 'stadholder' (regent) and prince of Orange (that is, a member of the Dutch royal family) Frederik Hendrik— provided reassurances and explained that the committee had to assess both the theoretical merits of Galileo's proposed method and its practical application, and that that would take time [60].

Reael called Galileo's stabilizing platform solution impractical and he thought it implausible that his sailors could operate such a complex device, since they were …

'rude people, men only superficially acquainted with mathematics and astronomy
... and who still find insuperable the problem of using your discovery on a moving
ship, continually being tossed about' [61].

But to show their goodwill, the Staten Generaal offered Galileo a gold chain (a necklace) valued at 500 guilders [62]. However, his acceptance of this generous gift was expressly blocked by (the Catholic) Pope Urban VIII, prohibiting Galileo to accept a gift from a Protestant government. Meanwhile, Van den Hove sent Diodati a lengthy letter on 27 April 1637 aimed at reassuring him that Morin de Villefranche had not been provided with anything crucial to enable him to scoop Galileo [63]. Nevertheless, the entire enterprise became mired in misfortune. The official letter from Reael containing the initial response of the Staten Generaal did not reach Galileo until 23 June 1637 [64]. His detailed response with instructions regarding the practical applications of his proposed instruments, returned to Reael on 22 August 1637, arrived on Reael's death bed and as such remained unopened on his desk [65]: De Groot's brother-in-law, councillor Nicolaes van Reigersberch (1584–1654), notified the former in a letter dated 25 October 1637 that Reael had passed away on 10 October:

'The loss of two children through the contagious sickness [the plague] had
plunged this good man into such profound melancholy, that he forgot all other
thoughts and even those that were very dear to his heart. Even a letter from
Galileo Galilei, which was given to him when he was still healthy, remained
unopened, which I mention so that the said Galileo may be informed of it' [66].

In his response to Huygens' 1637 missive, dated 28 February 1640 [67], almost three years later, Diodati refers to these and other unfortunate events that had happened in the meantime. He says that Galilei's proposal, as submitted to the 'Messeigneurs les Éstats Generaux' (their Lordships of the Staten Generaal) has been affected by several interruptions. These included the complete loss of Galileo's eye-sight during the past two years and the untimely demise on 17 August 1639 of Van den Hove, who had until recently been the sole survivor of the four commissaires tasked with researching Galileo's proposal of using Jupiter's moons as navigational aids. Beeckman and Blaeu had died earlier, in May 1637 and October 1638.

Nevertheless, Diodati expressed Galileo's keen wish to forcefully continue his pursuit of the construction of a timing device to determine longitude at sea, despite the string of bad luck the project has experienced. He suggested that the Staten Generaal appoint new committee members. In fact, Diodati asked Huygens Sr in exceptionally flattering language to take on this task himself, on behalf of Galileo, given the former's expertise in the subject at hand. The Staten Generaal indeed briefly considered convening a new committee, in 1640, but nothing materialized [68]. Despite Diodati's impassioned plea to Huygens, I have not uncovered any evidence of the latter's involvement since receiving Diodati's letter from 1640. One additional problem affecting sustained progress was that the poor quality of the lenses used in telescopes available in the Low Countries was insufficient for Galileo's purposes [69,70,71]: not only was the glass often of low quality, the most commonly used lenses were also seriously affected by both chromatic and spherical aberration, that is, by lenses failing to focus all colours in the same focal point, also depending on where on the lens a light ray would hit. Beeckman dedicated himself to polishing telescope lenses in order to facilitate much-needed improvements, but he passed away long before good-quality lenses became available to Dutch scientists.

3.2 Christiaan Huygens, inventor of the pendulum clock

3.2.1 A rising star in science

Following this early exchange of letters, the next time longitude determination is discussed in the *ePistolarium* corpus of 17th Century letters is 15 years later, in 1655. The young, 26-year-old Christiaan Huygens, son of Constantijn Huygens, is tasked by the Staten Generaal of the Dutch Republic to assess a proposed invention by Jan Kołaczek of Leszno (or Lissa; *Lat.*: Johannes Placentinus, Lesnensis; 1629/30–1683/7) [72], a professor of mathematics from Francfort (Frankfurt) an der Oder in Prussia (located on the present-day border between Germany and Poland), for the purposes of possibly patenting the idea. Christiaan had already caught the attention of a number of leading thinkers of the era; for instance, on 15 June 1646 Descartes described the young Huygens' promise as a scientist in a letter to David le Leu de Wilhem (1588–1658), the German adventurer, philosopher, and orientalist who had by that time settled in Leiden and married Constantia Huygens, Christiaan's aunt:

'Some time ago, Professor [Van] Schooten sent me a manuscript compiled by the second son of Mr. van Zuylichem about a mathematical invention he had been pursuing [73], also commenting that he had not quite solved the problem (which was not at all strange, given that he had sought something no one had ever found before); he had approached the problem in such a sophisticated manner that I am convinced that he will become an excellent practitioner of this type of science, where I know of hardly anyone who knows anything' [74].

Huygens' charge was a careful assessment of Placentinus' invention, which the latter had dedicated to the Electoral Prince of Habsburg. It purported to allow the '*determination of East and West*' using observations of the Moon, ...

'provid[ing] a way to find the longitude of places, both on land and at sea, at any time, day or night, and in this way, given the latitude of the location and having found its longitude, to determine the position of a ship as it is being swayed by storm and wanders back and forth, etc.' [75].

For Huygens Jr to be able to reach a decision on the suitability of Placentinus' method, in March 1655 he wrote [76] to the scholar and preacher Andreas Colvius (1594–1671) from Dordrecht, in the Dutch Republic, ...

'I expect, in turn, that you will send me manuscripts regarding the determination of longitude and whatever else you own from Galilei's legacy'.

Galilei's legacy which Huygens refers to included correspondence with Diodati, De Groot, Van den Hove, Reael, Alphonse (Alfonso) Pollot(to)—an Italian officer employed by the government of the Dutch Republic, with whom Galileo had exchanged letters about the determination of longitude around 1637—and Constantijn Huygens [77]. In 1622 Colvius, a personal friend of Beeckman, had accompanied Johan Berck (1565–1627), the first Dutch ambassador to Venice, as chaplain. He used this opportunity to meet Italian scholars and collect and copy numerous books and manuscripts, including Galileo's unpublished work *Del Flusso e Riflusso del Mare* (*Dialogue on the Ebb and Flow of the Sea*) in which Galileo compared the new Copernican system with the traditional Ptolemaic system. This manuscript was eventually published in 1632, bearing the title *Dialogo Sopra i Due Massimi Sistemi del Mondo* (*Dialogue Concerning the Two Chief World Systems*), a change made under pressure from the Inquisition. Upon Huygens' request, Colvius subsequently sent the young scholar his notes, as well as a manuscript written by Galileo, attached to a letter dated 23 March 1655 [78].

Placentinus provided tables of the maximum lunar elevation as well as of the constellations Leo (the Lion; specifically for the lion's tail) and Lyra, calculated for the Frankfurt (an der Oder) meridian (that is, due south) from April through June 1655, in addition to step-by-step instructions to determine one's longitude:

'First, measure during the months of April, May, and June of 1655 at your location the time at which the Moon reaches its highest elevation, either during the day as derived from the solar elevation or at night by determining star heights; such observations are not unknown to those who practice math, and also to sailors they are hardly a secret.

2°. Compare this time, observed at your meridian, with the time indicated in the table, and note the difference in the meridians between your location and that of Frankfurt [an der Oder], in hours and minutes.

3°. Convert the observed meridian difference into degrees and minutes away from the Equator, according to the second table, and you find the difference in longitude between Frankfurt [an der Oder] and your location.'

Huygens' assessment of Placentinus' invention was that the proposed method was wholly unusable since it violated basic astronomical principles, including the assumption that the Moon would traverse 15 degrees on the sky in an hour, just like the Sun. This is indeed incorrect; the Moon's orbit covers approximately half a degree less per hour than the Sun's (see also chapter 2.2.2). As such, Huygens referred to the proposed invention as '*Placentinus' nonsense*' [79]. He set out by stating that Placentinus' proposed approach had …

'by no means as good a basis as those of others before him, who have tried to achieve the same. Their [others'] inventions, although they were considered of

little or no use (because of major miscalculations that could arise from the smallest observational errors or imperfections in their Ephemeris tables), however, were theoretically well founded. But this discovery of Johannes Placentinus is so far removed from offering any benefit or use, that it even sins against the foundations of astronomy, and it is nothing but a gross fallacy' [80].

Although most likely unintentional, this shows a character trait that comes back time and again in his letters, throughout most of his career: while he tends to be very polite, even flattering, to his correspondents, he clearly is a man who does not suffer (those he considers) fools easily nor does he have much patience for those he believes to be lacking in education or insight. He often comes across as arrogant and clearly considers himself the better scientist or even the better person. And that attitude includes his closest family members... For instance, in a letter of 3 September 1646 to his brother Constantijn Jr we read,

'I therefore respond now to your most recent letter in which you sent me that geometric problem. You will find its solution in my comments, which I drafted as soon as I received it. To tell you the truth, you clearly show that you are not as well practiced in algebra, at least not to the same extent as me...' [81]

Perusing the vast number of letters left by Huygens, it becomes abundantly clear that he must have disliked one person in particular, notably the French philosopher Descartes, whom he got to know fairly well as a family friend of his father's. The language Huygens uses in some of the letters in which he refers to Descartes would not pass editorial review even at the present time...

Not everyone agreed with Huygens' assessment of Placentinus' proposal, however. The German–British polymath Samuel Hartli(e)b (c. 1600–1662), for instance, was rather impressed by Placentinus' ideas:

'Placentinus Bohemus Professor Mathematicus upon the Oder at Frankford hath published a Booke De Longitudine which is thought the best that ever hath been written and should not have beene thus plainly discovered the States of Holland and others having set so great a price and reward for it, which belike this Professor was ignorant of' [82].

Although Hartlib suggests that Placentinus was not aware of the great interest in viable solutions to the longitude problem shown by the Dutch government, I think that this statement is incorrect. After all, Placentinus applied for a patent from the Staten Generaal, specifically introducing his ideas as *'[n]ew and careful research of the longitudes of places, by Dutchmen, French, English, and Spanish most desired.'* This clearly reflects his awareness of the importance afforded to solving the problem by a number of European governments at the time.

In response to Huygens' dismissive assessment, the Dutch Republic's executive branch decided to consult a number of other experts for a second opinion, including

Frans van Schooten, professor of mathematics at Leiden University [83], and Jan (Joan) Willemsz Blaeu (1598 or 1599–1673) [84], a printer, cartographer, and renowned map maker from Amsterdam. Huygens and Van Schooten subsequently exchange numerous letters—all in Latin—between March 1655 and November 1656; Van Schooten hence became *de facto* Huygens' mentor during the early stages of his career. Their exchange included a copy of the *Opere di Galileo Galilei* (*Works of Galileo Galilei*). This is important in the context of our analysis of the Dutch efforts to determine longitude at sea, because the latter work included ideas about pendulums.

3.2.2 Opportunistic competition

Christiaan Huygens is, in fact, universally credited with the invention of the pendulum clock, although he conceded, in a concept letter in Latin from 13 August 1657 [85], that he had seen Galileo's description [86]. Nevertheless, Huygens adopted a much simpler escapement mechanism than that proposed by Galileo. A number of contemporaries had employed 'pendulum timing devices' before 1657, including the Flemish chemist, physiologist, and physician Jan Baptist van Helmont (1580–1644), Galileo, the Italian astronomer and Jesuit priest Giovanni Battista Riccioli (1598–1671), Johannes Hevelius (whom we will encounter soon again), as well as the English mathematician, astronomer, and Bishop of Exeter Seth Ward (1617–1689) in collaboration with the English polymath Robert Hooke. In fact, Hooke became one of Huygens' main competitors, vying for the priority of invention on numerous occasions. His ground-breaking contributions to clock development and pendulum theory are the focus of Patterson's lucid exposé, *Pendulums of Wren and Hooke*, to which I refer the interested reader for further details [87].

Riccioli's recommendation of the application of the pendulum as a timing device in *Book 2* of his *Almagestum Novum* (*New Almagest*, 1651) was particularly influential in triggering experimentation with pendulums for timing purposes in both England and the Low Countries. Riccioli had verified experimentally that the period of a small-amplitude pendulum is constant to better than two swings out of 3212. He aimed to develop a stable pendulum with a period of precisely one second— at that time a newly conceived unit of time also known as the 'second sexagesimal division of the hour,' that is, the hour subdivided into 60 smaller units, each of which would in turn be subdivided into an additional 60 units—that would maintain its swing amplitude for at least 24 hours; extensive experimentation and careful counting of pendulum swings over periods of 24 hours by a team of nine of his Jesuit brothers eventually resulted in pendulums that were indeed stable and reliable with periods within 1.85% of the desired value. In April 1642, Riccioli and his associates kept a '*seconds pendulum going for 24 hours, counting 87 998 oscillations,*' indeed close to the 86 400 oscillations expected. In the following months, he repeated this experiment twice and managed to improve the accuracy to better than 0.69% [88,89]. Huygens' invention of the pendulum clock in 1656 was aimed at automating this process by adapting the existing clockwork mechanism to count the pendulum swings and to sustain its motion in the presence of dissipative forces (friction).

Competition in this area between English and Dutch experimentalists was intense at the time. Hooke even stated that …

'Dr. Wren, Mr. Rook, Mr. Ball & others made use of an Invention of Dr. Wrens for numbring the vibrations of a pendulum a good while before Monr. Zulichem [Christiaan Huygens] *publisht his'* [90].

In addition to Hooke, the Curator of the Royal Society of London, some of the main actors on the English stage referred to here were the famous architect Christopher Wren, the astronomer and mathematician Lawrence Rooke, and the astronomer William Ball (c. 1631–1691), first treasurer of the Royal Society. All were members of the so-called Gresham College group, a loose collection of English scientists who are regarded as the main precursor to the Royal Society of London, and which had been meeting in London and Oxford from approximately 1645 (see chapter 1). The 26th verse of the *Ballad of Gresham College* (1661), a poem entitled '*In Praise of the Choyce Company of Philosophers and wits who meete on Wednesdays weekly at Gresham College*,' leaves no doubt as to one of their main aims [91]:

The Colledge will the whole world measure,
Which most impossible conclude,
And Navigators make a pleasure
By finding out the longitude.
Every Tarpalling[4] *shall then with ease*
Sayle any ships to th'Antipodes.

The dating of these early efforts is also supported by a report from the horologist Thomas Reid (1746–1831) that he had seen a report of an observation of a solar eclipse, made with a pendulum-based timing device in Oxford in February 1656 [92]. However, these early experiments do not challenge the fact that Huygens first combined the pendulum—a metal ball suspended by a silk thread—with the clock mechanism [93].

On 12 January 1657, Huygens wrote [94] to Van Schooten that …

'one of these days, I invented a new type of construction for a time piece, which can be used to measure times so accurately that there is more than a little hope that this can be used to determine the longitude, at least as regards travel on the seas.'

Huygens' personal notes, collected in his *Adversaria*, imply that he completed a prototype of the pendulum clock during the last days of December 1656. This is corroborated by a letter [95] to the French astronomer and mathematician Ismaël Boulliau dated 26 December 1657, in which Huygens stated that '*yesterday, it was exactly a year ago that I made the first model.*' His invention was to derive the pendulum's motion from that of the clock, while basing the clock's regularity on that

---

[4] *'Tarpalling' is old spelling for 'tarpaulin;' one of its meanings is 'sailor.'*

of the pendulum. The mechanical clocks with which Huygens must have been familiar were regulated by slowly falling weights. In turn, these would turn the clocks' gears. However, the rate at which the weights descended was irregular, and hence the clocks were highly inaccurate.

Huygens applied Galileo's pendulum law—the period of a pendulum is approximately independent of its swing amplitude—and used it to develop a system that combined a pendulum with a weight-driven clock. In 1656, he discovered through experimentation that a swinging pendulum was not truly isochronous, unless the arc it described was not completely circular. Huygens worked out that if the amplitude of the pendulum's swing changed, the period of the swing would also change. Indeed, in *Horologium* (1658), we read the following passage:

'It is asserted with truth that wide and narrow oscillations of the same pendulum are not traversed in absolutely equal time, but that the larger arcs take a little longer which it is possible to demonstrate by a simple experiment. For if two pendulums, equal in weight and length, are released at the same time, one far from the perpendicular, the other only a little deflected, it will be perceived that they are not long in unison, but that of which the swings are smaller outstrips the other' [96].

'Yet as I have said, my time piece is less likely to an inequality of this kind, because all the vibrations are of equal amplitude. Nevertheless, it remains not entirely free from inequalities, although these are very tiny, and as is needful, I intend to pursue the matter' [97].

A number of drawings attributed to Huygens, as well as the clocks constructed by his master clockmaker, Salomon Hendriksz Coster (1620–1659) of The Hague, provide clear evidence that Huygens and Coster attempted to somehow alter the bob's path to overcome this problem, known as the 'circular error.' This latter error is, in essence, owing to the fact that the restoring effect of the gravitational force increases as the sine of the angle of swing; it causes a decrease in the period of oscillation with increasing amplitude. Huygens and Coster empirically designed clocks that had curved metal plates—'chops' or 'cheeks'—on both sides of the pendulum suspension, which were placed such that the pendulum thread wrapped around them over arcs of approximately 50° on both sides. Records from Huygens' *Oeuvres Complètes* (*Complete Works*) reveal experiments with curved chops done in May and June 1657 [98]. In notes dating as far back as 1659, Huygens calculated that *cycloidal* chops—forcing the bob of a simple pendulum to swing along a cycloidal path with a vertical axis equal to half the pendulum's length instead of the simple pendulum's circular arc—would render the pendulum isochronous. We will return to this idea in detail in chapter 4; specifically, in chapter 4.1.2 we will use modern calculus to derive the isochronicity of the cycloidal swings.

A cycloid is the path followed by a point on the circumference of a circle as that circle rolls along a straight line: see figure 3.9. Cycloids have been investigated by many of the leading thinkers throughout modern history. They were likely first studied by the German philosopher, theologian, jurist, and astronomer Nicolaus de Cusa (1401–1464), Cardinal Cusanus, in 1451. Many of the world's great

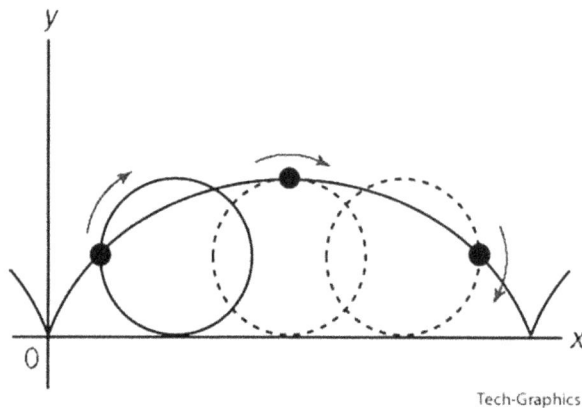

Tech-Graphics

Figure 3.9. A cycloid is the path followed by a point on the circumference of a circle as that circle rolls along a straight line. (Source: Wikidot/MathOnline; licensed under the Creative Commons Attribution-ShareAlike 3.0 License.)

mathematicians and physicists spent considerable effort on understanding their properties, including Charles de Bouvelles (Carolus Bovillus, 1471–1553; France), Gilles Personne de Roberval (France), Galileo (Italy), Evangelista Torricelli (1608–1647; Italy), Descartes (France), the Bernoulli family (in the second half of the 17th Century; originally from Antwerp, later resettled in the Swiss Confederacy), Pierre de Fermat (France), Gottfried Wilhelm (von) Leibniz (Germany), Wren (England), and Blaise Pascal (France). Indeed, by the mid-1660s, the cycloid was probably the most studied curve in the history of mathematics.

Earlier, in 1646, the young Huygens had corresponded with Mersenne about Galileo's achievements and insights into the pendulum. It is often reported [99] that, in 1644, Mersenne may have been the first to confirm experimentally that the pendulum's period of oscillation is inversely proportional to the square root of its length. He used his pendulum for timing purposes and is said to have specifically recommended the use of pendulums as timekeeping devices to Huygens during their 1646 correspondence. Through their correspondence, Huygens likely also learnt about work undertaken at that time to understand the properties of the cycloid. His own work on this problem was acknowledged by Pascal in 1658 and 1659. Huygens eventually published his geometrical proof that the cycloid was indeed the shape of the curve required for isochronicity in his *Horologium Oscillatorium* in 1673. We will return to the mathematical aspects of this derivation in chapter 4.

By the end of the 1660s, British investigations in support of the development of vibration theory had resulted in similar insights to those attained by Huygens. The English mathematician William Neile (1637–1670), as well as Christopher Wren, pursued 'rectification' (that is, determination of the length) of the semicubic parabola in 1657, followed by the cycloid in 1658, which John Wallis (1616–1703)—the English mathematician who is partially credited with the development of infinitesimal calculus and with introducing the symbol '∞' for infinity—published in his *De Cycloide et Cissoide* in 1659 [100,101] .

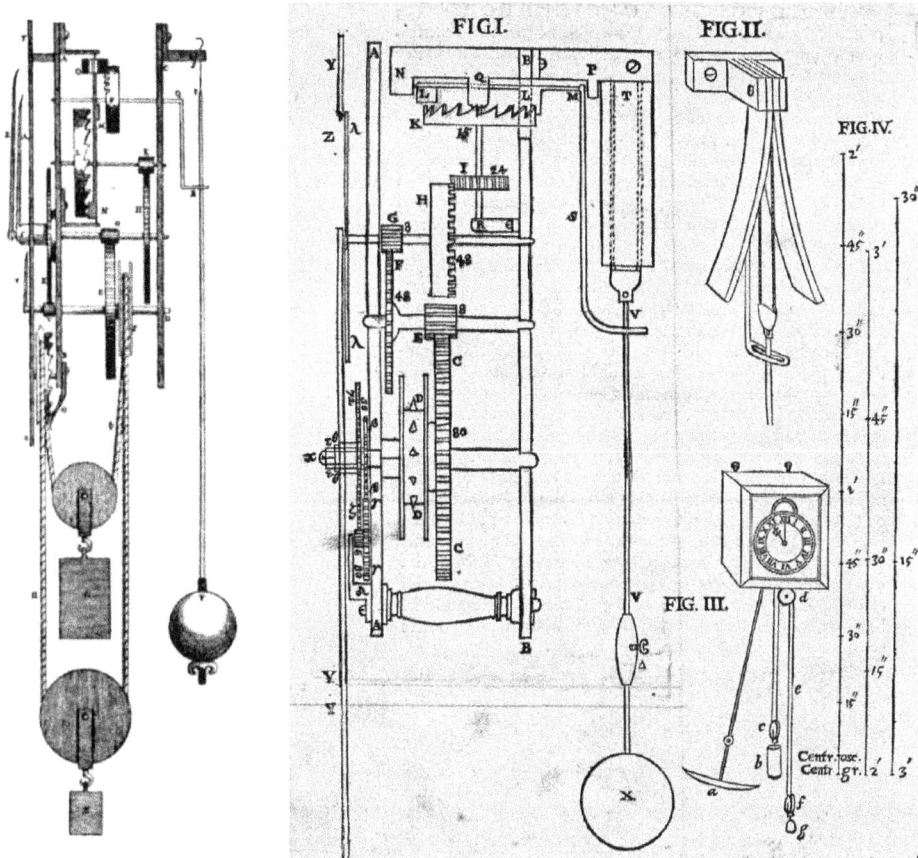

Figure 3.10. (*Left*) Earliest published drawing of Huygens' pendulum clock, from his 1658 treatise *Horologium* (*Opera Varia*, Vol. 1, S.1). (*right*) Plate from Huygens' *Horologium Oscillatorium* (1673), showing the curved chops in Fig. II.

The left-hand panel of figure 3.10 shows what has long been assumed to be the earliest surviving illustration of a clock by Huygens (published in 1658), which is in fact his second type of clock. The device is equipped with a vertical verge escapement and a carefully adjusted gear train to reduce the pendulum's amplitude, but no cycloidal chops. It beats half-seconds. The right-hand panel of the same figure represents a conventional, seconds-beating pendulum clock with isochronous chops from Huygens' *Horologium Oscillatorium* (1673). It is likely, however, that this latter was copied from the original design submitted with his patent application, as suggested by Benjamin Martin (1705–1782) in his manuscript *Newtonian Mathesis* (1764), reproduced in the left-hand panel of figure 3.11:

'*The construction for application of the pendulum is now somewhat different from what it was in the original invention of the pendulum clock by Mr. Christian Huygenius of Zulichem in Holland which he first described and published in a diagram cut in wood in the year 1657. And as this may be justly esteemed as one*

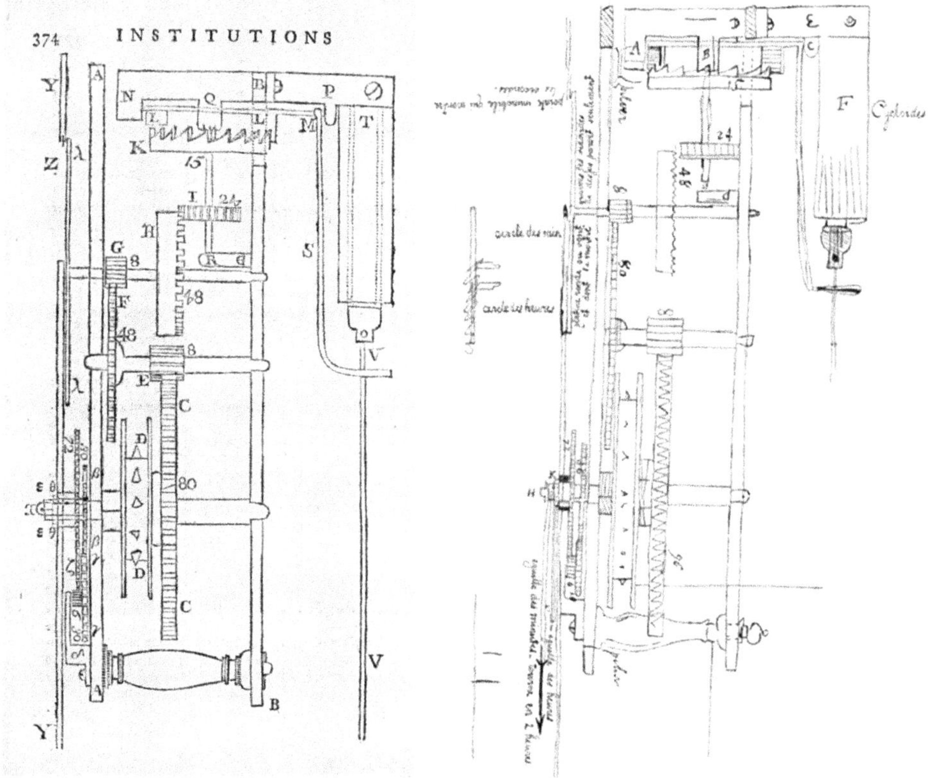

Figure 3.11. (*Left*) copy of Huygens' woodcut of 1657 published by Benjamin Martin in *Newtonian Mathesis* (1754). (*Right*) drawing of Huygens' original design in a letter to Canon Estienne (1669).

of the great curiosities of art and was never (that we know of) exhibited to the view of an English reader, we shall here present it, cut in wood exactly as the original.'

A number of elements in the right-hand panel of figure 3.10 suggest that this sketch is not a 1670s version of Huygens' pendulum clock. In the latter case, we would likely have been presented with a model that had undergone improvements to the extent one might have expected given the decade and a half that passed between his original patent applications and the publication of *Horologium Oscillatorium*. The boat-shaped pendulum bob (see Fig. III), the large-diameter escape gear wheel, and the seconds disc all raise red flags. If Huygens had wanted to include an up-to-date pendulum clock model representing the state of the art in 1673, he would most likely have added a spring-driven remontoire as well as the anchor escapement in common use by that time (see chapter 4). The clock's detailed description in *Horologium Oscillatorium* is much more reminiscent of the successes attributed to the original model than of some arbitrary, somewhat outdated later version.

There is additional evidence suggesting that a drawing of Huygens' first clock had been circulated before his publication of *Horologium Oscillatorium* [102]. This includes a French engraving of 1671, while a similar clock with a great wheel of 96 teeth is illustrated in a letter from Huygens to Canon Estienne (d. 1723) in 1669 [103]: see the right-hand panel of figure 3.11.

On 1 November 1658, Huygens recalled in a letter to his friend Pierre Petit (1594/ 8–1677), French physical scientist and instrument maker, …

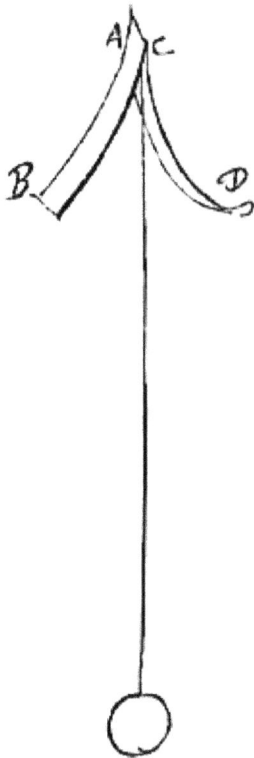

'at first I suspended the pendulum between two curved plates …, which by experiment I learnt … how to bend … And I remember having so well adjusted two clocks in this manner that in three days they never showed a difference of even seconds although in the meantime I often changed their weights rendering them heavier or lighter' [104].

He thus solved the most important problem underlying Galileo's earlier concept—which Vincenzo Galilei had been unable to overcome—by ensuring that the period of his clock's pendulum was truly constant. He did so through implementation of a pivot (the point of rotation) that caused the bob to swing along the arc of a cycloid instead of that of a circle. He made his clock's pendulum swing exactly once each second, allowing him to precisely regulate the motion of the clock's hands. The falling weight would drive the gears, while transferring just enough energy to the pendulum to balance the effects of air resistance and mechanical friction. This novel design almost instantly rendered clocks more reliable and accurate, improving their randomness from approximately half an hour a day for the existing mechanical clocks to better than one minute a day for Huygens' new device.

While Huygens was in Paris in late 1660, his younger brother Lodewijk took one of the new pendulum clocks with cycloidal chops on the first ever long-distance sea trial while on a diplomatic mission to Spain. Although Lodewijk reported with a heavy heart that pendulum clocks were rather useless on a pitching and rolling ship, Christiaan was apparently not so bothered by this news. He argued that the same storm his brother had sailed through had apparently also wrecked five ships of the Dutch merchant marine, and that he was more interested in his clocks' performance in reasonably temperate weather [105]. After all, he continued, in extreme weather the clocks could always be stopped to protect them from suffering damage and subsequently be reset at the next port of call. This shows a major flaw in his thinking, however: positional accuracy would seem most important after having been tossed about by strong winds...

3.2.3 Patent wars

It must have taken Huygens considerable time to experiment and perfect his timepiece before announcing it to the community of scholars—in fact, in the passage taken from his letter to Pierre Petit he refers to his experimental efforts. In response to the concerns expressed by the Flemish astronomer Godfried Wendelinus (1589–1667) in relation to performance challenges caused by seasonal variations, including temperature changes, Huygens writes, *'to me certainly it was not given to observe anything of this kind'* [106]. This implies that he had tested his new pendulum clocks, possibly for many months, prior to its public display. In addition, his patent application, which was filed on 14 June 1657, included a detailed description of the profile of the chops. It stands to reason that it would have taken him quite some time to determine the optimal shape—as also suggested by the passage from his letter to Petit—which implies that most of his experiments likely took place during 1656, culminating in a working prototype by Christmas day of 1656.

Huygens was, at least initially, keen to discuss his new invention with Van Schooten and other contemporaries, including the French mathematician Claude Mylon (1618–1660), member of the *Académie Parisienne* (the Parisian Academy). He proudly announced his invention to Mylon in a letter dated 1 February 1657, although without going into any detail [107]:

> *'The news that you tell me regarding the journey of Mr. Bulliaut [Boulliau] to these lands rejoices me very much ... I also want to show him one of my new inventions, which will be of the greatest benefit to astronomy, and I sincerely hope to apply it successfully in search of longitude. You may hear more about in a little while.'*

Huygens' invention of the pendulum clock must have rapidly made the rounds among the European scholarly community—even the Grand Duke of Tuscany had been provided with a copy by December 1657—given that Mylon responded to him, in a letter [108] dated 12 April 1657, that ...

> *'[y]our invention of a time piece is thought of very highly by all with whom I have discussed it. It will be even better if you can make sure that it is not affected by changes to either counterweight or spring. [109] With such a watch you can derive a true equation of time, as required to determine the longitude.'*

It appears, therefore, that by April 1657, Huygens—perhaps assisted by Coster—had managed to construct a clock that worked well enough to be copied and catch the attention of the intellectual community. This impression is corroborated by his response to a letter from Boulliau, in which the latter referred to a clock Huygens had shown him during his visit in April 1657 [110].

In that same letter to Boulliau, dated 26 December 1657 [111], Huygens was clearly pleased to announce the imminent conversion to pendulum control of a clock on a church steeple in a nearby village, *'near the sea, half a mile from here,'* likely the

Figure 3.12. The church of Scheveningen, after a drawing by Christiaan Huygens of 29 July 1658. (*Oeuvres Complètes de Christiaan Huygens*, Vol. **XVII**, p. VI)

church clock of Scheveningen (see figure 3.12). Huygens and Coster apparently modified the clock at Scheveningen by adding a pendulum of some 21 *pied* (or 21 Paris feet; approximately 493 cm) long and weighing 40–50 *livres* (pounds; one *livre* is equivalent to 403 grams) [112], although there are no contemporary records of the conversion, nor of any subsequent tests. They next modified a tower clock in Utrecht, using a rigid, 24 *pied*-long pendulum that weighed 50 *livres* and was equipped with six *pouce* (inch)-long suspension cords. Huygens included a number of sketches in his letter—reproduced in figure 3.13 [113]—which, he said, demonstrated the principle of pendulum clocks.

Note that the pendulums in his sketches do not include chops. Based on these sketches, it seems possible that Huygens may have produced his first pendulum-controlled clock by modifying an existing clock such as a spring-driven table clock, which were common in the early 17th Century. (Such a clock, characterized by swing amplitudes of order 30° but without having been equipped with chops would also be subject to significant deviations from isochronicity.) Also, note the gear configuration in the third sketch from the left, which shows the simple maintaining-power system now commonly referred to as a '*Huygens' endless chain*.' Huygens invented an early version of this mechanism, known as a 'maintaining-power' or 'going barrel,' around 1658 (based on its publication in *Horologium*; see the left-hand panel of figure 3.10) in order to keep a clock going while winding. During the latter operation, the driving force of the weight or main-spring is taken off the going barrel. Wheels **A** and **B** are the 'going-wheel' and the 'striking-wheel.' The system is weight-driven by two weights, **G** and **H**. When pulley **D** is pulled down by chain section **K**, the ratchet-pulley (**B**) runs under the click (**E**; shown in detail in the rightmost sketch), and the pulley **C**/weight **G** combination is pulled up without taking its pressure off the going-wheel, thus allowing continuous, smooth operation also while winding.

Despite the numerous technical details about his clock's operation, which he provided in his letter to Boulliau, Huygens did not explain why he introduced the

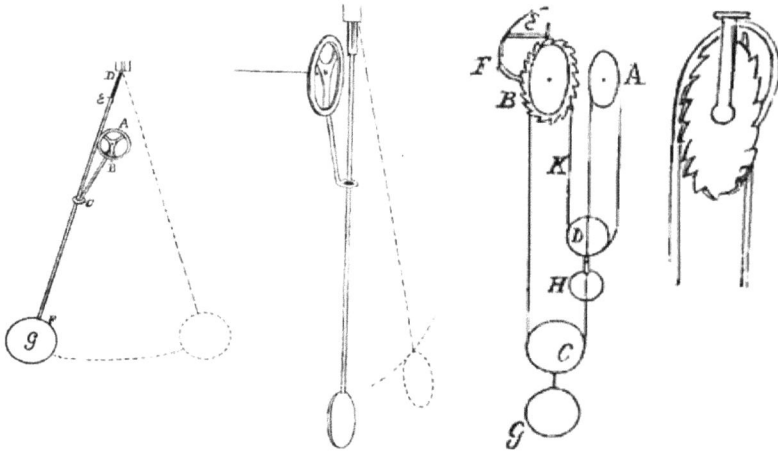

Figure 3.13. Huygens' sketches of pendulums and their application in mechanical clocks sent to Jean Chapelain of the *Académie Française* in a letter dated 28 March 1658.

crutch (shown as the rod between **B** and **C** in the leftmost sketch in figure 3.13). Emmerson speculates [114] that Huygens may have been concerned that, without it, the pivot (**D**) and balance cock (the bridge that carries the top pivot of the balance staff) would be subject to unwanted non-central (lateral) loads. He would also have discovered that if the pendulum string had to remain straight while receiving impulses from the escapement and without adding an exceedingly heavy bob, the pendulum had to be driven near the bob. Through addition of a rigid pendulum rod, needed to accept the impulse torque or the lateral impulse force, the crutch could be shortened and the pendulum arc could consequently be reduced. However, replacement of the silk thread by a pendulum rod would have rendered his pendulum a compound rather than a simple pendulum, as we will see in chapter 4.1.3. Alternatively, he may have introduced the crutch to aid the transfer of torque from the spring and the escapement to the pendulum itself, despite the increased slip associated with this configuration [115].

Almost simultaneously to these developments, in June 1657, Huygens contacted Samuel Carolus Kechelius ab Hollenstein ('Kechel;' 1611–1668), Bohemian mathematician and an astronomer loosely affiliated with Leiden University, to discuss the merits of pendulum clocks—and the nature of their mechanical components—for accurate timekeeping [116]. Through Boulliau's correspondence, we know that Kechel kept careful observational records of a series of solar and lunar eclipses, using either a water clock or counts based on the swings of a single pendulum to accurately determine the eclipse durations. Huygens admitted that he was inspired by Kechel's pendulum-based timekeeping efforts, given that he specifically told him that …

> *'[t]he stimulus for this discovery was provided by those pendulums, which you have been using for quite a few years already. When I noticed that because of the remarkable consistency of their swings they would be particularly suitable for*

accurate time keeping, I began to wonder whether it would be possible to turn their swings into a continuous movement so that they can be used to eliminate the tedium of counting' [117].

Despite having engaged in this initial period of a relatively open exchange of letters and information, Huygens soon realized that he could patent his invention. This realization led to more cautious, almost secretive correspondence. Both the national and provincial governments, the Staten Generaal [118] and the Staten van Hollant ende Westvrieslant [119], awarded patents for the exclusive development, manufacturing, and sale of pendulum clocks in their respective territories for periods of, respectively, 21 and 20 years to Huygens and Coster in June and July 1657. Upon Coster's death in 1659, the *privilege* to make these so-called Royal 'Haagse' (Hague) clocks was transferred to the clockmakers Claude Pascal of Geneva (Swiss Confederacy; d. 1671 or 1672) and Severijn Oosterwijck of Middelburg (Dutch Republic; 1637?–1694). The patents specified that these timepieces had certain characteristics, including a different movement mechanism from anything used previously, leading to more accurate timekeeping that was not affected by changes in weather or mechanical faults.

From the time of completion of his first working prototype until filing for a patent, Huygens continued to make improvements and refinements to his timepiece, in particular, improvements to maintaining power, development of an optimal gear train, a pendulum crutch with flexible suspension, and isochronous chops. These efforts required professional training and lengthy experimentation. Although Huygens filed his patent applications jointly with his master clockmaker Coster, there is historical confusion as to whether the latter was also involved in the construction of the prototype pendulum clock [120]. Instead, it may be more likely that the original design was Huygens' own and that Coster was appointed as the scientist's commercial partner: in the patent award, the committee appointed by the Staten Generaal of the United Netherlands refers to the invention as *'done* [gepractiseert] *by Huygens,'* with permission for Coster to copy (*'naer te maeken'*). On 19 October 1658, the clockmaker and surveyor Johan van Kal (Jan van Call) from Batenburg, near Nijmegen, also in the Dutch Republic, was—jointly with Coster—also given the *privilege* to apply pendulums to clocks [121]. Van Kal came highly recommended by Huygens himself:

'There is in Nijmegen a clockmaker named Master Jan Cal, who is an honest man and a renowned maker. He will visit you an bring on my behalf a description of these clocks, which I have issued [Horologium]. *Should there be any turret clocks that need to be made or changed to the new invention, I wish that he be .* *employed, rather than someone sent by [Rotterdam clockmaker Simon] Douw. Because I have informed him of the best method, which through experience has already proven to be good in several turret clocks; whereas Douw's has nowhere yet been put in working order'* [122].

Huygens and Coster intended to completely dominate the new pendulum clock market. The improvement in timekeeping accuracy achieved by Huygens' new pendulum clock design was so significant, however, that other clockmakers could not idly stand by. To the significant distress of Huygens and Coster, a year after their patent applications had been granted—in August 1658—the Staten Generaal awarded [123] an almost identical patent to the Rotterdam clock and watchmaker Simon Douw (c. 1620–1663), according to Huygens '*under false pretences.*' Huygens had somehow heard of Douw's patent application and proceeded to (apparently) falsely accuse Douw of plagiarism. He alleged [124] that Douw had secretly examined the inner workings of the Scheveningen public pendulum clock on 15 April 1658. He formally attested to this allegation and had it legalized by Hermanus de Coninck, notary public, in the presence of a well-respected witness, Frans van Schooten, Huygens' long-time mentor.

Huygens and Coster therefore jointly sued Simon Douw to prevent him from selling his competing system, evidently mostly concerned about their potential income and not so much worried about their intellectual priority. In this context, we get to see Huygens from a different side again. In letters to John Wallis [125], to his cousin Willem Pieck [126], and to his brother Lodewijk [127], as well as in a lively exchange with Van Schooten [128], dated throughout September and October 1658, he calls Douw's behaviour '*shameless and criminal*' and the perpetrator himself '*a stupid and shameless man.*' However, in their patent resolution of 8 August 1658, the assessors of the Staten Generaal rightly pointed out that Douw had corrected for the presence of irregular escapement forces in all Huygens–Coster systems using a *spring* remontoire with a single vertical beam escapement, and hence that Douw's pendulum clock was fundamentally different from that of his competitors, which was equipped with a *weight* remontoire. (The gear train of remontoire clocks contains a secondary spring or weight, which drives the clock's escapement; it is rewound regularly by the mainspring or the main weight, thus exerting a steady force on the escapement and ensuring the clock's smooth operation.) Douw's (re-) design was much better suited for marine purposes than Huygens', an insight the latter did not appreciate for a number of years as he continued to experiment with weight-driven pendulums. Nevertheless, until his death in September 1663, Douw remained tight-lipped about his intentions regarding the use of his clocks at sea.

But even their expert witness, Van Schooten, did not fully appreciate Douw's fundamental contribution: the latter had realized that Huygens' remontoire created a train discontinuity. Douw was so confident of the validity of his own improvement to the pendulum clock system pioneered by Huygens that he challenged Van Schooten to a bet for 1000 guilders—a bet which Douw won. Devastatingly for Huygens and Coster, the plagiarism allegation was also refuted by a number of professional clockmakers. The controversy [129] and legal wranglings [130,131] continued well into 1659, without any clear resolution. Meanwhile, the presiding Court forced the plaintiffs into a humiliating settlement by December 1658: one-third of all future profits from their pendulum clock would have to be shared with Douw, who was also given permission to sell clocks of his own design in The Hague.

With Huygens' invention apparently out in the open, opportunistic competition caused him significant difficulties in monetizing his work. His unpublished but ubiquitous 1657 clock caused a serious conundrum, however. Without a license signed off on by Huygens, any other clockmaker who wanted to jump onto the bandwagon and share in this profitable new environment had no other choice but to either directly copy ('plagiarize') or re-invent a Coster clock already at their disposal. However, some clockmakers managed to independently design their own pendulum-based timepieces. Among the latter, the most noteworthy partnerships are those of the astronomer Johannes Hevelius of Danzig (1611–1687) in collaboration with the clockmaker Wolfgang Günther [132] and of the brothers Giuseppe and Matteo Campani, who were given the exclusive manufacturing right for silent clocks by Pope Alexander VII in 1656. Clockmaker Johann Philipp Treffler (1625–1698) of Augsburg (Germany) apparently copied Coster's design in 1658 to construct a pendulum clock for Grand Duke Ferdinand II de'Medici of Tuscany (Italy). Crucially, however, he added a fusee—a cone-shaped pulley surrounded by a helical groove, wound with a cord or chain attached to the mainspring barrel, now known as a German or 'Augsburg' fusee (see chapter 4)—to Coster's design, which represented a major improvement in its own right.

Historians generally assume that Huygens permitted Coster to provide training on pendulum clocks to foreign clockmakers, who would then become licensees. One of the latter was John Fromanteel (1638–1692?), son of the illustrious London clockmaker Ahasuerus Fromanteel. On 3 September 1657, Coster drew up a Notarial Act or 'Contract' with the younger Fromanteel before The Hague public notary Josua de Putter and witnesses, enabling the young apprentice to learn about Huygens' new pendulum clocks [133]. In 1658, once John Fromanteel had returned to London with his new skills, father and son Fromanteel made the first English pendulum clocks using a verge and a short pendulum. This development has led historians to wonder about the proper context in which this transfer of skills had occurred. All available evidence suggests that neither Huygens nor Coster ever intended to freely distribute the design of their short-pendulum clock, their most profitable design. Yet, it appears that the Fromanteels managed to acquire their skills and license, as we probably should interpret the Contract [134], without any monetary promises. Perhaps to divert attention of would-be competitors from their most prized design, in *Horologium* Huygens published his first pendulum clock design in detail, albeit of the so-called 'long pendulum' with its complex 'OP gearing' (see chapter 4), which was applied to limit the pendulum's swing amplitude.

It is highly likely that Huygens published his pendulum-controlled clock design in *Horologium* to establish his priority of invention. By that date, September 1658, Huygens had ascertained that the 'circular error' associated with the operation of anisochronous pendulums was more dominant for larger swing amplitudes. He wrote, specifically,

'With large arcs the swings take longer, in the way I have explained, therefore some inequalities in the motion of a timepiece exist from this cause, and, although it may seem to be negligible, when the clocks were so constructed that the movement of the pendulum was somewhat greater [than at present] I have used an appliance as a remedy for this also.'

His drawing in *Horologium* shows that to reduce the pendulum's amplitude while maintaining sufficient leverage for the verge to release the escapement crown wheel, Huygens had introduced a pirouette, that is, a contrate wheel (a gear wheel with teeth set perpendicularly to its plane) and a pinion wheel, between the crutch and the verge. The appliance referred to in the preceding citation most likely consisted of a pair of (possibly cycloidal) chops, although this is not indicated as such in the treatise. It appears that Huygens was not content that his chops worked properly, however,

> *'At the present time, certainly, this method is not the cure,'*

so for a while at least the Huygens–Coster clocks did not include these devices; instead Huygens opted for small swing amplitudes:

> *'Therefore, by rendering all the swings short, ... individual times are distinguished by no remarkable difference. ... doubling the driving weight does not thereby accelerate the movement of the pendulum or alter the working of the time piece, which was not so in all others hitherto in use'*[5].

However, Huygens seems to have had a number of changes of heart regarding his application of chops. Both in November 1658 [135] and in October 1659 [136] his notes suggest that he reconsidered introducing chops. In December 1659 he had gone back to experiments to ascertain their performance [137], as we learn from his correspondence with Van Schooten [138].

Construction of the pendulum clock included in Huygens' *Horologium Oscillatorium* is generally attributed to Isaac Thuret (c. 1630–1706), *Horloger Ordinaire du Roi* (clockmaker to the French Royal court) and clockmaker of the *Académie Royale des Sciences*. In the autumn of 1655, when Huygens spent 4–5 months in France, mostly in Paris, he would have had ample opportunity to meet Thuret, who was then rapidly developing into France's greatest 17th-Century clockmaker. Although Thuret is not mentioned in Huygens' correspondence prior to 1662, this could very well be explained either by lost correspondence (it is unknown how complete Huygens' body of surviving letters is in the *Oeuvres Complètes*), by Huygens' attitude that the clockmaker did not have anything to do with the actual invention [139], or equally by the need for secrecy. After all, Huygens was in direct competition with the Parisian clockmakers, as we will see below, and intellectual piracy was rife at the time, conditions Huygens appears to have been all too familiar with, given his complaint to this effect published in *Horologium*.

In June 1658, Boulliau promised [140] Huygens that he would try to gain access to the Parisian market for his friend. However, Boulliau subsequently reported [141] that Pierre Séguier—Chancellor of France from 1635—'*ne vouloit pas faire crier apres luy tous les maistres horologeurs de Paris*': he did not want to alienate the master clockmakers of Paris, and hence Huygens was not given the French patent— and the important royalties from licenses he so desired—despite three attempts to

---

[5] It appears that Huygens did not realize that the recoil of the verge escapement to some extent masked the effect of changing the driving torque

secure approval. His patent application in Italy was also refused, on the grounds that he had plagiarized Galileo, irrespective as to whether or not that was actually true. Competition but also cooperation with French clockmakers dominates part of Huygens' correspondence. For instance, in a letter of 12 April 1662 to his brother Lodewijk Huygens, who was conveniently based in Paris, Christiaan asked, ...

> 'how are these Thuret clocks made, for which my father pays 10 or 12 pistoles [gold coins] and [which he] prefers to his own? If we could know the form it could be used to instruct the clockmakers here ...' [142]

This passage is an enquiry about a specific type of clock with which Huygens and his colleagues in the Dutch Republic, under Huygens' patronage, were in direct competition. Despite Huygens' apparent respect for Thuret, the latter was also not shielded from Huygens's wrath for his direct plagiarism of other constructions, however, as we will discuss in detail in chapter 5.1.1.

Following a number of letters in 1664 among Huygens' family members that involve sending clocks for repair to Thuret, in 1665 Huygens received a letter from Jean Chapelain (1595–1674), French poet and literary critic who was a founding member of the *Académie Française* (the French Academy), in relation to a French patent eventually awarded to the Dutch inventor, ...

> 'that excellent clockmaker Monsieur Thuret, of whom you yourself have told me much good, visited me yesterday and asked me to offer you his services for the construction of clocks to be used on ships and for their sale and distribution' [143].

Huygens agreed to this request two weeks later [144]. In 1666 Huygens moved to Paris himself. He likely continued to employ Thuret. Yet, despite his praise for the workmanship of Thuret, the accuracy of Huygens' timepiece still required significant improvements before it could be used for the purpose it was originally invented for: longitude determination.

References and notes

[1] Heidegger M 1998 On the Essence and Concept of φύσις in Aristotle's *Physics* B, 1; *Pathmarks* ed W McNeill (Cambridge: Cambridge University Press) pp 183–230

[2] Galilei G 1632 *Dialogo Sopra i Due Massimi Sistemi del Mondo* (*Dialogue on the Two Chief World Systems*) (Florence: Giovanni Batista Landini)

[3] Quoted in: Reston J Jr (ed) 1994 *Galileo, a Life* (New York: Harper Collins) p 9

[4] Hawking S 1988 *A Brief History of Time* (New York: Bantam Books) p 179

[5] Drake S 2003 *Galileo at work: his scientific biography* (New York: Dover) pp 19–21

[6] *Ibid.* p 20

[7] Brown J R 2011 *The Laboratory of the Mind: Thought Experiments in the Natural Sciences* (London: Routledge)

[8] Stuart M T 2015 Thought experiments in science *PhD Thesis* University of Toronto pp 23–4

[9] Note that the provenance of this thought experiment has recently been called into question; Palmieri P 2005 'Spuntar lo scoglio più duro': did Galileo ever think of the most beautiful thought experiment in the history of science? *Stud. Hist. Phil. Sci.* **36** 223–40

[10] Sobel D 1995 *Longitude: The True Story of a Lone Genius Who Solved the Greatest Scientific Problem of His Time* (London: Walker Books) chapter 4

[11] In his *Discorsi e Dimostrazioni Matematiche Intorno a Due Nuove Scienze (Discourses and Mathematical Demonstrations Relating to Two New Sciences,* 1638), he claimed that the pendulum was isochronous for all arcs less than 180°, although that claim seems overly optimistic today.

[12] Drake S 1978 *Galileo at Work: His Scientific Biography* (Chicago, IL: University of Chicago Press) p 399

[13] Palmieri P 2009 A phenomenology of Galileo's experiments with pendulums *Brit. J. Hist. Sci.* **42** 479–513

[14] Drake S 1970 Renaissance music and experimental science *J. Hist. Ideas* **31** 483–500; Drake S 1975 (Jan–June) The role of music in Galileo's experiments *Sci. Am.* **232** 98–104

[15] Drake S 1994 Theory and practice in early modern physics *Trends in the Historiography of Science* ed K Gavroglu, Y Christianidis and E Nicolaidis (New York: Springer) pp 15–30

[16] Galilei G 1892–1909 (reprinted 1929–1939 and 1964–1966) *Le Opere di Galileo Galilei Edizione Nazionale* ed A Favaro (Florence: Barbèra)

[17] Drinkwater Bethune J E 1830 *The Life of Galileo Galilei, with Illustrations of the Advancement of Experimental Philosophy* Project Gutenberg e-Book p 84 www.gutenberg.org/files/43877/43877-h/43877-h.htm

[18] Drake S and Drabkin I E 1969 *Mechanics in sixteenth-century Italy* (Madison WI: University of Wisconsin Press) p 245

[19] Palmieri P 2009 *op. cit*

[20] MacLachlan J 1976 Galileo's experiments with pendulums: real and imaginary *Ann. Sci.* **33** 173–85

[21] Koyré A 1966 *Études Galiléennes* 1st edn 3 vols (Paris 1939)

[22] Galileo G 1638 *Discorsi e Dimostrazioni Matematiche Intorno a Due Nuove Scienze (Discourses and Mathematical Demonstrations Relating to Two New Sciences), Edizione Nazionale* vol **VIII** 277–8

[23] Palmieri P 2009 *op. cit.*

[24] Palmieri P 2009 *op. cit.*

[25] Emmerson A 2010 *Things are seldom what they seem—Christiaan Huygens, the pendulum and the cycloid* www.antique-horology.org/piggott/rh/images/81v_cycloid.pdf (App B) [accessed 11 July 2016; 23 August 2017]

[26] In his *Discorsi* (1638) Galileo refers to his pendulums as '*repeating their goings and comings a good hundred times by themselves*'; Blackwell R J 1986: transl. Huygens C 1673 *The Pendulum Clock or Geometrical Demonstration Concerning the Motion of Pendula as Applied to Clocks (Horologium Oscillatorium)* (Ames, IA: Iowa State University Press) 19

[27] Drinkwater Bethune J E 1830 *op. cit.*

[28] Galileo G 1638 *op. cit* vol **VIII** 276

[29] Galileo G 1638 *op. cit.* vol **VIII** 128–9, 277–8

[30] Palmieri P 2009 *op. cit.*

[31] Naylor R 1974 Galileo's simple pendulum *Physis* **16** 23–46; Galileo G 1638 *op. cit.*

[32] Balliani G. B 1638 *De motu naturali gravium solidorum et liquidorum*

[33] Drake S 1969 Galileo's 1604 Fragment on Falling Bodies *(Galileo Gleanings XVIII) Brit. J. Hist. Sci.* **4** 340–58

[34] de Grijs R and Vuillermin D 2017 *Measure of the Heart: Santorio Santorio and the Pulsilogium* Hektoen Int'l Winter 2017, Moments in History http://hekint.org/measure-of-the-heart-santorio-santorio-and-the-pulsilogium/; arXiv:1702.05211

[35] Levett J and Agarwal G 1979 The first man/machine interaction in medicine: the *pulsilogium* of Sanctorius *Med. Instrum.* **13** 61–3

[36] Sanctorius S 1631 *Methodi vitandorum errorum omnium qui in arte medica contingunt* (Genevae: P. Aubertum) p 289

[37] Mersenne M 1634 *Les méchaniques de Galilée* 7th addition p 77. *The privilége du roy* suggests that 30 June 1634 was the date on which the printing was completed: cf. Mersenne M *Correspondance* **IV** pp 76–77, 207–12

[38] Mersenne M *Correspondance* **V** 132; cf. **VI** 242

[39] Blundeville Th. 1594 *M. Blundeville His Exercises, Containing Eight Treatises* (London: William Stansby) 6th edn 1622 p 390

[40] Drake S 1978 *Galileo at Work: His Scientific Biography* (Chicago, IL: University of Chicago Press) p 419

[41] 1637-04-13: Huygens, Constantijn—Diodati, Élie; Worp J A 1892–1894, *De Briefwisseling van Constantijn Huygens* (http://resources.huygens.knaw.nl/retroboeken/huygens/#page=0&source=19; Rijks Geschiedkundige Publicatiën) ('s Gravenhage: Martinus Nijhof) No 1542

[42] Frisius G 1530 *De Principiis Astronomiae Cosmographicae (On the Principles of Astronomy and Cosmography—with Instruction for the Use of Globes, and Information on the World and on Islands and Other Places Recently Discovered)* (Antwerp: Johannes Grapheus)

[43] Pogo A 1935 Gemma Frisius, his method of determining longitude *Isis* **22** i–xix and 469–85

[44] Friendly M, Valero-Mora P and Ulargui J I 2010 The First (Known) Statistical Graph: Michael Florent van Langren and the 'Secret' of Longitude *Am. Stat.* **64** 174–84

[45] This idea was also championed by the English astronomer and mathematician Lawrence Rooke (1622–1662), one of the founders of the Royal Society of London, who proposed to treat the Moon's irregular surface as gnomons on a sundial. *Dictionary of National Biography* (reprinted, Oxford: Oxford University Press 1949–1950) **17** 209–210; Ward J 1740 *Lives of the Professors of Gresham College* (London) pp 90–95; Ronan C A 1960 Lawerence Rooke (1622–1662) *Notes and Records of the Royal Society* **15** pp 113–8

[46] Davids K 2011 Dutch and Spanish Global Networks of Knowledge in the Early Modern Period: Structures, Connections, Changes *Centres and Cycles of Accumulation in and around the Netherlands during the Early Modern Period* ed L Roberts (Zurich: Lit Verlag GmbH & Co. KG Wien) pp 38–39; Gould R T 2013 *The Marine Chronometer, Its History and Development* (Woodbridge: Antique Collectors' Club Ltd) p 12

[47] de Waard C 1939–1953 *Journal tenu par Isaac Beeckman de 1604 à 1637, publié avec une introduction et des notes par C. de Waard*, 4 vols (The Hague: Martinus Nijhoff) **4** 236; following accepted practice, I will henceforth refer to this collection of Beeckman's notebooks as the *Journal*. For details, see van Berkel K 2013 *Isaac Beeckman on Matter and Motion—Mechanical Philosophy in the Making* (Baltimore, MD: Johns Hopkins University Press) Note 10 (Introduction) p 188

[48] *Ibid.* **4** 236–237; cf; Garcia S 2004 *Élie Diodati et Galilée: Naissance d'un réseau scientifique dans l'Europe du XVIIe siècle* (Florence: Leo S. Olschki)

[49] *Ibid.* **4** 241–244; Drake S 1978 *Galileo at Work* p 374; 1637-08-15: Galilei, Galileo—Staten Generaal, *Oeuvres Complètes de Christiaan Huygens* **III** No 673d

[50] 1636-06-06: Galilei, Galileo—Reael, Laurens Laurensz.; *Ibid.* **III** No 673c

[51] 1636-08-15: Galilei, Galileo—de Groot, Hugo; No groo001/2712

[52] 1636-09-20: de Groot, Hugo—Galilei, Galileo; No groo001/2763

[53] *Journal* **4** 245–51

[54] *Ibid.* **4** 253

[55] *Ibid.* **3** 229–230. In 1614, Beeckman had suggested that the Moon might be used as a celestial clock ('hemelse klok'): *Journal* **1** 33–34; *Journal* **4** 62

[56] Galilei G 1892–1909 *op. cit.* **17** 46–9, 96–105

[57] *Ibid.* **17** 39–41

[58] Ariotti P E 1972 Aspects of the Conception and Development of the Pendulum in the 17th Century *Arch. Hist. Exact Sci.* **8** 329–410 specifically p 368; Galilei G 1892–1909 *op. cit.* **16** 96–9

[59] *Journal* **4** 261–2

[60] *Ibid.* **4** 267–8

[61] Galilei G. 1892–1909 *op. cit.* **17** 116–117 (22 June 1647); Ariotti P E 1972 *op. cit.* 329–410 369

[62] *Journal* **4** 267

[63] *Ibid.* **4** 270

[64] *Ibid.* **4** 258–9

[65] Galilei G. 1892–1909 *op. cit.* 96–105, 174–5

[66] *Journal* **4** 282

[67] 1640-02-28: Diodati, Élie—Huygens, Constantijn; Worp J A *op. cit.* No 2318

[68] Mersenne M *Correspondance de Mersenne* **IX** 172

[69] Keil I 2003 *Von Ocularien, Perspicillen und Mikroskopen, von Hungersnöten und Friedens- freuden, Optikern, Kaufleutern und Fürsten: Materialien zur Geschichte der optischen Werkstatt von Johann Wiesel (1583–1662) und seiner Nachfolger in Augsburg* (Augsburg) p 40, citing a letter of Gassendi to Galileo dated 19 January 1634

[70] *Ibid.* 48, citing a letter of Galileo to Diodati dated 6 June 1637

[71] Letter from René Descartes to Diodati, 13 April 1637, in: Worp J A *op. cit.*

[72] Omodeo P D 2016 Central European Polemics over Descartes: Johannes Placentinus and his academic opponents at Frankfurt on Oder (1653–1656) *History of Universities* **XXIX/1** ed M Feingold (Oxford: Oxford University Press) pp 29–64

[73] The young Huygens had attempted to find proof that a suspended cord or chain does not attain a parabolic shape, and how one should apply pressure on a mathematical cord (that is, a weightless cord) to achieve such a shape. See his letters of 1646 to Marin Mersenne in *Oeuvres Complètes de Christiaan Huygens* **I** No 14 (pp 27–28) and No 20 (p 34)

[74] 1646-06-15: Descartes, René—le Leu de Wilhem, David; *Oeuvres Complètes de Christiaan Huygens* **I** No 9 (pp 14–6)

[75] 1655: Placentinus, Johann; *Ibid.* **I** No 215 (p 320); appendix to 1655-03-04: Huygens, Christiaan—Staten Generaal; *Ibid.* **I** No 214 (pp 318–9)

[76] 1655-03: Huygens, Christiaan—Colvius, Andreas; *Ibid.* I No 217 (p 322)

[77] Galilei G 1892–1909 *op. cit.* **3** 123–191; with specific reference to letters from 1718

[78] 1655-03-23: Colvius, Andreas—Huygens, Christiann *Oeuvres Complètes de Christiaan Huygens* **I** No 218 (p 323)

[79] 1655-03-04: Huygens, Christiaan—Staten Generaal; *Ibid.* **I** No 214 (pp 318–9)

[80] *Ibid.*

[81] 1646-09-03: Huygens, Christiaan—Huygens, Constantijn Jr.; *Ibid.* **I** No. 11 (pp. 18–9)

[82] *The Hartlib Papers, Ephemerides*, 1655, Part 2 (February–21 April 1655) 29/5/15A-28B: www.hrionline.ac.uk/hartlib; this probably refers to an earlier version of the second edition of Placentinus' *Geotomia*.

[83] 1655-03-08: Staten Generaal—Huygens, Christiaan; *Oeuvres Complètes de Christiaan Huygens* **I** No 216 (pp 320–1)

[84] 1655-03-27: van Schooten, Frans—Huygens, Christiaan; *Ibid.* **I** No 220 (p 325)

[85] 1657-08-13: Huygens, Christiaan—unknown; *Ibid.* **II** No 400 (p 46). This incomplete letter is found on the reverse of a concept of letter No. 399 (Huygens Christiaan—de Sluse René François).

[86] Note that it is sometimes erroneously alleged that Huygens was unaware of Galileo's contributions to clockmaking; e.g Applebaum W (ed) 2000 *Horology Encyclopedia of the Scientific Revolution: from Copernicus to Newton* (New York: Garland Publ. Inc.) Sec IV D

[87] Patterson L D 1952 Pendulums of Wren and Hooke *Osiris* **10** 277–321

[88] Meli D B 2006 *Thinking with Objects: The Transformation of Mechanics in the Seventeenth Century* (Baltimore, MD: Johns Hopkins University Press) pp 131–4

[89] Heilbron J L 1999 *The Sun in the Church: Cathedrals as Solar Observatories* (Cambridge, MA: Harvard University Press) 180–1

[90] Robertson J D 1931 *Evolution of Clockwork* (London: Cassell & Co. Ltd) 168

[91] Howse D 1980 *Greenwich time and the discovery of the longitude* (Oxford: Oxford University Press) 15; for the full Ballad of Gresham College, see https://en.wikisource.org/wiki/Ballad_of_Gresham_College [accessed 16 February 2017]

[92] Reid T 1826 *Treatise on Clock and Watch Making—Theoretical and Practical* (Glasgow, Edinburgh, London: John Fairbarn) 184; Robertson *op. cit.* 129

[93] Note that Richard Harris of London is credited to have converted a church clock to employ a pendulum in 1642: Reid Th 1826 *A Treatise on Clock and Watchmaking* (Edinburgh) 179

[94] 1657-01-12: Huygens, Christiaan—van Schooten, Frans; *Oeuvres Complètes de Christiaan Huygens* **II** No 368 (p 5)

[95] 1657-12-26: Huygens, Christiaan—Boulliau, Ismaël; *Ibid.* **II** No 443 (p 109)

[96] Edwardes L E 1977 *The Story of the Pendulum Clock* (Altringham: John Sherratt & Son, Ltd) 51 (transl)

[97] Huygens C 1657 *Oeuvres Complètes de Christiaan Huygens* **XVII** 17–20

[98] *Ibid.*

[99] This information is frequently referred to in the literature, although it appears that there are no primary sources in support of this notion. See also Koyre A (ed) 1992 *Metaphysics and Measurement* (London: Taylor & Francis) p 100

[100] 1673-06-12: Wallis, John—Huygens, Christiaan; *Oeuvres Complètes de Christiaan Huygens* **VII** (No 1948 App II to No 1946) p 309

[101] 1673-10-18: Brouncker, William—Huygens, Christiaan; *Ibid.* **VII** No 1962 (pp 344–5)

[102] Whitestone S and Sabrier J C 2008 (December) The Identification and Attribution of Christiaan Huygens' First Pendulum Clock *Antiquarian Horology* 31/2 www.antique-horology.org/Thuret/Default.htm

[103] 1669-09-07: Huygens, Christiaan—Estienne; *Oeuvres Complètes de Christiaan Huygens* **VI** No 1759 (pp 490–2)

[104] 1658-11-01: Huygens, Christiaan—Petit, Pierre; *Ibid.* **II** No 546 (pp 270–4)

[105] 1660-12-18: Huygens, Christiaan—Huygens, Lodewijk; *Ibid.* **III** No 823 (pp 209–11) Lodewijk's letter to Christiaan Huygens is missing.

[106] Huygens C 1658 *Horologium* (The Hague) p 1

[107] 1657-02-01: Huygens, Christiaan—Mylon, Claude; *Oeuvres Complètes de Christiaan Huygens* **II** No 370 (p 7)

[108] 1657-04-12: Mylon, Claude—Huygens, Christiaan; *Ibid.* **II** No 382 (pp 22–3)

[109] Mylon continues to encourage Huygens, saying that '*I am glad that you are continuously perfecting your new clock and have no doubt that you will make it work just as well at sea as in your chambers, and that changes from dry to wet will not alter its accuracy anymore than changing the clock's weights.*'—1657-05-18: Mylon, Claude—Huygens, Christiaan; *Ibid.* **II** No 388 (p 29)

[110] 1657-04-27: Huygens, Christiaan—Boulliau, Ismaël; *Ibid.* **II** No 387 (p 28)

[111] 1657-12-26: Huygens, Christiaan—Boulliau, Ismaël; *Ibid.* **II** No 443 (pp 108–10)

[112] 1658-03-28: Huygens, Christiaan—Chapelain, Jean; *Ibid.* **II** No 477 (pp 156–62)

[113] *Ibid.*

[114] Emmerson A 2010 *op. cit.*

[115] *Ibid.*

[116] Although the addressee is not referred to by name, Kechel was at that time the only person known to have engaged in the astronomical eclipse (and other emphemeris) observations that were discussed. He had published a pamphlet on this topic: *Eygentlicke afbeeldinge der Drie Sonnen, dewelcke verschenen zijn, Ao. 1653 den 14/24 Jan. alhier binnen Leijden, ende op den Toren van de Academie waergenomen door Sam. Car. Kechel van Hollenstein* (1653).

[117] 1657-06: Huygens, Christiaan—Kechel, Samuel; *Oeuvres Complètes de Christiaan Huygens* **II** No 392 (pp 34–5)

[118] 1657-06-16: Staten Generaal—Coster, Salomon; *Ibid.* **II** No 525 (pp 237–8)

[119] 1657-07-16: Staten van Hollant ende Westvrieslant—Coster, Salomon; *Ibid.* **II** No 526 (p 239)

[120] Whitestone S and Sabrier J C 2008 *op. cit.*

[121] 1657-06-16: Staten Generaal—Coster, Salomon; *Oeuvres Complètes de Christiaan Huygens* **II** No 524 (pp 236–7); Huygens C 1658 *Ibid.* **XVII** 78; 1658-10: Huygens, Christiaan—Pieck, Willem; *Ibid.* **II** No 532 (pp 247–8) Note 3

[122] 1658-10: Huygens, Christiaan—Pieck, Willem; *op. cit.*; transl van Leeuwen P 2013 Jan van Call and the age of the pendulum clock in the Netherlands *Proc. Antiquar. Horol. Soc.* **34** 33–41

[123] 1658-08-08: Staten Generaal—Douw, Simon; *Ibid.* **II** No 527, 528 (p 240)

[124] *Horologium*, appendix V http://gallica.bnf.fr/ark:/12148/bpt6k778667/f86

[125] 1658-09-06: Huygens, Christiaan—Wallis, John; *Oeuvres Complètes de Christiaan Huygens* **II** No 512 (pp 210–1)

[126] 1658-10: Huygens, Christiaan—Pieck, Willem; *Ibid.* **II** No 532 (pp 247–8)

[127] 1658-10-18: Huygens, Christiaan—Huygens, Lodewijk; *Ibid.* **XXII** 782

[128] 1658-10-04, -05, -13, -18: exchange between Huygens, Christiaan and van Schooten, Frans; *Ibid.* **II** No 523, 531, 534, 535 (pp 235–6, 246, 249–50, 251, respectively)

[129] 1659-02-13: van Schooten, Frans—Huygens, Christiaan; *Ibid.* **II** No 587 (pp 352–3)

[130] 1658-12-16: Raden van Hollandt, Zeelandt ende Vrieslandt—Staten van Hollant ende Westvrieslant; *Ibid.* **II** No 555 (p 288)

[131] 1658-12-17: Staten van Hollant ende Westvrieslant—Douw, Simon; *Ibid.* **II** No 557 (pp 291–2)

[132] Hevelius J 1673 Machinae Coelestis *De Horologiis* Book 1 **XVII** 360–72

[133] Kats F 2005 Compilation of the Coster–Fromanteel notarial act; www.antique-horology. org/_editorial/costerfromanteel/ [accessed 22 August 2016]

[134] Dereham W 1696 *The Artificial Clockmaker* (London: James Knapton)

[135] 1658-11-01 Huygens, Christiaan—Petit, Pierre; *Oeuvres Complètes de Christiaan Huygens* **II** No 546 (pp 270–4)

[136] Huygens C 1659 *Ibid.* **XVII** 84 (note 4)

[137] Huygens C 1659–1661 L'horloge à pendule à arcs cycloïdaux de décembre 1659 et son réglage par le poids curseur de 1661 *Ibid.* **XVII** 97–9

[138] 1659-12-06 Huygens, Christiaan—van Schooten, Frans; *Ibid.* **II** No 691 (pp 521–2)

[139] Although Chapelain comments in a letter to Huygens of 20 August 1659 about a clockmaker of Paris ('de notre') *'who has tried to rob you of your claim.'* This may refer to Isaac Thuret. 1659-08-20: Chapelain, Jean—Huygens, Christiaan; *Oeuvres Complètes de Christiaan Huygens* **II** No 655 (pp 467–9)

[140] 1658-06-13: Boulliau, Ismaël—Huygens, Christiaan; *Ibid.* **II** No 490 (pp 183–4)

[141] 1658-06-21: Boulliau, Ismaël—Huygens, Christiaan; *Ibid.* **II** No 492 (pp 185–6)

[142] 1662-04-12: Huygens, Christiaan—Huygens, Lodewijk; *Ibid.* **IV** No 1004 (pp 109–10)

[143] 1665-03-13: Chapelain, Jean—Huygens, Christiaan; *Ibid.* **V** No 1352 (pp 267–8)

[144] 1665-03-26: Huygens, Christiaan—Chapelain, Jean; *Ibid.* **V** No 1361 (p 281)

Chapter 4

The importance of high-precision timekeeping

In addition to his numerous other projects, Huygens continued to innovate and improve his pendulum clock. In this chapter, we will explore the developments towards construction of a practically viable timepiece for longitude determination at sea which occurred during two distinct phases in Huygens' life. The first phase, which started in 1657 with Huygens' and Coster's patent applications, covers Huygens' discovery of the isochronism of the cycloidal pendulum, his implementation of cycloidal chops in his clock designs, his subsequent development of a conical pendulum, which led to an exploration of the centre of oscillation of a compound pendulum, and his innovative use of a sliding weight to vary the period of his compound pendulum clock. This will take the narrative through to 1661. The next decade, from approximately 1662 to 1672, represents the second stage. During those years, Huygens worked on stabilizing his clocks onboard inherently unstable ships. He also worked out in detail how ship-board clocks could be used to determine longitude at sea. This second phase culminates with the final preparation of his treatise, *Horologium Oscillatorium Sive de Motu Pendulorum*—henceforth *Horologium Oscillatorium*. Chapter 5 will continue with the final phase, from his invention of a spring balance in 1674 to his death in 1695, a period of sea trials and increased competition.

During the entire period from the invention of the pendulum clock in 1657 to his final sketches of practical marine clocks, Huygens pursued a detailed assessment of those factors which may affect a clock's regularity at sea. He considered what the effects might be owing to changes in humidity and temperature, such as those one would encounter when sailing from temperate to tropical climates and back. In addition, he considered the effects which might be due to changes in counterweight, pendulum length, or spring balance, including their dependence on geographic location—that is, under different gravitational strengths given the slightly flattened shape of the Earth—and because of the ship's motion and the effects of the Coriolis force, which takes into account the Earth's rotation. Meanwhile, towards the end of this period, the *Vereenigde Oost-Indische Compagnie* (VOC; the United or Dutch East India Company), took note of his efforts. Its board was persuaded to issue a number of resolutions in support of taking one of Huygens' clocks on a journey to the Cape of Good Hope, the southernmost point of the African continent.

4.1 *Horologium* (1658) and beyond

As we saw in chapter 3, the initial efforts of Huygens and his master clockmaker Coster to commercialize their newly invented pendulum clock proceeded all but smoothly. The available evidence suggests that Huygens was fully intent on licensing his invention and drawing an income of royalties, but the fact that he had published a blueprint of his pendulum clock in *Horologium*, in 1658 (see figure 3.10), made it difficult to secure the exclusive commercial rights. However, we also realized that this blueprint was of the rather imprecise 'long pendulum,' a model of the pendulum clock that would be adequate for domestic use but not to determine longitude at sea.

4.1.1 Technical improvements

By early to mid-1657, around the time that Huygens and Coster filed their patent applications, we know from Huygens' correspondence with Claude Mylon that he was exploring the development of a more precise, spring-driven maritime pendulum clock, most likely in secrecy:

> '*I am glad that you are continuously perfecting your new clock and have no doubt that you will make it work just as well at sea as in your chambers, and that changes from dry to wet will not alter its accuracy anymore than changing the clock's weights*' [1].

Meanwhile, across the English Channel, Robert Hooke also investigated the potential of a spring-driven pendulum for use in marine clocks [2]. Given that spring-driven clocks could be regulated more reliably than the common weight-driven pendulum in use at that time, in 1658–1660 Hooke explored the possibility that his spring-driven watches might be eligible for any of the awards issued for 'finding the longitude.' However, the legendary rewards appeared to no longer exist on either the European continent or in England [3]. At this stage, he had not yet attempted to secure an English patent for his spring-regulated clocks; in 1660, a patent application was drafted, backed by Christopher Wren, the Scottish statesman Robert Moray, and the mathematician William Brouncker (1620–1684), 2nd Viscount Brouncker and first President of the Royal Society. Since Hooke had made significant progress in constructing reliably working timepieces, he could have made a fortune from patenting his invention. However, when he realized that, if awarded, the patent would allow anyone who improved his design to receive the royalties, he refused to continue:

> '*To which Clause I could no waies agree, knowing 'twas easie to vary my Principles an hundred waies, and it being ... facile Inventis addere.*'

A second attempt to patent Hooke's clock design was made in late 1663 or early 1664. At a meeting with Robert Boyle—one of the pioneers of the modern experimental scientific method—as well as Moray and Brouncker, conditions for a patent for Hooke's pendulum clock would be worked out. However, once again, the proposed conditions implied that all further improvements would likewise enjoy claims for patents, which again led to Hooke's refusal to continue with the process.

Despite this intense competition with his counterparts in England, or perhaps because of it, Huygens was clearly determined to develop a pendulum that would be able to deal with the rolling and pitching conditions buffeting ships at sea. He was most interested in using a *going barrel* (see figure 4.1), but without a fusee. Indeed, Huygens was very pleased with his isochronal pendulum's lack of sensitivity to power variations in a going barrel, but he realized that implementation of a fusee would cause difficulties in maintaining power during winding. This is reflected in a passage in *Horologium*, where he also refers to his use of springs rather than weights:

'I have indeed seen in the workshop of him whose labours I first employed for these constructions, completed clocks which go, not by weight, but by force of a spring. In this kind of work up to the present time, the differing power of the spring when wound up and when wound down was equalized by a fusee, round which was coiled a gut line; now these are disused. For the teeth are brought together with the barrel itself in which the spring is enclosed ... I pass over clocks of this kind which have been contrived to sound the hours by one and the same motor, either a weight or a spring, which serve also for turning the hand of the timepiece, since all these have nothing to do with my invention except as occasioned by the opportunity it presents' [4].

Figures 4.1 and 4.2 show the springbarrel–fusee arrangement typical of mechanical clocks in use elsewhere at the time of Huygens' innovations. The fusee's operation is based on the mechanical principles of 'levers.' Once charged, the spring tensions the gut line to the smallest radius of the cone. As the line unwinds, the fusee's cone shape causes the spring's radius to increase. In turn, the increasing leverage compensates for the reduction in power as the spring relaxes, so that the

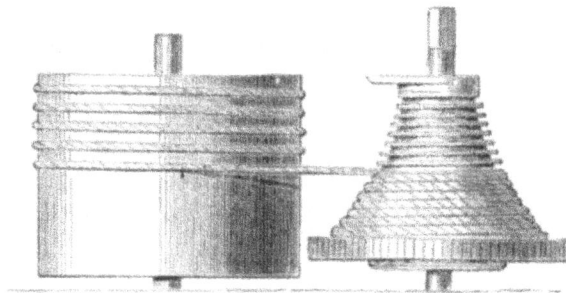

Figure 4.1. Drawing of a fusee, the cone-shaped pulley on the right, used in early mechanical clocks to even out the force of the *going barrel*, shown on the left, as it unwound. (Goodeve T M 1897 *The Elements of Mechanism* (New York: Longmans, Green & Co.)

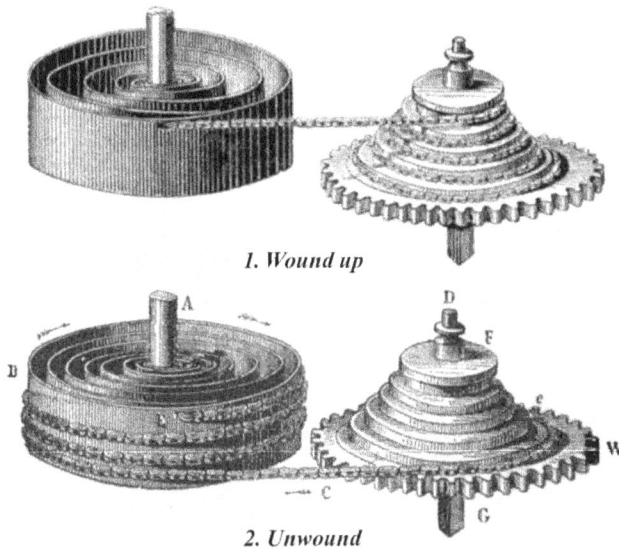

1. Wound up

2. Unwound

Figure 4.2. Fusee and *spring* barrel, showing operation. A: Mainspring arbour. B: Barrel. C: Chain. D: Fusee pivot. E: Attachment of chain to barrel. e: Attachment of chain to fusee. F: fusee. G: Winding arbour. W: Output gear. (Lardner D 1855 *The Museum of Science and Art* vol **6** (London: Walton & Maberly) figs 14 and 15.)

clock's gear train is subject to a constant force. As we saw in chapter 3, while Huygens proceeded with his marine pendulum clock development without fusees, the German clockmaker Treffler introduced a major innovation in the form of a short pendulum equipped with so-called German or 'Augsburg' fusee, which had been long in use by the German clockmakers' guild. Treffler's implementation of the Augsburg fusee is significant in the context of pendulum clock development, since it represents the first obvious difference in the pendulum clock's operation with respect to Coster's timepieces. Deviating from Coster's pendulum clocks, Treffler also employed the traditional German flanged (or ridged) springbarrel, connected by a gut line to a shallow-profile, deeply grooved brass fusee: see figure 4.3.

Treffler is thought to have copied Coster's new pendulum, presented as a gift by the Italian polymath and nobleman Tito Livio Burattini (1617–1681) to Ferdinando II de'Medici, Grand Duke of Tuscany, on 23 September 1657 [5–7]. Except for the pendulum clock model submitted in support of their patent application on 14 June 1657, it appears that Burattini's gift to Ferdinando II de'Medici may well be Coster's earliest timepiece which was constructed following Huygens' novel design. In a 1690 de'Medici inventory, it is described as '*having a short pendulum, in an ebony case with a wavy cornice*' [8]. Although records of this clock's location have been lost—and so may the clock itself—in late 1657 Ferdinando II de'Medici ordered that a copy be made by Treffler, his clockmaker.

This latter clock survives at the *Museo Galileo* in Florence [9]. It has a short pendulum suspended between chops, like all of Coster's clocks, but it also indicates seconds. It has been suggested that Treffler's pendulum clock preceded that of

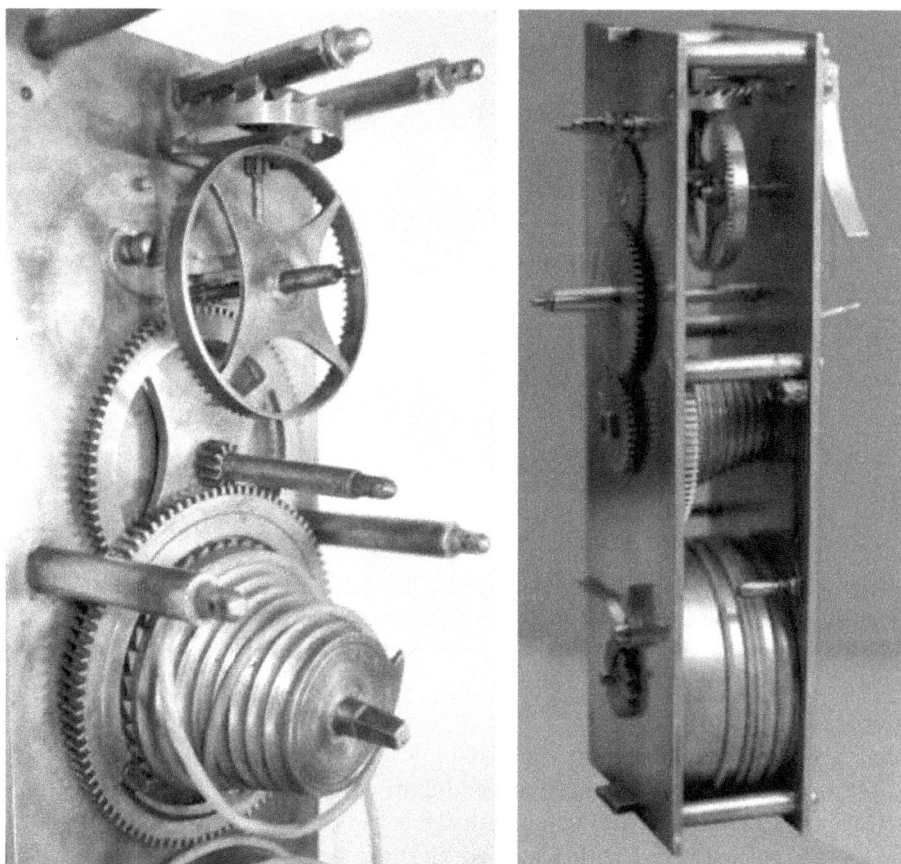

Figure 4.3. Treffler's original four-wheel gear train, (*left*) mounted on the front plate of the de'Medici clock, including a back-wound seven-turn Augsburg shallow-cone fusee; (*right*) gear train including the spring barrel. (Photos © Andrea Palmieri, Master Horologist, Florence; reproduced with permission.)

Huygens [10], a claim made by Treffler himself and supported by his patron, the Grand Duke of Tuscany, but this seems unlikely [11]: while Treffler added the Augsburg fusee, he probably copied Coster's short-pendulum spring-driven gearing. In addition, the crutched verge and curved chops Treffler had implemented were, in fact, Huygens' intellectual property, not Treffler's. Instead, Treffler most likely used Burattini's gift timepiece as his working model, although he constructed his own, unique component parts and pursued his own design [12].

Around the same time, Huygens discovered that the isochronal chops he had developed for his pendulum clock worked very well in stable conditions on land, but they led to changes in pendulum length when even slightly tilted—which, in turn, would lead to changes in the swing period. By the end of 1657 or perhaps early in 1658, this led to his development of a half-second pendulum clock with four-wheel 'OP gearing,' a crutched verge, and a vertical escapement. The advantage of this gear type—shown in Huygens' first sketch from February 1658 and reproduced in figure 4.4—is that it reduces the pendulum's swing amplitude to an arc where the

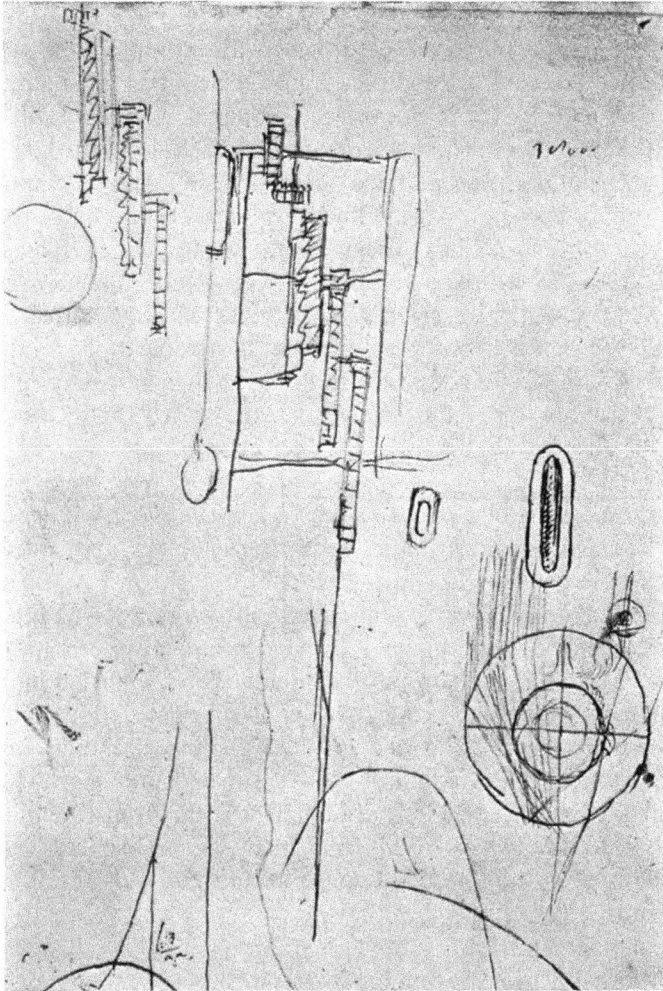

Figure 4.4. Huygens' first sketch of his 'OP gearing,' where the relevant gear wheels are high-lighted through hatching; February 1658. (Huygens, C 1662 *Oeuvres Complètes de Christiaan Huygens*; acknowledgement: Leiden University.)

circular and isochronal paths remain very similar. It is likely that Huygens intended to develop this second clock into a marine clock [13], although the effects of pitching and rolling still had to be dealt with. Clearly, his first priority was to reduce variation in his clocks' performance. He started by testing a version powered by descending weights, but he quickly realized that a radically new approach was needed to develop an accurate timepiece for use at sea.

4.1.2 The conical pendulum and the development of cycloidal paths

One attempt at radically overhauling his pendulum clock design commenced in November 1659, when Huygens developed a conical or 'circular' pendulum clock whose bob, instead of swinging, rotated at constant speed in a circle about a vertical axis, with the cord tracing out a cone. Independently, although a few years later [14], Hooke also studied the conical pendulum but in the context of (multiple) planetary orbits in our solar system [15]. This followed earlier suggestions from around 1637 by the English astronomer Jeremiah Horrocks (1618–1641) that the Moon orbited the Earth in an elliptical path and that comets also followed elliptical orbits, a worldview he supported by analogy to the movements of a conical pendulum [16]. Although this new conical clock design is not mentioned in Huygens' *Oeuvres Complètes*, we know that he had it constructed, because he obtained a new value for the constant of gravitational acceleration (in other words, for free fall) '*ex motu conico penduli*' [17], that is, from the movement of a conical pendulum.

Both Huygens and Hooke continued to improve their respective conical pendulums. Hooke worked on his clock through the summer of 1666. Gunther describes how, on 13 June 1666,

> '*Mr. Hooke exhibited a new contrivance of a circular pendulum applicable to a watch, and moving without any noise, and in continued and even motion without any jerks. ... He was desired to show the use of it in a watch, which he said the President [of the Royal Society] had already given order for*' [18].

Hooke's study of both the regular and the circular pendulums led him to conclude independently that both timekeepers would run isochronously for orbits of different sizes only if the paths of the bobs followed a paraboloid. Although this shows that Hooke was capable of carrying through a complex mathematical argument to demonstrate a mechanical result, his argument was far from convincing. He may indeed have heard of Huygens' result, but given the problematic reasoning, it is likely that this was not an attempt at reproducing Huygens result but rather an independent effort [19].

Despite the mathematical inconsistencies of Hooke's theory, Brouncker reported a successful practical demonstration on 18 July 1666:

> '*The circular pendulum applied to a clock being inquired after, the President affirmed, that he had made trial of one, and observed the motion of it for four days, in which time it had gone so equally with his pendulum clock, that after these four days were elapsed, he found it only to have gone one minute too fast*' [20].

Indeed, the excellent regularity of Hooke's design was later confirmed by William Derham (1657–1735) in his treatise *The Artificial Clockmaker* (1734):

> '*[T]he motion of the pendulum being as regular as the vibrating one, was contrived by Mr. Hooke, to give warning at any moment of [its] circumgyration, either when it had turned a quarter, half, or any lesser or greater part of [its] circle. So that here you had notice not only of a second, as in the pendulum vibrating seconds, but of the most minute parts of a second of time, by which means it was made very useful in astronomical observations*' [21].

Following further improvements made by Hooke, on 21 February 1667 the English naturalist Phil(l)ip Skippon (1641–1691) wrote a letter to his fellow naturalist John Wray (Ray; 1627–1705), stating that …

'*It is somewhat difficult for me to explain in Writing the new way of the Pendulum. There is the common Vibration that Hugenius [Huygens] invented in watches, and Mr. Hooke hath to that added a Circular Motion; the weight at the end of one Vibration is turned off by a kind of Spring, which makes the Motion circular*' [22].

The Royal Society's records reveal that, on the same day,

'*Mr. Hooke produced a circular pendulum so contrived, that its motion should be equal, whatever weight was appended to it. He affirm[ed] that he knew the demonstration of it, was ordered to give it in writing at the next meeting. He was ordered likewise to compare the motion of this circular pendulum with a clock*' [23].

The Royal Society's Fellowship returned to the matter at their next meeting, on 28 February 1667:

'*The circular pendulum designed for an equal motion with unequal weights being again spoken of, the President affirmed that though the inventor Mr. Hooke had demonstrated, that the bullet of the circular pendulum, if it can be always kept rising or falling in a parabola, will keep its circular motion in the same time; yet he had not demonstrated, that the diameter of the parabola from the point of contact in the curve to the vertex of the diameter is equal to that portion of the curve from the said point of contact to the vertex of the same curve, plus half the latus rectum*[1] *or plus double the focus of the parabola*' [24].

Hooke had thus invented the *parabolic* pendulum, which he derived from the circular (conical) pendulum. Huygens did not publish his own calculations pertaining to parabolic clocks until 1673, in *Horologium Oscillatorium*, while his derivation of the isochronicity of parabolic surfaces did not see the light until 1669. Just as for the simple, swinging pendulum, the conical pendulum's bob is also kept on an isochronous path using a curved plate whose shape is determined by the theory of *evolutes*[2].

Although evolutes were first described by the ancient Greek geometer and astronomer Apollonius (c. 262 BCE–c. 190 BCE) in Book V of his *Conics* (Κωνικά), Huygens is often credited with their first application, in 1659, based on mechanical rather than mathematical principles. The editors of Huygens' *Oeuvres Complètes* tentatively identified [25] his use of the evolute of the parabola for the suspension of a conical pendulum in an early sketch from 1664 (see the left-hand

---

[1] The *latus rectum* is the line segment through a focus of parabola, perpendicular to the major axis, which has both endpoints on the curve.
[2] The term *evolute* has its etymological origin in the participle *evolutus* (*Lat.*; unrolled) of the verb *evolvere* (to unroll).

drawing of figure 4.5)—although the curves are not identified as parabolas. This was after his visit to London in 1663, and after he had discussed the idea of a conical pendulum with Wren [26]. Huygens pursued development of the conical pendulum for marine timekeeping purposes after Hooke's demonstration of his prototype at the Royal Society in 1667. The Dutch scholar seemed to have been rather taken by the concept, given his excited comments in a letter to his brother Lodewijk of 4 December 1667 [27], which was accompanied by the sketch reproduced in the right-hand panel of figure 4.5.

Consider figure 4.6. The black curve **A–G** is the evolute. It can be 'unrolled' by fitting a 'thread' (*filum*) to each tangent point while maintaining the length of the total curve. The free extremity of the black curve will move outwards along the red curve— the *involute*—from **A** through **A1**, **A2**, ... to **A6** as the thread unwinds, a curve Huygens always referred to as '*that drawn by the unrolling [descripta ex evolutione]*'.

Huygens' theory of evolutes and their relationship with the corresponding involutes was based on mechanical experiments, that is, it followed naturally from his studies of the pendulum. The complete theory—a product of his mathematical response to this physical challenge—preceded a more rigorous, calculus-based description, which he would develop a few decades later. Modern calculus defines the evolute as a curve that is, at any given time, the locus of the centres of rotation or curvature of its involute. In Part III of *Horologium Oscillatorium* (1673), Huygens proved that the tangents to the evolute (which he used to mechanically derive the involute) and the normals to the involute (from which he derived the shape of the evolute mathematically) are the same: indeed, he showed that both approaches— the mechanical and the mathematical—are equivalent.

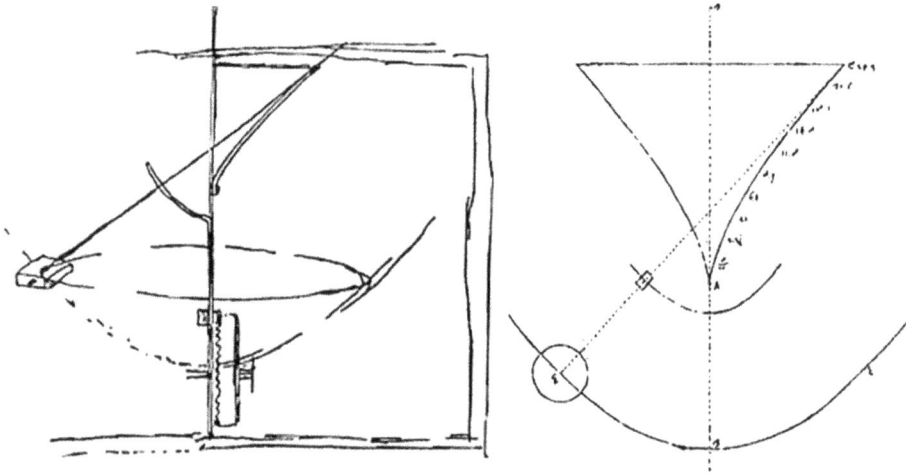

Figure 4.5. (*left*) Early sketch from 19 September 1664 (*Codices Hugeniani, Manuscript B* p 160; reproduced courtesy Leiden University Library) showing Huygens' ideas for a conical pendulum. To ensure isochronous rotation, Huygens appears to have applied the evolute of the parabolical path of the pendulum bob. (*right*) Huygens' explanation of the conical pendulum (*Codices Hugeniani, Manuscript C* p 203; reproduced courtesy Leiden University Library) dated 5 September 1667. The figure suggests that his conical pendulum clocks were also equipped with a sliding weight (see chapters 4.1.3 and 4.1.5).

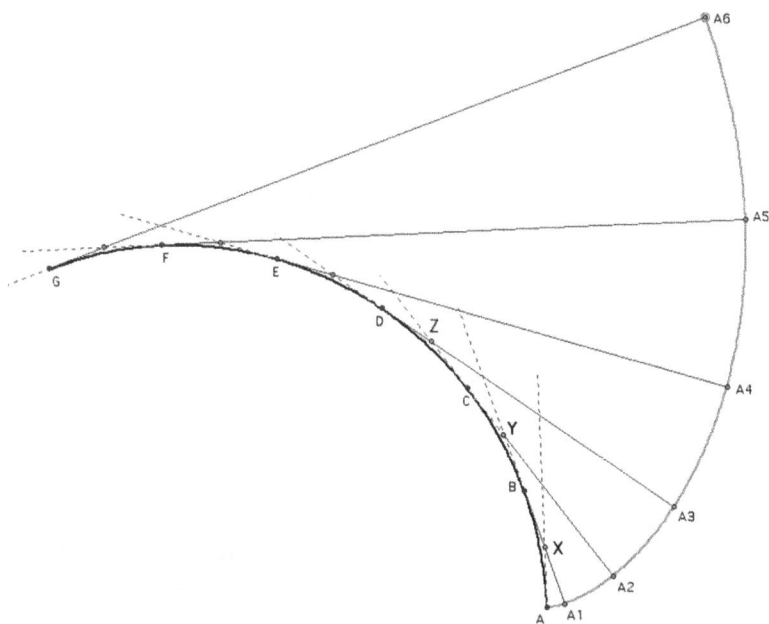

Figure 4.6. Graphical representation of the construction of an involute (red) from an evolute (black). (Source: The Math Images Project, www.mathforum.org; available under the GNU Free Documentation License, Version 1.2. Credit: Chengying Wang, Swarthmore College.)

How does this all relate to the development of a stable pendulum clock? As we saw in chapter 3, an isochronous pendulum bob must be moving along a cycloidal path. Marin Mersenne and others—including Galileo himself, although he had readily dismissed his observations (as we saw in chapter 3)—had already observed that simple pendulums are not fully isochronous. Instead, their period depends on the swing amplitude; isochronicity is only approximated for small amplitudes. This was independently realized by Hooke. He had invented an early predecessor of the anchor escapement in the second half of the 1650s [28] to secure isochronism of land-based pendulum clocks by employing a heavy, simple pendulum with a rigid suspension. Such pendulums, which came to be known in England as 'Royal Pendulums' [29], would have only small swing amplitudes and, hence, they would be approximately isochronous. We will return to this innovation in chapter 4.2.3.

As the amplitude increases, the period increases slightly as well, in the same manner as how $\sin \theta$ deviates from θ for increasing but small angles θ (particularly if θ is expressed in radians). In December 1659 Huygens analysed this deviation from isochronicity by working out the shape of the curve along which a frictionless particle will slide down under the influence of gravity during the same time span, irrespective of its starting point [30]:

'*What ratio does the time of a minimal oscillation of a pendulum have to the time of perpendicular fall from the height of the pendulum?*' [31]

Based on geometrical arguments, reproduced in chapter 4.2.2, he showed that a pendulum's period only depends on its length. However, the implied isochronicity resulted from an approximation he had made in his analysis, that is, he had represented the pendulum bob's swing arc by a parabola instead of the circle it describes in reality. Indeed, Huygens knew and demonstrated that the circular arc of a simple pendulum bob is not isochronous. After all, from his experiments he already knew that the period of a pendulum also depends on the swing amplitude. This realization led to his next question: he wondered which path the pendulum bob had to trace for his approximation to apply to all swing amplitudes and, hence, render the pendulum isochronous. The answer, of course, is a curve whose tangent is drawn using the same method that was then used for drawing the tangent to a cycloid [32]. The obvious next step was to determine what shape the clock's chops should be so that, as the pendulum suspension thread wrapped and unwrapped around them, the bob would follow a cycloidal path.

We will return to his derivation below, but here we will first consider the theory of evolutes that resulted from these considerations. As it happened, Huygens had fortuitously been studying cycloidal curves for two mathematical challenges issued by Blaise Pascal in 1654, namely, the problems to determine the area of any segment of a cycloid and the centre of gravity of any such segment, as well as the volume and surface area of the solid of revolution formed by rotating the cycloid about the x axis. His calculations at that time had shown that a body falling from *any* point along a cycloid will reach the bottom in the same amount of time, and the ratio of this time to the time of free fall from rest along the cycloid's axis is $\pi{:}2$,

> 'On a cycloid whose axis is erected on the perpendicular and whose vertex is located at the bottom, the times of descent, in which a body arrives at the lowest point at the vertex after having departed from any point on the cycloid, are equal to each other; and these times are related to the time of a perpendicular fall through the whole axis of the cycloid with the same ratio by which the semicircumference of a circle is related to its diameter' [33].

Huygens' notes of 20 December 1659 show that he numerically correlated the coordinates of a cycloid with the angular displacement of the pendulum thread. By the following summer, he had determined that his clock's chops should be semi-cycloids, congruent with the bob's path, a conclusion he expediently announced to his contemporaries. This led to a frenzy of activity by December 1661, since the cycloidal pendulum became the focus of numerous investigations as a possible device to establish a standard of length. (The conclusion was published in 1673 as Proposition VI in Part III of *Horologium Oscillatorium*.)

As we saw above and in chapter 3, Huygens' British competitors had almost simultaneously realized the need for cycloidal paths to make their pendulum isochronous, yet Huygens claimed to have derived this property of isochronous pendulums first—despite not having published his insights until 1673. Naturally, this led to protests from across the English Channel. Brouncker wrote to the Royal Society's first Secretary Oldenburg on 18 October 1673,

'It is very sure, that Mr. William Neil [Neile] *had in the year 1657 found out and demonstrated a Streight line equal to a Paraboloeid; and did then communicate and publish the same (though not in print) to my self and others, who used to meet at Gresham Colledge, and it was there received with good approbation; and the same was, presently afterwards, otherwise demonstrated by my self and others: And therefore ancienter than that of Monsieur Heurat* [Hendrik van Heuraet, 1633–1660?],[3] *which (as it seems) is not pretended to have been done before the year 1659; and ancienter too than that of Sr. Ch. Wren, finding a Streight line equal to a Cycloid in the year 1658; and by him admitted so to be. Nor ought it at all to prejudice Mr. Neil, that M. Heuraet's was somewhat sooner abroad in print, than that of M. Neil, (though both in the same year 1659;) since it is well known to many of us, that Mr. Neil's was done before. Otherwise M. Hugens* [Huygens], *by the same reason, will grant the precedency to Heuraet, of that which he now claims to be his own invention (that Rectifying the Parabolical Line and Squaring the Hyperbolical Space do mutually depend on each other:) for this was published in print by M. Heuraet (or M. Schooten* [Frans van Schooten] *for him) in the year 1659, and not by M. Hugens till now, 1673: And yet M. Hugens thinks, the may well claim that invention to be his own, because he now tells us, that he found it out about the end of the year 1657, and did (some time after) communicate it privately to some friends. And whereas, he doth suppose, that this invention of his might give occasion to that other of Heuraet, we may also as well suppose, that he might have taken such occasion from hearing of M. Neil having done the like, (for this had been then commonly known for a great while:) Or might have taken occasion (as well as Mr. Neil) from that of Dr. Wallis ... or from that of Sr. Ch. Wren having found a Steright equal to another Curve the year before: ... It was easie (for M. Heuraet, or M. Hugens, or any other,) to infer, That, if we can Rectifie the one, we may Square the other, & vice versa. But from whence soever M. Heuraet had it; we may, as before, reasonably conclude, that Mr. Neil had it before him: And M. Hugens is a person of that ingenuity, that, when he shall better consider of it, he will (I doubt not) be of the same mind'* [34].

In any case, claims of precedence aside, Huygens needed to find the curve that would 'unwind' to form this cycloid. In other words, he had to derive the evolute to the isochronous path traced by his pendulum bob: see figure 4.7. From this mechanical basis, the problem simplified to the need to meet three basic requirements:

1. Each trace from the bob's cycloidal path to the pendulum's pivot point, that is, the shape of the pendulum's cord or chain, must be tangent to the curvature of the suspension chops;
2. The pendulum's cord must be perpendicular to the cycloid's arc at each point of contact;
3. The pendulum's length is constant.

---

[3] Hendrik van Heuraet (1633–1660?) was a Dutch mathematician and physician; he is known as one of the founders of the integral.

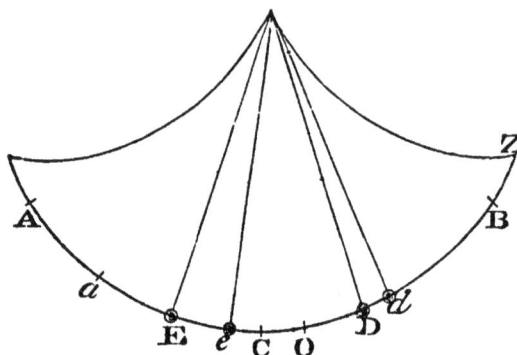

Figure 4.7. Derivation of the shape of the chops using the theory of evolutes, adopting a cycloidal path, **A–BZ**, for the pendulum bob. (Newton I 1687, *Philosophiæ Naturalis Principia Mathematica* Proposition XXV, Theorem XX p 304.)

The evolute must thus be a curve characterized by the same base, height, and length as the pendulum's cycloidal path; in other words, the cycloid is its own evolute. We will show this now on the basis of modern calculus. The cycloid is given by

$$x = a\,(\theta - \sin\theta); y = a\,(1 - \cos\theta),$$

where a is the cycloid's height and θ the running parameter, which for a pendulum is its swing amplitude. By definition, a curve's evolute is the envelope of the normals to its centres of curvature, so that the cycloid's evolute is given by

$$x = a(\theta - \sin\theta); \; y = a(\cos\theta - 1),$$

which is simply a shifted version of the original cycloid.

In practice, and in relation to Huygens' design, for small angles θ of the pendulum from the vertical direction, $\cos\theta \approx 1$, while the periods, P, of a conical and a swinging pendulum of the same length, L, are the same and largely insensitive to changes in θ. In other words, the period of rotation is approximately independent of the driving force. That is, such conical pendulums are isochronous, just like ordinary simple swinging pendulums. It is straightforward to determine P.

Figure 4.8 shows a conical pendulum of length L with a bob of mass m revolving—in the absence of friction—in a circle at a constant speed v at a fixed suspension angle from the vertical direction θ. (If the angle θ were not fixed, the bob would be moving on the surface of a sphere, not along an isochronous paraboloid.) The force, T, exerted from the bob to the pivot point—that is, the tension in the string—can be decomposed into horizontal and vertical components, $T\sin\theta$ and $T\cos\theta$. The bob is additionally pulled down by its weight, mg, where g is the usual gravitational acceleration.

Although these concepts will be familiar to anyone with even a cursory knowledge of modern physics, the notions of centrifugal motion and the centripetal force were new to Huygens and his contemporaries. In fact, Huygens coined the term

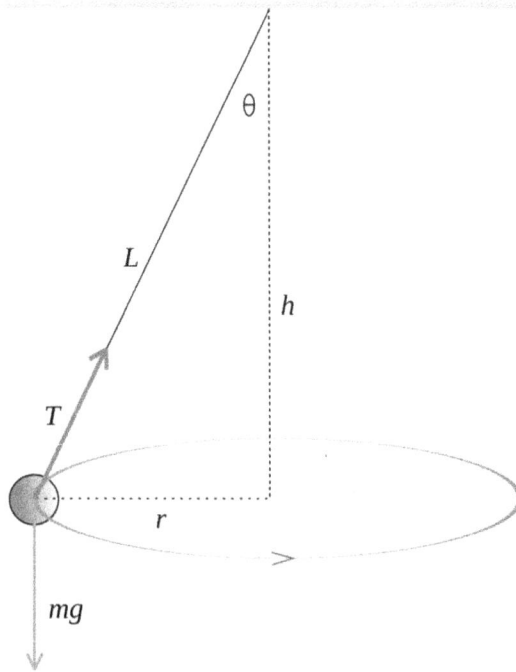

Figure 4.8. Conical pendulum geometry. θ: Suspension angle. r: Radius of bob's circular motion at constant speed v. h: Vertical suspension height. L: Pendulum length. T: Tension force acting on the bob; mg: Gravitational force. (Source: CosineKitty/ Wikipedia, available under the Creative Commons Attribution-ShareAlike License.)

'centrifugal force' in his treatise *De Vi Centrifuga* in 1659 [35], which was published posthumously in 1703. In modern physics, use of the notion 'centrifugal force'—which refers to the tendency of an object following a curved path to fly outwards, away from the curve's centre—is frowned upon. The object's outward motion results from its inertia, that is, its tendency to resist any change to its state, either at rest or in motion. A careful assessment of the forces at work reveals that the actual force that counteracts the centrifugal motion is the *centripetal* force, which keeps the object moving at uniform speed along a circular path. I will hence refer to centrifugal *motion* and centripetal *force*. This latter concept—*vis centripita*—was coined by Newton in his discussions of gravity in his manuscript *De Motu Corporum in Gyrum* ('*On the motion of bodies in an orbit*'), which he sent to Edmond Halley, the English scientist who lent his name to Halley's comet, in November 1684. Newton's second law—acceleration is produced when a force acts on a mass—implies that the horizontal component of the tension in the string imparts a centripetal acceleration on the bob towards the centre of its circular path, while there is no acceleration in the vertical direction:

$$T \sin \theta = \frac{mv^2}{r};$$
$$T \cos \theta = mg.$$

Note that the first of these equations represents the centripetal force and is equivalent to what is now known as Newton's second law of motion in quadratic form: Huygens' derivation was equivalent to that of Newton's but it preceded the latter by several decades. We can now eliminate T and m, and by realizing that during one period P the bob travels the full circumference of a circle with radius r, we derive

$$\frac{g}{\cos\theta} = \frac{(2\pi)^2 r}{P^2 \sin\theta},$$

which can be rewritten to derive the period of rotation,

$$P = 2\pi\sqrt{\frac{r}{g\tan\theta}} = 2\pi\sqrt{\frac{L\cos\theta}{g}}.$$

The period of rotation of a conical pendulum is thus independent of either the velocity or mass of the bob; it only depends on the suspension angle θ for a given pendulum length. Since $\cos\theta \approx 1$ for small angles θ, it follows that the conical pendulum is approximately isochronous. Note that for $\cos\theta = 1$ the latter equation reduces to the expression of the period for simple, swinging pendulums.

Huygens' original sketch of the conical pendulum is reproduced in the left-hand drawing of figure 4.9 [36]. Moving along a circular path, the pendulum traces a conical surface, **ABE**. To achieve a period of one second for a suspension angle **ABG** of 45°, Huygens calculated that the cord length needed to be $11^8/_{11}$ inches [37]. For any arbitrary angle, and therefore for any arbitrary height **H**, he derived that the bob will remain at constant height for a period of $\sqrt{\mathbf{HA}}/\sqrt{\mathbf{GA}}$, where the radius of the circular path travelled in a period of one second, **GA** = **BG**. Therefore, for fixed angles, the period of a conical pendulum scales directly with the height of the bob,

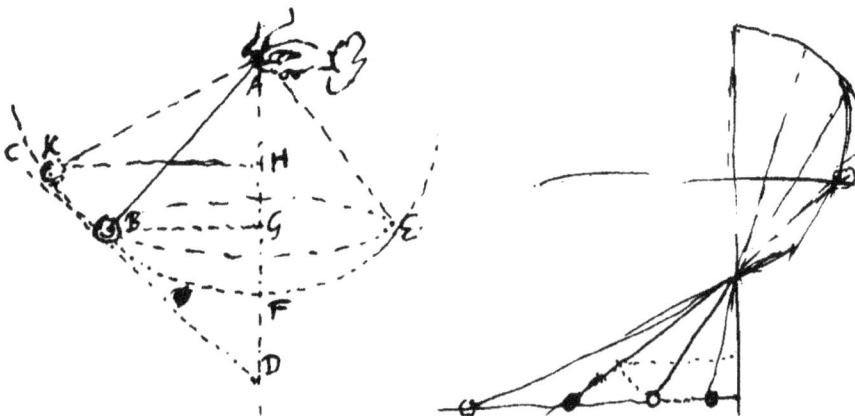

Figure 4.9. (*left*) Huygens' original sketch of the conical pendulum (Leiden University Library, HUG 26 f.7r). (*right*) Preliminary drawing for the conical pendulum clock while maintaining the bob at constant height. (Huygens C 1659 De Vi Centrifuga, *Oeuvres Complètes de Christiaan Huygens* **XVI** pp. 255–301; Leiden University Library, HUG 10, f.87r.)

HA. In turn, this implies that if the bob can be maintained at the same height, irrespective of the pendulum's length and, hence, the angle with the vertical direction, all periods will be the same: see Huygens' preliminary sketch in the right-hand panel of figure 4.9.

Huygens next proceeded to examine the forces exerted through both the string's tension and the gravity on the bob, drawing on the mechanics first worked out by the Flemish mathematician, physicist, and engineer Simon Stevin (1548–1620). He concluded that the tension does not depend on the pendulum's length nor on the centripetal force, but only on the bob's weight (that is, m times g) and the angle of the pendulum with respect to the vertical. The principles resulting from his geometrical considerations allowed Huygens to construct his first stable, conical pendulum clock. Most importantly, the period can be maintained stably if the bob is kept at the same height throughout the clock's operation. This implies that, as the angle of the pendulum's suspension changes, the cord length must be adjusted—as shown in the right-hand sketch of figure 4.9.

Huygens implemented this crucial aspect in his final conical clock design: see figure 4.10. This design, which includes the date of the invention, '*Inventum die 5 Oct. 1659*,' shows a circularly moving weight **A** attached to a cord that passes over a pulley to a weight **E**, which—together with the chain **CD**—counterbalances the forces exerted on **A**. This arrangement allows for changes in the suspension angle— and hence in the balance of the cord's tension and the bob's weight—to be counteracted by providing more or less slack in the cord. In turn, this allows the clock to maintain an equilibrium configuration, but crucially *at the same height*, so that the period of rotation and, as a consequence, the clock's stable performance do not change.

Note that the process leading to a new equilibrium is rather complex. If the weight **A** is too heavy for its present configuration, it will tend to fall, thus leading to a reduction in the cord's angle with the vertical. The idea is that the bob will continue to move inwards until it reaches a new balance between the tension in the cord and the weight of the bob. However, in reality the excess weight of the falling bob will raise the counterweight **E**, as well as a number of links of the chain beyond **D**, which in turn increases the tension in the cord [38].

Crommelin discussed the conditions for isochronous operation of this timepiece [39], which Huygens did not address in his own notes. I will now develop the arguments following Crommelin. We start from the conical pendulum's period of rotation we just derived. The condition that needs to be satisfied for isochronicity is $\Delta P = 0$, that is, $\Delta(L \cos \theta) = 0$, for all L and θ. As we saw above, all circles that the weight A (note that A represents the *weight*, not the mass of the bob; that is, it includes the local gravitational pull) could describe must lie in the same horizontal plane. The tension felt by the pendulum's cord is given by

$$s = \frac{A}{\cos \theta} = \frac{LA}{L \cos \theta},$$

so that

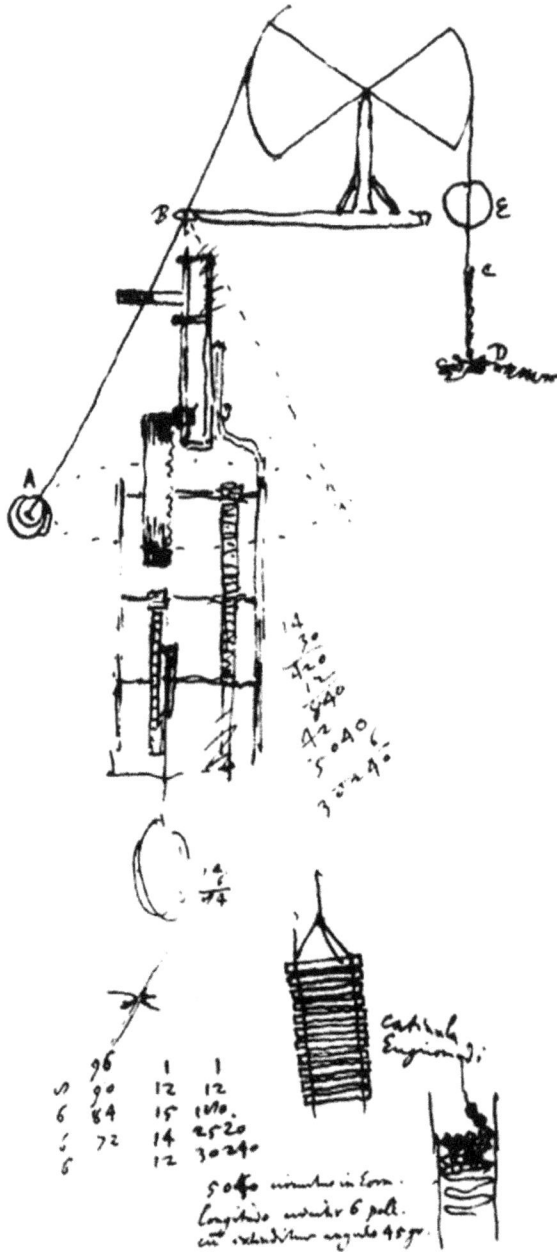

Figure 4.10. Huygens' conical pendulum clock design, including an adjustable string or chain. (Huygens C 1659 *De Vi Centrifuga, Oeuvres Complètes de Christaan Huygens,* **XVI**, pp. 255–301.)

$$\Delta s = \frac{A\Delta L}{L \cos \theta} = \Delta(\varepsilon + pb),$$

where ε is the weight suspended on the right-hand side of the pendulum (see figure 4.10), p that of one centimetre of chain length, and b the effective chain length expressed in centimetres, that is, the length of the chain contributing to the overall weight on the pendulum's right-hand side. It thus follows that

$$\frac{A\Delta L}{L \cos \theta} = p\Delta b,$$

but $\Delta L = \Delta b$, so that we have the condition for isochronous operation,

$$p = \frac{A}{L \cos \theta}.$$

Crommelin attempted to construct a conical pendulum such as that sketched by Huygens in 1659 [40]; he found that his reconstructed pendulum clock ran adequately but not perfectly, suffering a loss of 16 seconds per 24 hours in approximately constant-temperature conditions.

Huygens employed his new insights into the motions of conical pendulums to obtain a new value for the gravitational constant, g, based on measurements of his clock's period. He discussed the similarity between the gravitational and centripetal forces at the beginning of *De Vi Centrifuga*:

'Furthermore, whenever there are two bodies of equal weight each one held by a cord, if they have the same conatus [effort, tendency] *due to accelerated motion, whereby they would pass through equal spaces in the same time, receding along the extension of the cord: we assert that the same tension is felt on their cords be they drawn downwards or upwards or in whatever direction ... And it ought to be measured by the initial motion, taking an arbitrarily small span of time ... Now let us see what and how much* conatus *there is in bodies bound to a cord or to a rotating circle as they recede from the centre'* [41].

In essence, here he equates the tension in a cord with the weight of—that is, the gravitational force acting on—the bob. Since *'the same tension is felt'* due to both gravity and the centrifugal motion (or the centripetal force), measuring the latter can provide insights into the former. Indeed, we already saw that, given a fixed angle and a fixed length, the period of a pendulum is entirely determined by the value of the gravitational constant. Following a number of experiments, he eventually reached the conclusion that the height from which he needed to drop a ball for it to hit the ground in one second—measured with his conical pendulum clock with a pendulum length of 9½ inches—was 15 feet 7½ inches [42], so that $g = 9.81$ m s^{-2}—very close to the value commonly adopted today for northern European latitudes.

Huygens next proceeded to improve his conical pendulum, as evidenced in chapter 5 of *Horologium Oscillatorium*, where he describes a parabolic conical pendulum that indeed allows robustly isochronous operation. His proposition VI discusses the improved pendulum design, but evidence of its operation was not disseminated until the publication of *De Vi Centrifuga* in 1703. Crommelin has pointed out, however, that that description originated from that latter manuscript's editors—Burchard de Volder (1643–1709) and Bernard Fullenius (1640–1707)—and not from Huygens himself [43]. With Huygens having stumbled upon the invention of the parabolic conical pendulum in 1660 or 1661, we do not learn anything about its operation until a letter from Huygens to his brother Lodewijk from 4 December 1667:

'I am presently working on the construction of another kind of clock, or even two different kinds, of one has a pendulum that rotates around…' [44],

while in a letter to his brother Constantijn of 12 October 1668, he announced,

'I have here a circular movement of my new invention, which runs very well and quietly' [45].

For regular conical pendulums and different elevation angles, the rotating bobs describe paths on the surface of a sphere. The idea at the basis of the parabolic conical pendulum concept is that its bob would follow a paraboloidal surface. Huygens invented an ingenious way to ensure that the pendulum bob would indeed describe a paraboloid for any given L and θ: see figure 4.11 for the technical details [46]. The axis of the parabolic conical pendulum consisted of a metal rod **OX**. A metal strip shaped like a cubic parabola, **ED**, was attached to the rotation axis at point **E** and maintained in position by a third metal strip, **XD**. The pendulum's metal components are coloured red in figure 4.11. When rotating around its pivot point, **O**, the bob, **A**, would be found following a circular path anywhere on the three-dimensional surface **FOF′** while rotating around the axis **OX**. The pendulum's cord, **DA**, with weight A, is shown in figure 4.11 under the assumption that the pendulum is in operation and the bob has been raised by the centripetal force; at rest, the cord would describe the curve **DEO**. In other words, the pendulum cord is at all

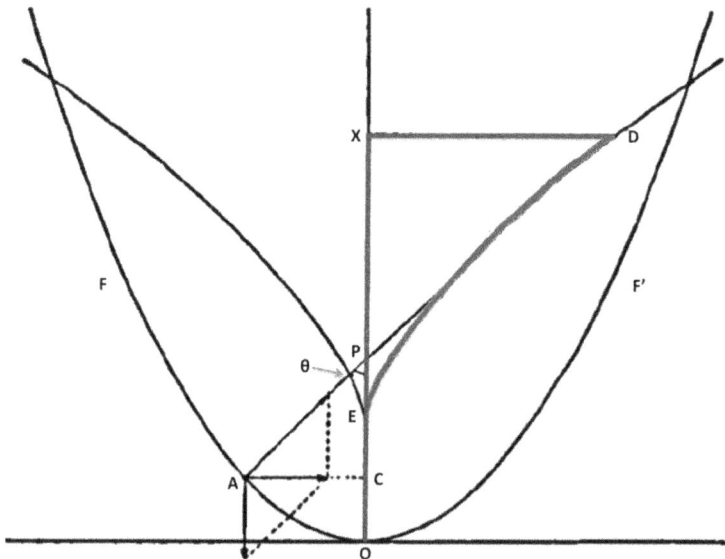

Figure 4.11. Technical justification of the parabolic conical pendulum as an isochronous timepiece.

times oriented perpendicularly to the cubic parabola's envelope and tangent to the metal strip extending from **D** towards **E**. Isochronicity of the pendulum's motion can now be demonstrated very easily. We can assume, for mathematical simplicity, that the pendulum's bob is suspended on its cord from point **P**; this is mathematically equivalent to the practical set-up of the device. The pendulum's length hence becomes **AP**. The height of the pendulum is $L \cos \theta = $ **CP**, which is simultaneously the 'subnormal' of the parabola **FOF'**. A parabola's subnormal is conveniently constant for any angle θ and any point p on the curve, so that $L \cos \theta = p$ and the pendulum's period is hence independent of θ.

4.1.3 Compound pendulums

From a mathematical perspective, Huygens' conical clock experiments were a great success. After all, they had allowed him to explore the centripetal force for the first time and to determine an accurate value for the gravitational constant. However, Huygens suggested that to construct a practical pendulum, one also needed to know its 'centre of oscillation,' a physical property that was later renamed the 'moment of inertia' by the influential Swiss polymath Leonhard Euler (1707–1783). This realization was driven by the notion that realistic pendulums do not have all of their mass concentrated at the end of a massless string. The question Huygens needed to answer in order to develop an accurate timepiece was, *What is the length of a simple pendulum that oscillates with the same period as a realistic compound pendulum?*

Compound pendulums had already been subject of intense experimentation and discussion for a few decades by that time. By the 1640s, the concept of the compound pendulum was already well established and its properties had attracted the attention of some of the leading contemporary mathematicians. In fact, in 1647 the relationship of the compound pendulum's period of oscillation to that of the simple pendulum was the subject of a heated debate between Personne de Roberval and Descartes. Both mathematicians were attempting to establish the quantitative, rather complex law of motion for a compound pendulum, a pendulum that does not have its weight and motion concentrated in a single point co-located with its bob. From our modern perspective, we now realize that both mathematicians approached the problem from complementary angles. At the time, however, the debate deteriorated into a war of words. History shows that both men were right: Descartes' insistence to consider only the absolute value of the inertia forces was correct, while Roberval was also correct in insisting that gravity should also be considered [47].

Returning now to Huygens' conundrum, to make significant progress on the development of his timepiece, he needed to determine the distance from the compound pendulum's pivot point to its centre of mass for comparison with the simple pendulum. His early attempts at addressing this question date back as far as August to November 1661 [48]; his definite solution was published in *Horologium Oscillatorium*, where he started by explaining the idea behind his investigation:

'*The investigation of the centre of oscillation or movement was one proposed by the most learned Mersenne, and it was famous amongst the geometers of the day;*

indeed this problem was attempted by me when I was still a boy, along with many others, and I have kept the letters Mersenne sent me; and neither should we exclude the work of Descartes recently published, which contains answers to these things proposed by Mersenne. Moreover, he wanted to find the centre of oscillation as I shall do here, for sectors of a circle, which was either suspended from the vertex of the angle or from the middle of the arc, and disturbed from the side; likewise for circular segments, and for triangles suspended either from a vertex or from the middle of the base. These problems are reduced to the case of the simple pendulum, that is, a weight hanging from a string that has been found at some distance, so that the oscillations are made in the same time as that of the object suspended in the manner indicated' [49].

Huygens' working notes [50] show that at first he considered two dimensionless weights, **B** and **C**, swinging together on an inflexible, massless rod **AC**: see figure 4.12 (left). Next to this compound pendulum, he drew a simple pendulum, **HP**, with the same oscillation characteristics (that is, reaching the same swing amplitude over the same period). It follows that the speeds of **B** and **C** will be directly proportional to that of **P** at corresponding points of their swings. The speed of **P** can be calculated from the square root of the small height **QP** through which it falls to **K**, which is proportional to the heights **BO** and **CS**, that is, the heights through which **B** and **C** fall towards **E** and **D**. However, **B** and **C** are not bodies in free fall, but they are linked by a fixed rod. To determine their speeds, Huygens proposes a thought experiment in which he makes **B** and **C** collide with identical bodies **G** and **F**. He then inserts a plane at an inclination of 45° off of which the latter two imaginary bodies would be reflected; each will reach a height corresponding to the square of its

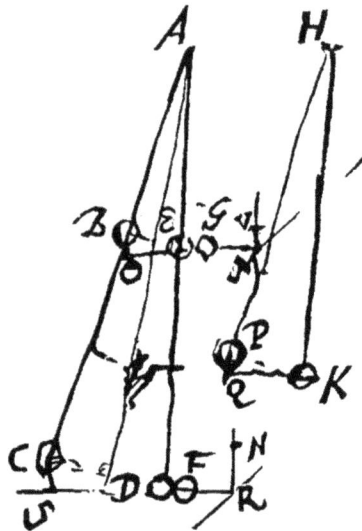

Figure 4.12. (*left*) Huygens' two-bob pendulum, showing bobs **B** and **C** attached to the same pendulum rod, compared with (*right*) the simple pendulum. (Huygens C 1661 *Oeuvres Complètes de Christiaan Huygens*, **XVI**, Part IV pp 415–427.)

velocity. In turn, he expresses this concept as a function of the height **CS** and the ratio of the distances from the bob to **A** to the height of the centre of oscillation, **HK**. Having reasoned through the arguments thus far, Huygens now invokes the important principle that the centres of gravity of **G** and **F** will rise to the same heights as that of the compound pendulum at the beginning of its swing (and not higher or lower).

At this point, Huygens cannot proceed with his analysis based on geometrical arguments alone, because the masses cannot be represented geometrically. Without access to modern calculus techniques, Huygens resorts to algebraic manipulation, where **HK** is the unknown parameter. Let us now retrace his steps. Assume that the distances **HK**, **AB**, and **AD** are given by x, b, and d, respectively, and B and D represent the weights of the bobs. Then we can obtain an equation for the centres of gravity before and after the swing, that is,

$$\frac{Bb + Dd}{(B + D)d} \quad \text{and} \quad \frac{Bb^2 + Dd^2}{(B + D)xd}.$$

The unknown pendulum length **HK** then follows directly:

$$x = \frac{Bb^2 + Dd^2}{Bd + Dd}.$$

Rigid rods can thus be represented by a sequence of individual oscillators. His solution formed the basis for Huygens' subsequent analysis of more complex systems [51]. Building on Galileo's law of free fall and the Italian scientist's exploration of swinging bodies on inclined planes, the basic premise of Huygens' approach was that when the centre of gravity of a system falls from rest and then moves upwards again to the same equilibrium position, it starts and terminates at the same height, irrespective of the changes in the relative position within the system itself. Indeed, Huygens' first hypothesis states that

> *If some weights begin to move under the force of gravity, then it is not possible for the centre of gravity of these weights to ascend to a greater height than that found at the beginning of the motion.*

Huygens clarifies a number of times that the real meaning of this hypothesis is that bodies cannot '*uniquely by virtue of their own weight*' rise to a greater height than that from which they started their fall. This is important, as was also already realized by Galileo in his *Discorsi*:

> '*Indeed, if those builders of new machines who tried in vain to produce perpetual motion* [motum perpetuum] *had known how to use this hypothesis, they would have easily seen their errors and would have understood that this is in no way possible through mechanical means* [mechanica ratione].'

In other words, a quantity of work cannot be done without a corresponding compensation, *perpetual motion is impossible*. Huygens proceeds by introducing a second hypothesis:

In the absence of air, and with the removal of all other known impediments to motion, ..., the centre of gravity of a disturbed pendulum travels through equal arcs in falling and rising.

He clarifies that ...

'*[t]his has been demonstrated for a simple pendulum in Proposition 9 of* 'On the fall of heavy bodies' [Horologium, *Part II*]*. Experience declares that the same should also be held for composite pendulums, inasmuch as, whatever be the shape of the pendulum, it is equally fitted to continue its motion, unless to some greater or less extent it is impeded by the air present.*'

Next, in his Proposition III, Huygens specifies that ...

[i]f weights of certain magnitudes all descend or ascend, through any unequal intervals; then the distances of descent or ascent of these, multiplied by their respective weights, give a sum equal to that produced by the distance of descent or ascent of the common centre of gravity, multiplied by the sum of all the weights.

In modern parlance, this implies $H = \sum_i m_i r_i / \sum_i m_i$, where H is the height of ascent or descent over which the centre of gravity has travelled, m_i are the masses of the compound pendulum's segments, and r_i are their respective heights of up- or downward travel.

Then, in Proposition IV, he states that the removal of the constraints between the bodies or their parts does not affect the equivalence between height of ascent and descent, that is, these constraints do not perform work:

If a pendulum is composed of a number of weights, and it is released from rest, then any part of the whole oscillation is carried out by the weights together; and thus again it is understood that the individual weights of the pendulum, without a common bond, convert the acquired speed by rising up for as long as they are able to ascend; from this fact, it follows that the common centre of gravity of all the weights returns to the same height that it had at the start of the oscillation.

He proceeds to prove this proposition elegantly. I reproduce his proof here to allow the reader to appreciate its mathematical beauty; refer to figure 4.13 for details [52].

The pendulum is composed of some number of weights **A**, **B**, **C**, and the rods or surfaces to which they are attached are considered weightless. [*Note that until Newton's publication of his* Principia *in 1687, 'mass' and 'weight' were often used interchangeably; in this instance, 'weightless' would have been 'massless' in modern physics parlance.*] The pendulum is suspended from the axis drawn through point **D**, which is understood to be perpendicular to the plane shown here. The centre of gravity **E** lies in the same plane as the weights **A**, **B**, and **C**; and the line **DE** from the centre is inclined to the

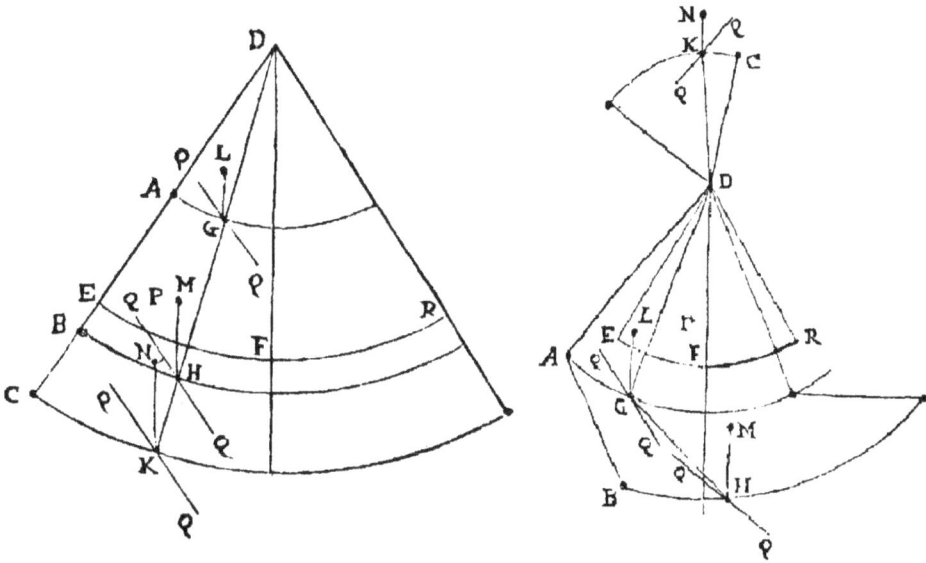

Figure 4.13. Huygens' sketches in support of his derivation of the centre of oscillation for a compound pendulum (proposition IV). (Huygens C 1673 *Horologium Oscillatorium* **IV**.)

perpendicular line **DF** by the angle **EDF** to which line the pendulum is obviously attracted all the time. From this position the pendulum is released and completes some part of an oscillation, so that the weights **A**, **B**, and **C** arrive at points **G**, **H**, and **K**. Henceforth, with the common rod abandoned, it is understood that the speeds acquired can be converted to upward motion (since this can be done by the masses striking certain inclined planes such as **QQ**) and rise for as long as they are able, truly to **L**, **M**, and **N**. When they arrive at these points, the common centre of gravity is point **P**. I say that this point **P** is at the same height as point **E**. For in the first place it is agreed that point **P** cannot be higher than point **E**, from the first assumed hypothesis.

But we will show that neither can it be lower. For, if it is possible, let **P** be lower than **E**, and imagine the weights to fall again from the same heights to which they ascended, which are **LG**, **MH**, and **NK**. Hence indeed it is agreed that they have acquired the same speeds needed to ascend to these heights, that is, these speeds acquired from the motion of the pendulum from **CBAD** to **KHGD**. Whereby, if with the said speeds the weights are now re-attached to the rod or surface, and they continue the motion began along the arcs. The question can now be resolved of what happens, if they are considered to continue rather than to rebound from the inclined plane **QQ**, as before being attached to the rod, in the following manner, by which the restored motion of the pendulum continues, or equally, if the motion continues without any interruption. Thus as the centre of gravity of the pendulum **E**, by descending and ascending, runs through equal arcs **EF** and **FR**, and hence it returns to the same height **R** as **E**. Moreover, if the centre of gravity **E** is placed higher than the corresponding centre of gravity **P** for positions **L**, **M**, and **N**, then **R** will be lower than **P**: and hence the centre of gravity of the weights at **L**, **M**, and **N** will be less in height in the descent than the height in the ascent, which is absurd. Therefore the centre of gravity **P** is not lower than **E**. But neither is it higher. Therefore it necessarily must be equal in height. *Q.e.d.*

We already know from Galileo's experiments that the vertical distance covered by a free-falling heavy body starting from rest is proportional to the square of the velocity attained during its fall, v_f. Therefore, we have $v_f = \sqrt{2gH}$ for each compound pendulum segment, so that

$$\sum_i m_i r_i = \frac{\sum_i m_i v_i^2}{2g}.$$

If the falling motion is constrained rather than free, we can only detect the individual velocities by measuring the individual heights of free ascent. However, armed with that information, we can simply apply Galileo's relation to the heights of ascent and express the final velocities using the same law:

$$\sum_i m_i r_i' = \frac{\sum_i m_i v_i'^2}{2g}.$$

From the equivalence of the height of ascent and descent, we can now derive the centre of gravity,

$$\sum_i m_i v_i^2 = \sum_i m_i v_i'^2.$$

Thus Huygens' result implies that for a system of bodies jointly moving under the influence of gravity, the sum of the products of their masses multiplied by the squares of the final velocities is the same, irrespective of the nature of the motion, that is, constrained or free. In his Proposition V, Huygens concludes,

> For a given pendulum composed of any number of weights, if the individual weights are multiplied by the square of their distance from the centre of oscillation, and the sum of the products is divided by that product, which is the sum of the weights multiplied by the distance of the common centre of gravity of all the weights from the axis of oscillation; then the quotient arising from this is the length of the simple isochronous pendulum, or the distance between the axis and the centre of oscillation of the composite pendulum.

In essence, Huygens used an equivalent approach to the conservation of energy to define the centre of gravity of a compound pendulum in terms of the modern notion of its moment of inertia. This theoretical breakthrough allowed him to develop his understanding of the use of pendulum clocks equipped with weights sliding along the pendulum rod to adjust for—or, indeed, measure—geographic differences. None of the previous versions of his pendulum clock—neither the simple pendulum nor that equipped with cycloidal chops—required integration of sliding weights to keep it going precisely. After all, their bobs could be adjusted easily to calibrate their timekeeping to agree with the stars, with the solar time (that is, corrected for the equation of time), or with a second clock.

The idea of using sliding weights came from an unexpected direction. Around the same time that Huygens completed his calculations allowing him to properly adjust mean clock time to solar time, he announced the addition of sliding weights to his clock. This would make its regulation by the table of the equation of time easier and more accurate. While considering the importance of developing an accurate timepiece, Huygens was struck by the idea that the length of a simple, isochronous 'seconds' pendulum—a little longer than 39 inches, making seconds pendulums non-portable—which beat with a precisely calibrated period, might be employed as a universal standard of measure. Together with Wren, he suggested to his counterparts at the newly established Royal Society of London that one-third of that length would be a 'clock foot' (*pes horarius*). The 12 founders of the Society first met on 28 November 1660. Following a recruitment drive to increase the number of Fellows, the record of their next meeting, held on 5 December 1660, states that …

> '*[i]t was ordered that Mr Wren be desired to prepare against the next meeting for the Pendulum Experiment. …*' [53]

Huygens was the first foreign scientist to visit the Royal Society and work with its Fellows in 1661. This eventually led to his election as a Fellow of the Royal Society in 1663, although he does not seem to have greatly appreciated the honour:

> '*If you are a Fellow of the Royal Society, you should not think that you have to be extraordinary, because I see that everyone is admitted [to the Fellowship] easily. Two days ago [54], I was also admitted, without having noticed that I am now smarter than before my admission, in any field*' [55].

The idea he initially proposed to the Royal Society's Fellowship in 1661, that is, exploring the use of the pendulum to define a universal measure of length, was taken up enthusiastically, leading to numerous experiments with simple as well as cycloidal pendulums, many of which were pursued independently of Huygens' direct influence. Unfortunately, however, Huygens' idea did not work as well in practice as it had appeared in theory. In addition, pendulums made of different compounds were observed to behave differently [56].

English efforts to arrive at an all-encompassing theory of vibration, supported by the Royal Society, included investigations of simple harmonic motion, the behaviour of conical pendulums and vibrating springs, *ad hoc* applications of isochronism, and generalizations of the law of the conservation of the 'living force' (*vis viva*). The latter was an early formulation of the law specifying the conservation of a system's energy. Rival camps vehemently argued that the *vis viva* was either kinetic energy ($\sum_i m_i v_i^2$) or momentum ($\sum_i m_i v_i$). These arguments pitted famous contemporary physicists such as Leibniz against the likes of Isaac Newton and Descartes. Hooke led the English efforts to establish a viable theory of vibration. In support of this goal, he performed a series of demonstrations over the better part of two decades [57], on the circular pendulum (1666), the inclining pendulum (1666), the parabolic pendulum (1667), which we discussed earlier in this chapter, the spring pendulum

(1678), and the simple pendulum (1687). The Royal Society also asked Hooke to verify Huygens' new rules on the properties related to a pendulum's centre of oscillation [58], which the latter had announced in his correspondence with Moray in the early 1660s.

4.1.4 Geographic variations

At the time of their proposal, Huygens and Wren did not realize that, for the same length, the pendulum's period is affected by its geographic location. Of course, with the advantage of having access to the insights from modern physics, this dependence follows naturally from the pendulum period's dependence on the inverse square root of the gravitational constant (see chapter 3.1.1). The leading scientists of the time would soon be confronted with geographic differences in the behaviour of the pendulum clock, however. Between 1671 and 1673 the French astronomer Jean Richer (c. 1630–1696) undertook a number of experiments and measurements in Cayenne (French Guyana) at the request of the Parisian *Académie Royale des Sciences*, which had been established in 1666 partially to undertake scientific expeditions. Within three weeks of the *Académie*'s first meeting, on 11 January 1667, the French astronomer Adrien Auzout (1622–1691) proposed a scientific expedition to Madagascar [59], which was then expected to become the new headquarters of the recently established French East India Company [60]. Olmsted has pointed out that Azout, a leading scientist who was actively involved in the foundation of the *Académie* as well as in planning the Observatoire de Paris (Paris Observatory), ...

> '*had discerned with remarkable insight the great scientific possibilities of such a voyage. Moreover, the ends of the expedition as he conceived them were exclusively scientific. ... A considerable gulf separates these lucid proposals from the confused ideas about scientific voyaging current in scientific circles in Paris and in London at the time*' [61].

The members of the *Académie* agreed that the expedition's main purpose would be to test Huygens' proposals for solving the longitude problem using three of his pendulum clocks, commissioned at royal expense, although the research was eventually done in Cayenne, not in Madagascar.

Around the same time, the first reliable ephemerides of Jupiter's moons had been published, which finally allowed contemporary scholars to pursue accurate longitude determinations, at least on land [62, 63]. Simultaneous observations from a number of different locales of the eclipses of Jupiter's satellites by their parent planet allowed accurate determination of one's geographic position; excellent measurements were obtained in Paris, Florence, Rome, and Danzig (Gdansk) [64]. Therefore, in early 1669, upon the arrival in Paris of Giovanni Domenico Cassini, the new Director of the Observatoire de Paris, the records of the *Académie* reveal that they ...

'began to discuss sending observers under the patronage of our most munificent King into different parts of the world to observe the longitudes of localities for the perfection of geography and navigation' [65].

By March 1670, the high-profile French Minister and driving force behind the establishment of the French East India Company, Jean-Baptiste Colbert, wrote that the *Académie* would send assistant Jean Richer, accompanied by a certain M. Meurisse, both of whom had been made *'mathematicians designated to go to Cayenne to make astronomical observations of utility to navigation'* [66] in January of that year, to the East Indies (that is, initially not to Cayenne) ...

'to make various astronomical observations in connection with others which are to be made here [in Paris]*, and to test the clocks which have been constructed for the determination of longitude at sea'* [67].

Although Richer and Meurisse went to La Rochelle (France) in preparation for their sailing, when their ship set sail, neither Richer nor Meurisse were on board. Apparently, Huygens' clocks did not arrive in time for the ship's departure, as attested in a letter received by Colbert:

'Mr. Richer's clocks arrived here after the departure of the Persian squadron [to Madagascar]*. I think it will be decided to send him to Acadia. It is a voyage from east to west during which he will be able to make his experiments; and, if he returns in time, it will be possible for him to embark on* Le Breton [for the East Indies] *in October'* [68].

The *Académie* was intent on testing a competing method for longitude determination, proposed by the French astronomer and cartographer Jean Deshayes (Deshaies; d. 1706), on the same journey to Acadia (Canada, then known as Nouvelle France) in May 1670. The territory had just been returned to France by the English under the terms of the Treaty of Breda (31 July 1667), which marked the end of the Second Anglo–Dutch War (1665–1667). Nothing is known of Deshayes' family background, although it is clear that in 1668 he was clearly recognized as a promising scholar. After all, Colbert—King Louis XIV's right-hand ministerial adviser—selected him to validate a new, as yet secret method of calculating longitude at sea involving measurements of lunar eclipses and the ephemerides of Jupiter's satellites. (A year after his arrival in Québec, Canada, in 1685, he established a geographic measurement system in Canada based on triangulation and successfully confirmed the difference in longitude of Québec with respect to that of Paris, that is, 72° 13′ [69], using this lunar eclipse method.) Meanwhile, Colbert received a letter from La Rochelle from his protégé Jean Talon (1626–1694), Count d'Orsainville, the first Intendant of Justice, Public Order, and Finances in Canada, Acadia, and Newfoundland, announcing that ...

'I have already had the honour to write you that Richer will embark with his clocks on the Saint Sébastien *for Acadia. Deshayes will also sail on the same*

vessel with the instrument that he has made in Paris. It is to be hoped that from the contact of these two men, who are embarking on good terms, knowledge will result with which you may be satisfied' [70].

Huygens himself had high expectations of Richer's 1670 expedition; his excitement is palpable in a letter to his colleague, the German natural philosopher and London-based diplomat Henry Oldenburg (née Heinrich Oldenburg; c. 1619–1677) dated 22 January of that year:

'*The voyage to test our method to determine eastern and western longitudes has not yet departed, but will do so in a month. There is little reason to doubt its success after the successful recent voyage [under the command of] du Beaufort to Candia [Crete], where we determined the longitudes of several places along the Mediterranean Sea, both on our way out and on the return*' [71].

(We will return to these Mediterranean Sea trials in chapter 4.1.8.) However, in the same letter, he also conceded that …

'*errors made during long voyages are [expected to be] larger, so that we shall see more clearly the usefulness of this invention.*'

Prior to leaving La Rochelle, Richer set one of the clocks at exactly 12 o'clock at the precise moment when the Sun passed through the local meridian (that is, when the Sun had reached the point due south). Having sailed to a certain point in the Atlantic Ocean, he was meant to set the other clock at 12 o'clock sharp, again at the precise moment when the Sun passed through that location's local meridian. The difference in time between both clocks would indicate the difference in longitude between La Rochelle and Richer's location at sea.

The *Saint Sébastien* sailed on 1 May 1670, but the ship soon encountered inclement weather. Unfortunately, before conditions could improve, both of Huygens' clocks had stopped working, and neither was restarted during the voyage. One of the clocks stopped during the storm, but Richer did not restart it immediately, despite his clear instructions to the contrary. The second clock followed suit and stopped some 10–12 hours later. Again, contrary to instructions, Richer simply discontinued the tests completely. As a result, the clocks rapidly deteriorated in their mountings and eventually crashed onto the ship's deck. Huygens, upon hearing of this failure, accused Richer of incompetence. In short, Huygens concluded that the voyage had been a disaster, which no one at the time denied. He pursued the lack of cooperation he had experienced and in February 1671, Huygens complained to the *Académie*'s first secretary, the French natural philosopher and cleric Jean-Baptiste du Hamel (Duhamel; 1624–1706), that …

'*Richer's handling of the clocks had been bad throughout the voyage. For want of a little oil, properly applied, the clocks had been needlessly damaged and afterwards more or less ruined; for want of attention to the written instructions provided, they had not been started again after the storm so that they might be observed during the balance of the voyage. In short, the want of success on this occasion, as far as I can judge, stems more from the carelessness of the observers than from the failure of the clocks'* [72].

He continued to complain by saying *'qu'on s'est fort peu appliqué à bien faire réussir cette experience* (*that they have done very little to ensure a successful outcome*),' because of the *Académie*'s apparent lack of interest, but there are no records as to whether he ever received a response.

By September 1670, preparation commenced for a second scientific expedition, this time to Cayenne, which eventually left La Rochelle on 8 February 1672. Richer was initially meant to take with him a new, remodelled marine clock designed by Huygens, but this plan was abandoned because of time constraints (the clock had not been completed by the time the ship set sail because Huygens had been ill). Huygens reacted, perhaps unnecessarily, hostile; he had been unhappy and rather vindictive in his blame of Richer for the failure of the earlier trial of his marine clocks, which made him say that he was glad that the clock was not ready in time for Richer to test it. Huygens wrote to Oldenburg that, even if his remodelled marine clock had been ready for a new French expedition to the West Indies, he would not have sent it along with Richer:

'I think I would rather go one day myself on some little trip to see to the success of this invention, because I realize that much depends on the diligence of those who are committed to it and in whom I am not very satisfied until now' [73].

Nevertheless, Richer's observations of a lunar eclipse and of Jupiter's moons resulted in the determination of the longitude of Cayenne with respect to Paris, which was however three minutes too large.

At the same time, the astronomers at Paris Observatory proceeded with their appointed task to map France more precisely than ever before[4]. Importantly, to compare new observations with earlier records, it was crucial to confirm the geographic locations of the historical observatories used to obtain those earlier records. Most important among the latter were the observations made by Tycho Brahe between 1576 and 1597 at his Uraniborg observatory in Denmark, located on the island of Hven in the Øresund, the body of water between Denmark and present-day Sweden.

The French astronomer and geodesist Jean Picard (1620–1682), founding member of the *Académie*, proposed in 1669 that he travel to Uraniborg to precisely re-measure its geographic location using the emphemeris tables published by Cassini in Paris. Brahe's measurements were based on the Uraniborg meridian, while Cassini's related to the Paris meridian, so careful mapping from one to the other was required. Over a period of eight months in 1671–1672, Picard, his assistant

---

[4] The French King, Louis XIV, is famously said to have complained that he was losing more territory to his astronomers than to his enemies.

Estienne Villiard, the Danish mathematician Rasmus Bartholin (1625–1698), the young Danish astronomer and mathematician Ole Rømer (1644–1710), and Anders Spole (1630–1699), then professor of mathematics at Lund University, used eclipses of Jupiter's moons to determine that the difference in the longitude of Uraniborg with respect to that of Paris was 10° 32′ 30″ [74]. These new measurements and Rømer's observations of the eclipses of Jupiter's moons (see figure 4.14)—whose deviations from pre-calculated times led him to suggest, in 1676, that the speed of light is finite—contributed greatly to Cassini's revision of the ephemeris tables for Jupiter's satellites. Eventually, in the long-delayed [75, 76] 1693 revision of the tables, accurate eclipse times were given for Paris Observatory [77], which set the stage for accurate longitude determinations elsewhere on Earth [78]. Halley, for

Figure 4.14. Ole Rømer at his transit instrument (from Horrebow P 1735 *Basis Astronomiæ ...*, Ch. **VIII**. Copenhagen) showing on the left-hand side a version of Huygens' pendulum clock. Rømer's discovery in 1676 of a delay in the arrival time on Earth of the light signals coming from the eclipse of Jupiter's moon Io was made possible by virtue of the accuracy of one of Huygens' clocks.

Figure 4.15. Drawing from Richer's *Observations astronomiques et physiques faites en l'isle de Caïenne* (Paris, 1679). Most of the astronomical instruments used by Richer are shown, including one of Thuret's pendulum clocks. (Artist: Sebastian Le Clerc; available under the Creative Commons Attribution-ShareAlike License.)

instance, converted Cassini's tables for use in London in 1694, to which he added a detailed commentary [79].

In Cayenne (shown in figure 4.15), meanwhile, one of Richer's important tasks was to calculate the length required for a seconds pendulum and compare this with the length obtained in Paris. Although this observation was not considered particularly important from the outset, Richer wrote in his report,

> *'One of the most important observations I have made is that of the length of the seconds pendulum, which has been found shorter in Cayenne than in Paris. For the same measurement marked on an iron rod in the former place in accordance with the length found necessary to make a seconds pendulum was transported to France and compared with the Paris measurement. The difference between them was found to be $1^1/4$ lignes* [2.8 mm], *by which the Cayenne measurement falls short of the Paris measurement, which is 3 feet, $^{183}/_5$ lignes. This observation was repeated during ten whole months, when no work passed without its being carefully performed several times. The vibrations of the simple pendulum which was used were very short and remained quite perceptible up to 52 minutes, and were compared with those of an extremely good clock whose vibrations indicated seconds'* [80].

Indeed, these results were also deemed very important by Richer's contemporaries. Newton highlighted them in his *Philosophiæ Naturalis Principia Mathematica* (*Principia*; 1687), stating that Richer's pendulum clock—which kept perfect time in Paris—went slow in Cayenne:

> *'Now several astronomers, sent into remote countries to make astronomical observations, have found that pendulum clocks do accordingly move slower near the Equator than in our climates. And, first of all, in the year 1672, Mr. Richer took notice of it on the island of Cayenne; for when, in the month of August, he was observing the transits of the fixed stars over the meridian, he found his clock to go slower than it ought in respect of the mean motion of the Sun at the rate of $2^m 28^s$ a day'* [81].

Richer is generally regarded as the first person to have observed a difference in the force of gravity as a function of location on Earth, although other scientists had earlier made similar suggestions. For instance, Hooke did not subscribe to the idea that the pendulum may be a reliable standard because of differences in the Earth's gravity in different locations, at different altitudes—which had already been raised as a concern by the English philosopher and scholar Francis Bacon in his *Novum Organum Scientiarum* (*New Organon*) in 1620—in addition to differences in the local climate and in atmospheric conditions; these concerns prompted Boyle and Brouncker in 1661 to propose that a pendulum clock should be taken up the Pico Tenerife, the tallest mountain of the Canary Islands, presumably to test the effects of varying air pressure on the stability of pendulum operation. The first volume of the Royal Society's *Register Book* records as one of the main questions '*propounded and agreed upon to be sent to Teneriffe by the Lord Brouncker and Mr. Boyle*' on 2 January 1661,

> '*4, Try by an hour-glass, whether a pendulum-clock goeth faster or slower on the top of the hill than below.*'

In addition, in his monograph *Observationes diametrorum solis et lunae apparentium ... Cum tabula declinationum solis ... Huic adjecta est Brevis dissertatio de dierum naturalium inaequalittate et de temporis aequatione* (Paris, 1670), the French abbot and scientist Gabriel Mouton (1618–1694) attempted to quantify these differences. He published a value for the length of a seconds pendulum in Lyon that differed significantly from that measured at Paris. From the variations of this length with latitude, he proposed to deduce the length of the terrestrial meridian, that is, the size of the Earth. In turn, he suggested to take a fraction of this latter measurement—the minute of arc or *mille* (*milliare*)—as the universal unit of length.

Nevertheless, leading scientists such as Picard were unconvinced, insisting that the length was constant and independent of latitude. Picard persisted in his opposition to the reality of latitudinal variations even in the face of careful observations made by a dedicated, *Académie*-sponsored scientific expedition to Gorée, off the coast of present-day Senegal in West Africa, the Cape Verde islands, and the West Indies [82]. Huygens was undecided until well into 1687, commenting that he could provide a theoretical rationale for both the reality and the absence of latitudinal variations [83]. However, despite his initial reservations as regards Richer's competence, the latter's observations led Huygens to suggest that this geographic variation would be owing to the centripetal force from the Earth's rotation. This is indeed a latitude-dependent effect.

4.1.5 Sliding weights

Returning now to Huygens' 'sliding weights' analysis of the compound pendulum in the late 1650s and early 1660s, he has provided us with a straightforward way to determine where along a rod we would need to place a fixed weight so that the resulting compound pendulum has a period of exactly two seconds. This is equivalent to the statement that we can determine the distance from the point of suspension to the centre of oscillation so that its length is the same as that of a simple seconds pendulum. A small weight added to the pendulum's centre of oscillation will not cause any change to its period. We can calculate where to place a weight to result in a precise change in period. That is the key principle behind the sliding weight (*poids curseur*), which Huygens first wrote about to Moray on 30 December 1661 [84]:

> '*Since some time now [depuis quelques temps] I have found how to adjust very precisely the time of my clock by applying a small sliding weight to the copper rod of the pendulum.*'

Despite Huygens' confident tone, the phrase '*depuis quelques temps*' appears to be somewhat exaggerated. His 1660 revision of *Horologium* did not discuss the principles underlying his theory of the centre of oscillation. It thus appears that although his notes and calculations from the second half of 1659 indicate that Huygens had started to work on the problem, he did not solve the main difficulties until some time in 1661. At that point, he pursued calculations that led to a table listing the settings of a sliding weight for gains over 24 hours of up to 2 minutes by five-second intervals [85].

These first results allowed Huygens to account to some extent for the discrepancies uncovered by the Royal Society's experiments, but it took until 1664 before he managed to explain why the size of the bob made a difference. By that time, he had successfully developed the mathematical background to apply the basic principle to a range of bob shapes, including a sphere. He could not yet locate the precise centre of oscillation of a pendulum made of a fine wire and a round bob: for a fixed length of wire the distance from the geometrical centre turned out to depend on the bob's radius [86].

4.1.6 Theoretical insights inform practical applications

In the context of the development of new science in early modern Europe, Huygens was a typical practitioner, adept at deriving and proving his results in theory, supplemented by practical design sketches where viable. To him, his pendulum clock represented a practical interface between the philosophical and mathematical worlds on the one hand and physical reality on the other—indeed, between theory and practice. This attitude allowed him to increase his clocks' accuracy quite significantly in a carefully orchestrated, step-wise manner: each step forward aimed at reaching a new, more tightly constrained tolerance level. This interplay between theory and practice facilitated Huygens' drive for perfection, reducing the practical tolerance of his timing devices to fractions of seconds and inches. Huygens thus set the standard for future attempts at achieving improved mechanical precision.

Although Huygens was an excellent theoretician, he left the final construction of his designs to qualified clockmakers. This required him to establish productive working relationships with these latter professionals, something that did not come easy to him. This was largely a problem caused by his inability or, perhaps, unwillingness to recognize the unique expertise they brought to their collaboration, and to which he would otherwise not have had access, along with their skill. Think in this context of his famous dispute with the Parisian clockmaker Thuret, which we will discuss in detail in chapter 5.1.1; a dispute caused by this same type of arrogance would also develop from his collaboration with Alexander Bruce, as we will see below.

Translation of theoretical considerations into practically viable instruments is not always as straightforward as it may appear at first sight. One important obstacle which had to be dealt with was the unavoidable friction owing to air resistance and gearing. The obvious solution to the friction problem is to make frictional forces irrelevant by using sufficient mass in one's construction. A heavier bob allows the pendulum to keep moving smoothly until it gets its next power boost from the clock's escapement mechanism. In turn, this increases the pendulum's quality factor (*Q factor*), rendering the pendulum's motion less dependent on any mechanical errors passed on, thus leading to increased accuracy. However, a heavier bob also requires more energy to be transferred from the clock's power source, as well as more friction and wear. In addition, the pendulum rod—as well as the bob itself—must be thin so as to allow the pendulum to slice through the air, while the pendulum must also be enclosed to avoid perturbations by drafts. The requirement for sufficiently massive pendulums that were also thin (unlike Huygens' original pendulums) resulted in the tapered, lens-shaped discs in use on present-day pendulum clocks. In addition, the silk threads used for the pendulum rod in Huygens' design were extremely light and strong, with little stretch and high resistance to rot, and they also minimized friction at the pivot point.

Despite all of these improvements, theoretical breakthroughs, and practical inventions, one important aspect had not yet been resolved. Pendulum clocks, particularly those featuring the latest advances, work very well on stable surfaces, but not on the rocking and pitching platforms expected at sea—one intrinsic difficulty we have seen already: as soon as a pendulum is tilted from the vertical direction, its length changes, and therefore so does its oscillation period.

4.1.7 Huygens' rivalry with Alexander Bruce

Earlier in this chapter, we already encountered Richer's unsuccessful tests of Huygens' clocks at sea, but his voyages were not the first marine tests. Instead, the first viable marine pendulum resulted from a collaboration between Huygens and the Scottish inventor Alexander Bruce, Second Earl of Kincardine, in November and December 1662 [87–89]. In 1660, Bruce had joined King Charles II in The Hague before his triumphant return on 29 May of that year to England, a period known as the *Restoration* (or the *Stuart Restoration*, referring to the family name of the British monarchy), in which he reclaimed the thrones of England, Scotland, and Ireland for the monarchy after more than a decade of republican rule. It is likely that Huygens' new pendulum clocks were discussed at that time, leading to Bruce's interest in their application as potential Longitude Clocks.

Huygens and Bruce became involved in a collaboration that was at times awkward and sometimes hostile: Huygens felt that Bruce had elbowed his way into a field that the Dutch scientist had developed. Nevertheless, Bruce implemented a number of innovations into the pendulum clock design that preceded Huygens' own efforts. In 1661, when Huygens visited him in London, Bruce demonstrated his new double-fork, 'F'-shaped crutch for the first time (see figure 4.16 [90]), which had been designed—likely by the Fromanteels in London—to avoid the rigid-body rotation allowed by a single crutch. The known existence of a table clock with an offset winder as fusee dating from before 1662, fitted with an English dial plate and signed 'Severyn Oosterwyck Hague' suggests that Bruce added a fusee to his marine pendulum clock well before Huygens adopted the same approach [91]. This can be taken as evidence of the London origin of Bruce's original marine timepiece, which he showed to Huygens in 1661.

Both men continued to develop their own marine pendulums. Bruce spent the period from March to December 1662 in The Hague, when he commissioned the clockmaker Severijn Oosterwijck (before 1637–c. 1694) to construct two marine clocks to his own, spring-driven design [92]. At that time, Huygens was still pursuing rectangular, weight-driven designs—or perhaps he simply depicted his clock design as such in order to confuse and divert competitors. After all, it had become increasingly clear already by the early 1660s—at least to Hooke and the elder Fromanteel, who is thought to have made Bruce's prototype marine pendulum to Hooke's design—that spring-driven remontoires were much more practical on choppy seas than their weight-driven counterparts. On his return voyage to England, one of Bruce's Oosterwijck clocks was badly damaged, to the point that it was no longer useful for accurate timekeeping at sea. The London clockmaker John Hilderson was entrusted with making a copy, which in turn was taken on Captain Robert Holmes' (c. 1622–1692) voyage to West Africa in 1663–1664. We will return to Holmes' voyages later in this chapter.

Meanwhile, during the summer of 1662, Huygens' wrote a number of enthusiastic letters to both his brother Lodewijk and to Moray, referring to a …

> 'small pendulum clock … which works sufficiently well to serve for [the purposes of] longitude determination, and which, once I have given it a push, continues to move without stopping in my room, where it is suspended from 5 foot long ropes, but I have yet to test it on water, for which we should requisition a reasonably sized vessel to [allow us to] sail on choppy seas, something I do not know when I could achieve it' [93].

Upon hearing the news, Constantijn Huygens Sr, their father, appears to have become overly excited, since on 9 November 1662 Huygens asked his brother Lodewijk to urge their father to tone down his enthusiasm:

> 'I am not as advanced with the invention of [a method to determine] longitudes, as it seems that you believe, and I wish that Father does not talk about it until I have ensured that it is useful. Mr. Brus [Bruce], who has returned to Scotland, will undertake a sea trial whose success I look forward to, because it is of great importance for these affairs' [94].

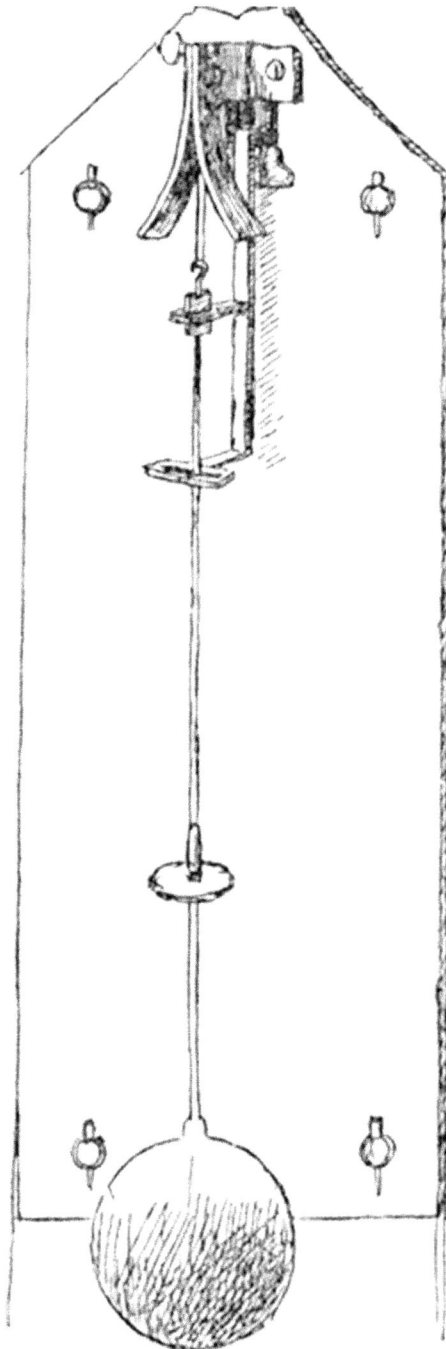

Figure 4.16. Bruce's 'F'-shaped crutch, developed in 1661; drawing from Huygens' 1664 patent application.

Indeed, he was still working on the longitude problem, but meanwhile Bruce had agreed to undertake sea trials, supported by the Royal Society, to test the clocks' reliability and accuracy. Moray, Brouncker, and Hooke were all present at Bruce's earliest sea trials, in 1662 [95, 96]. Hooke was clearly impressed:

> 'The Lord Kincardine did resolve to make some Trial what might be done, by carrying a Pendulum Clock to Sea; for which End, he contrived to make the Watch Part to be moved by a Spring instead of a Weight; and then making the Case of the Clock very heavy with Lead, he suspended it, underneath the Deck of the Ship, by a Ball and Socket of Brass, making the Pendulum but short; namely, to vibrate half Seconds, and that he might be better inabled to judge of the Effect of it, he caused two of the same Kind of Pendulum Clocks to be made, and suspended them both pretty near the middle of the Vessel, underneath the Deck; thus done, having first adjusted them to go equal to one another, and pretty near to the true Time; he caused them first to move parallel to one another, that is, in the Plane of the Length of the Ship, and afterwards he turned one to move in a Plane at Right Angles with the former; and in both these Cases it was found by Trials made at Sea, at which I (i.e. Dr. Hook) was present, that they would vary from one another, though not very much, sometimes one gaining and sometimes the other, and both of them from the true Time, but yet not so much but that we judged that they might be of very good Use at Sea, if some farther Contrivances about them were thought upon, and put in Practice. This first Trial was made in the Year 1662; whereupon, these being found to be able to continue their Motion without stopping, several other Clocks of this Nature were made and sent to Sea, by such as should make farther Experiment of their Use.'

The success of these first sea trials also raised concerns regarding national security issues. Abraham Hill (1633–1721), a British merchant, wondered aloud *'whether such arcana may be divulged, and so become of as much advantage to foreigners as to ourselves'* [97].

In his letter to Lodewijk of 14 December 1662 [98], Huygens wrote, clearly unhappy, that Bruce claimed intellectual rights to some of the marine clock's design, as well as a share in any profits resulting from its commercialization, but he did not go into details. Although Huygens eventually but reluctantly agreed to pay him an equal share, Bruce demanded more. He went so far as to dismiss Huygens' contributions as simply an obvious extension of earlier work, which was highly insulting from Huygens' perspective. In addition, Huygens felt cornered by the regulation that a foreigner could not apply for an English patent [99], leaving the way open for Bruce to proceed without any regard for Huygens' intellectual contributions. Mediation by Moray on behalf of the Royal Society [100] salvaged the situation to some extent. The result was that Bruce agreed to share the patent with the Royal Society, which acted both on behalf of Huygens as well as in its own right [101].

Publicly, Huygens' approach to this issue seemed to have consisted of simply ignoring Bruce's claims: even when Bruce and the Royal Society were granted an

English patent on 13 March 1665, Huygens deferred any discussion of their clock's design or Bruce's contribution, of which the most important aspects were the use of a steel ball encased in a brass cylinder as the marine pendulum's pivot and the application of the 'F'-shaped crutch. Both innovations were included in Huygens' patent drawings of November 1664 [102]. Nevertheless, we have to wait until the publication of *Horologium Oscillatorium* in 1673 to unveil some of the mystery surrounding the Huygens–Bruce marine clock design [103]:

> '*Instead of a weight they had a steel strip wound in a spiral, but the force of which the wheels were turned 'round, just as is commonly employed in those small watches that are wont to be carried about. So that the clocks could endure the tossing of the ship, he* [Bruce] *suspended them from a steel ball enclosed in a brass cylinder, and extending downward the arm of the crutch that sustains the pendulum's motion (the pendulum, by the way, was a half-foot in length) he doubled it to resemble the form of an inverted letter F; namely, lest the pendulum's motion wander out in a circle with the danger of stoppage*' [104].

Huygens had already considered a number of different ways in which to reduce friction and suspend a clock from a pivot so that the ship's motion would not affect the clock's operation [105]. Bruce's idea of using a steel ball inside a brass cylinder turned out to be highly effective and stable in tests performed in The Hague. Huygens shipped two clocks featuring this new design to Bruce at the Royal Society in London in early 1663. They were eventually sent on sea trials to Lisbon, Guinea, and into the Atlantic Ocean, under the command of Holmes on board the English Royal Navy frigate *H.M.S. Reserve* [106]. The first results were obtained between 28 April and 4 September 1663. They exceeded expectations, confirming that both clocks ran highly reliably, to within a few minutes over 24-hour time intervals, and allowed longitude determinations which were within a few minutes of arc of those determined by independent means—which explains the bitter disappointment Huygens felt following Richer's failed tests supported by the French *Académie* a decade later. However, Samuel Pepys (1633–1703), Clerk of the Acts to the Navy Board, politician, and diarist, questioned the accuracy of Holmes' measurements:

> '*The said master* [the captain of Holmes' ship] *affirmed, that the vulgar reckoning proved as near as that of the watches, which, added he, had varied from one another unequally, sometimes backward, sometimes forward, to 4, 6, 7, 3, 5 minutes; as also that they had been corrected by the usual account*' [107].

Nevertheless, given their early success, Moray and Brouncker offered their backing to Huygens and Bruce to market their clocks in the Dutch Republic, France, Spain, Sweden, and Denmark. They had previously attempted to patent Hooke's spring-driven pendulum clock, but we have already seen that Hooke aborted these efforts in 1660 because of the restrictive terms associated with the proposed application. Their new efforts to turn the Huygens–Bruce clocks into commercial success did not work out either [108].

Figure 4.17. Frontispiece of Thomas Spratt's *History of the Royal Society of London* (London, 1667). A garlanded bust of King Charles II stands upon a pedestal between William Brouncker and Francis Bacon.

Figure 4.17 is a reproduction of the frontispiece of Thomas Spratt's *History of the Royal Society of London* (London, 1667), of which Spratt, later Bishop of Rochester, was a co-founder. The engraving was designed by the well-known English writer and diarist John Evelyn (1620–1706) and produced by the Bohemian etcher Wenceslaus (Václav) Hollar (1607–1677). It is meant to show the scene at the founding of the Royal Society, including Bruce's (triangular) Longitude Clock as well as a Tall Clock, possibly Seth Ward's 'Lawrence Rooke' commemorative clock, made by Ahasuerus Fromanteel in 1662–1663.

Holmes' second voyage in 1664 on the *H.M.S. Jersey* proved to be a crucial confirmation of the promise of the Huygens–Bruce marine clock design. Holmes had set sail due west from St. Thomas off the coast of Guinea, heading out some 800 leagues (approximately 4500 km) into the Atlantic Ocean before revising their course northeastwards to the coast of Africa. When, after a few days the ship's stores and, particularly, their fresh water supply began to dwindle, Holmes' senior crew recommended that they reroute towards Barbados. Their traditional piloting methods placed them some 100 leagues from Cape Verde—a 3–4 day run—but longitude determinations based on Holmes' clock suggested that they were only some 30 leagues out of port. Holmes decided to trust his own measurements and continue their predetermined course; they reached the islands the next afternoon [109]. Huygens' clock had thus proved its potential as a marine navigation aid; the traditional method had been unable to detect that the ship had drifted with the ocean currents some 80 leagues eastwards...

Obtaining details about the voyage after its completion turned out to be troublesome, however: upon his return to English shores, Holmes was immediately escorted to the Tower of London as a prisoner, allegedly for unwarranted hostile actions against foreign interests [110]. Indeed, although Holmes' orders, signed by King James II of England, were to *'promote the Interests of the Royall Company'*—that is, the *Royal African Company*, which had been established to trade along the west coast of Africa—and to *'kill, take, sink, or destroy such as shall oppose you'*, he had been too successful; more successful even than anticipated by the most unreasonable (that is, commercial and greedy) expectations.

James's predecessor, King Charles II, had expressly promoted pursuing economic dominance, to the detriment of the Dutch. He clearly hoped that a combination of English naval force and state-sanctioned piracy would cripple the Dutch Republic financially and force the Staten Generaal to agree to a favourable peace settlement. Instead, Holmes' actions, which consisted of having captured Dutch forts on the West African coast—including the Dutch base at Gorée, Anta Castle on the Gold Coast, and the principal Dutch base of Cape Coast Castle near El Mina on 1 May 1664—and half a dozen ships (including the West Indiaman *Brill* on 27 December 1663 and the *Goulden Lyon* on 28 March 1664), led to the Second Anglo–Dutch war. This had significant repercussions back in London, and so Holmes was made a scapegoat and charged with exceeding his orders. However, it was not his military prowess that landed him in the Tower of London, but the greed of the Company's governors—Holmes' ill-gotten takings in merchandise were far behind the Company's rather unrealistic materialistic expectations, which led to him being committed to the Tower on both 9 January and 14 February 1665.

His imprisonment was cut short, however, by the Dutch declaration of 22 February 1665—which was interpreted by the English as a declaration of war—that they would retaliate against British shipping [111]. Holmes' leadership and naval expertise were hence sorely needed again. Meanwhile, Moray, acting on Huygens' behalf, tracked down another of the *H.M.S. Jersey*'s officers by early March 1665, who provided enough insights to suggest that the original data needed updating [112]. However, Moray kept the information about the ship having drifted on account of the ocean currents to himself until 27 March 1665 [113].

Huygens, nevertheless enthralled by the success of this latest voyage, proceeded forcefully. He published an account of the voyage and its scientific success in the 23 February 1665 issue of the *Journal des Sçavans* [114] and offered his clock for public sale. He offered anyone interested in purchasing his clocks detailed instructions regarding its regulation and use, collected in his *Kort Onderwys aengaende het gebruyck der horologiën tot het vinden der lenghten van Oost en West* (1665) [115], which was translated into English in 1669, titled *Instructions Concerning the Use of Pendulum-Watches, for finding the Longitude at Sea*. It contained the first known table of the equation of time, although Huygens used conventions that differ from those commonly applied today: he computed the equation of time as a positive number, setting his zero value at 11 February. To reconcile his values with the position of the 'mean' Sun, one should subtract the annual average from his values, corresponding to 14 min 25 s [116].

The Staten van Hollant en Westfrieslant (the Provincial States of Holland) issued a 15-year patent for his marine clock on 16 December 1664 [117]. A series of letters to Moray at the Royal Society and Chapelain at the French *Académie* in the spring of 1665 show that this prompted him to seek the same recognition from the French and English governments [118]. While his patent applications were being considered, he halted his plans to publish his new treatise, *Horologium Oscillatorium*, which contained full details of his clock's design and which he had been working on since making the first revisions to its precursor, *Horologium*, in 1660. Clearly, if he wanted to make a profit from his work, particularly now that his first marine clock looked like it might sustain rigorous sea trials and form the core of a practical and, most importantly, profitable method of longitude determination, he should not publicly release the clock's design details so as not to jeopardize the exclusive license he was after. In addition, he expected to soon be able to add more detailed and systematic data in support of the clock's seaworthiness [119].

4.1.8 Mediterranean sea trials

In 1667, Huygens requested that the VOC allow him to proceed with sea trials, but the VOC's insistence that he supervise them personally was impractical [120]: he had been appointed to a key position within the nascent *Académie* in 1666, so that a prolonged absence would have been inconvenient. Therefore, the next sea trials took place in 1668–1669, on the Mediterranean. During his relocation from The Hague to Paris, Huygens had taken along two clocks with novel remontoires, which he wanted to have tested at sea [121]. However, Huygens and his Parisian clockmaker Thuret encountered significant difficulties while putting these remontoires to the test, so that Huygens was forced to order new clocks without this 'improvement.'

In March 1668, the *Académie* managed to secure passage for one of its aides, M. de la Voye-Mignot (c. 1619–1684), and one of Huygens' clocks on board a ship sailing the Mediterranean, but this first voyage was largely unsuccessful. De la Voye-Mignot also sailed on a second voyage, which was completed by October 1669. On that latter, more successful journey, the scientific expedition formed part of a fleet that sailed under the command of the Duc de Beaufort from Toulon (France) to Candia (Crete) to engage the Turkish fleet and relieve the port's Turkish siege. Longitudinal differences between Toulon, Crete, and several intermediate points were measured with encouraging accuracy. Although the expedition lasted until the autumn of 1669, Huygens received a number of updates along the way. In a letter to his brother Lodewijk, dated 11 May 1668, Huygens was clearly confident as regards his clocks' stable maritime operation:

'*I have received news about my clocks—which are with Mr. de Beaufort—that even during a major storm they did not stop at all. They have not yet been able to use them for longitude determination, but I expect [to hear about] their success in forthcoming letters*' [122].

Indeed, subsequent longitude determinations proceeded successfully, as highlighted by Huygens' enthusiastic account in a letter to Oldenburg of 4 September 1669:

'*I have received very good news regarding the experience of [measuring] longitudes using our clocks on the Mediterranean Sea, obtained onboard Mr. de Beaufort's ship, who has determined the differences in longitude of Candia, Chania* [Crete], *a specific Sicilian promontory, and several Mediterranean islands with respect to Toulon, which are quite consistent with two highly accurate Dutch maps*' [123].

Colbert received the expedition's final report in October 1669; it is no longer available, but we can deduce its contents from Huygens' comments in a letter to his brother of 17 October 1669:

'*Mr. Colbert has been sent an exact relation regarding longitude determinations using my clocks, which have seen great success on the Mediterranean Sea, where poor Mr. de Beaufort has pursued their testing. Today, I read about this relationship with great satisfaction, and henceforth there is no more reason to doubt the viability of the invention*' [124].

Despite these encouraging initial tests, Huygens really wanted to proceed with the long-range tests ordered by the *Académie*—Richer's voyages we encountered earlier in this chapter. Indeed, the scientific duties were formally given to Richer on 10 March 1670 [125], and not to De la Voye-Mignot, apparently because of the latter's intolerable behaviour following his earlier success and his proven lack of reliability [126]. In fact, Huygens referred to him as a *fripon*, that is, a cheat or a charlatan:

'*de la Voye has performed these trials at sea but [he is] a* fripon. *He recorded nothing during the first voyage he made, while he promised miracles. But later, when he was in Candia with Mr. de Beaufort, he made very good observations so that we now have the [exact] relations* [between longitude and the time shown by a precise pendulum clock].'

Huygens was keen that these planned long-range tests were done, arguing that the experiment would greatly benefit from the lessons learnt from the earlier voyages. More importantly, however, Oldenburg had made Huygens aware that '*personnes intelligences*' (*intelligent people*) in the Royal Society doubted that Huygens' clocks could be trusted to operate properly and accurately on a pitching and rolling ship;

they also contended that Huygens' pendulum would be highly sensitive to changes in the ambient air:

'We are very glad that, on the last trip to Candia, your pendulum clock has succeeded so well as to leave you little doubt about its successful use on long voyages, where the errors are larger. I do not know [however] whether our members will support new trials, mainly because there are intelligent people here who think that all approaches known to date and used by others are unable to maintain the machine in a perpendicular position; in addition, they have found, through carefully made observations, that [changes in] the air [have] such an influence on the clock that it will sometimes run ⅛ of an hour faster or slower [over the course of] one day' [127].

We learn more about these *personnes intelligences* in a letter from the English traveller and author Francis Vernon (1637?–1677) to Oldenburg, dated 25 February 1670, in which he provides an account of a house call to a seriously ill Huygens. One of the anonymous critics of Huygens' approach referred to by Oldenburg may have been the mathematician Nicholas (or Nicolaus) Mercator, also known by his birth name, Niklaus Kauffman(n) (c. 1620–1687), at least if we can rely on Huygens' own speculations. The following passage provides an account of Huygens' sentiments and refers to Fellows of the Royal Society:

'Here hee fell into a digression concerning their Judgement about some things wich hee had written & hee said you had intimated to him as if one of the Society in his experiments made about Pendulums had iudged them variable & subject to the alterations of weather. This person hee conjectured to bee Mister Mercator. However, hee said notwithstanding the great ability & capacity of that Person who made those experiments hee durst assure him that a Pendulum was a machine the equality of whose motions one might safely relye upon & if it did not appear soe to him the defect was either in the Artificer who made it or else that this Pendulum was without a cyclois wich corrects its anomalies or else hee said hee did not Putt weight enough & that hee repeated againe I believe in England they doe not hang weight enough to their Pendulums & soe the air governes their motions butt the great secret to master the air is to hang weight enough & use a cyclois of which the severall experiences of the Pendulums here in Paris have soe convinced mee that of that I make noe longer doubt' [128].

Mercator designed a large marine chronometer (*"twas of a foot diameter'* [129]) based on the pendulum principle for King Charles II, which directly led to his nomination, by Moray, for election as a Fellow of the Royal Society in November 1666. The so-called 'equation clock' was exhibited at the Royal Society in London in August–September 1666, as recorded by the diarist John Evelyn (1620–1706) in his memoirs: *'to the Royal Society, where one Mercator, an excellent mathematician, produced his rare clock and new motion to perform the equations ...'* [130] Mercator's clock was the first timepiece from which the Sun's equation of time could be extrapolated. It was constructed to Mercator's specifications by John Fromanteel (the younger), who modified an 'old clock' for this purpose [131]. It was subsequently presented to King Charles II, who clearly understood the principle and commended the mathematician for his invention but without providing any monetary compensation. The clock's present whereabouts are unknown; it may not have survived until the present. In his biographical collection *Brief Lives* (1683/ 1684), John Aubrey (1626–1697), Fellow of the Royal Society, noted with disappointment,

> '*Well! This curious clock was neglected, and somebody of the court happened to become master of it, who understood it not; he sold it to Mr Knib[b], a watch-maker who did not understand it neither, who sold it to Mr Fromantle (that made it) for £5, who asks now (1683) for it £200.*'

Mercator sold a significantly improved design to France in 1669, before he moved to Paris with his family himself in February 1682, at the invitation of Minister Colbert, to design the waterworks at the Palace of Versailles.

On his sick bed, Huygens had speculated that Mercator may have been one of the opponents to have raised doubts about his approach. The latter had independently proposed to the Royal Society a method to determine longitude using a pendulum clock. Huygens strongly condemned this direct competition [132]. To understand Huygens' outspoken response, we need to take a step back and consider the broader context of the Royal Society's efforts at understanding the natural world. Huygens had been corresponding with Oldenburg, who requested, in 1668, that Huygens should contribute some new insights on mechanics. Huygens responded by sending some of his latest work on the physics of impacts. As a result of this correspondence, Huygens learnt of direct competition on the subject of impact and momentum by Wren and Wallis.

Additional clashes caused Huygens to feel unfairly disadvantaged in publishing his latest achievements; given his notoriously inflammable character, this sometimes led to him voicing severe yet possibly unjustified criticism. Oldenburg managed these contentious issues well and ameliorated much of this feeling of resentment, but he could not prevent an acrimonious exchange of letters about mathematical aspects that both Huygens and James Gregory (1638–1675) had pursued independently around the same time—elliptic integrals for the simple pendulum, which Gregory had pursued between 1669 and 1672, but which were not published until after his death [133]. This thus shows that Huygens had become nervous about the impact of these controversies on his reputation, which also explains his attitude with respect to Mercator's new proposals.

It is with this broader background in mind that we can understand Huygens' disappointment in the failure of Richer's voyages. Huygens' initial sea trials had come to an end, with no further tests on the horizon. Despite significant effort, his proposed method of longitude determination remained unproven in practice—but not by fault of the clock, Huygens insisted [134].

4.2 From *Horologium Oscillatorium* (1673) to new long-range sea trials

At that stage, in 1671, he finally decided to pursue publication of *Horologium Oscillatorium*; although much of the manuscript was actually written in 1660 [135], it was eventually published 13 years later, in the spring of 1673. *Horologium Oscillatorium* is considered one of the masterpieces of the 17th-Century scientific literature. Huygens' planning had started early. Just over a year after the publication of *Horologium*, he had started planning a second edition of his original treatise. His aim was to collect all of the numerous discoveries that he had made since that original publication, and publicize the new designs he had developed as a result, in a single volume. In its eventual form, it became a treatise on the accelerated motion of a falling body using the bob of a pendulum clock as his basic tool. In practice, he used it to present his 1658 pendulum design, but with added cycloidal chops, a revised escapement, a discussion of the conical pendulum, as well as of an adjustable ('sliding') compound pendulum. He also summarized the results of the sea trials that had taken place in 1663, 1664, and 1668. However, in *Horologium Oscillatorium* Huygens attempted to go beyond the state of the art at the time; he introduced and illustrated his newest marine clock model, with a triangular suspension and a so-called 'Cardan' mounting consisting of an iron ball enclosed by a brass cylinder to facilitate free movement. This Cardan suspension is named after the Italian mathematician and physicist Gerolamo Cardano (1501–1576), who described it in detail. As we saw earlier in this chapter, such a construction had already been implemented by Alexander Bruce in his designs of the first Huygens–Bruce marine pendulum clocks, and it had been used occasionally since for clocks meant to be taken on board ships.

However, most of all *Horologium Oscillatorium* is an accomplished work of applied mathematics. Huygens included individual chapters which focussed on his mathematical analyses of the fall of heavy bodies and their motion along a cycloidal path—which is of course closely related to his development of the theory of the conical pendulum—the evolution of curves, and the centre of oscillation. This latter analysis included his newly developed theory of sliding weights, the design of a seconds pendulum as a universal measure of time, and his analysis of the vertical distance through which a body falls freely in a given time. In turn, this leads into an appendix containing his theorems regarding the centripetal force (that is, the centrifugal motion), although without providing mathematical proof. The results of Huygens' exploration of centrifugal motion found their way into Newton's *Principia* (1687). Huygens' proofs were eventually published posthumously in his edited collection *De Vi Centrifuga* (1703).

In this section, I will first return to Huygens' mathematical demonstration that the cycloid is a tautochrone and discuss the implications in modern physics language. I will also reassess his theorems about the centrifugal motion from a modern perspective. The remainder of this section deals with developments that saw the light of day following the 1673 publication of *Horologium Oscillatorium*, in particular the redesign of the pendulum escapement. In Chapter 5, I will discuss a renewed focus on spring-driven clocks which originated in the mid-1670s as well as Huygens' exploration of pendulums suspended from three wires, the so-called 'tricorn pendulums.'

4.2.1 The cycloid as a tautochrone

Although he had already discovered the isochronicity of cycloidal oscillations by 1661, Huygens returned to an in-depth exploration of tautochronic oscillations in the years immediately following the publication of *Horologium Oscillatorium*. In fact, according to the editors of his *Oeuvres Complètes*, he did not discover the general condition for isochronicity until well after that publication.

Here, I will first employ a modern, calculus-based approach to show that pendulums tracing cycloidal paths are indeed isochronous, provided that one can overcome the added friction introduced by the cycloidal chops. In chapter 3, I defined a cycloid as the path followed by a point on the circumference of a circle as that circle rolls along a straight line. As a function of time, t, this can be written mathematically as

$$x = a(t - \sin t); y = a(1 - \cos t).$$

In the context of our swinging pendulums, we simply replace the time t by the time-dependent swing amplitude, θ: see figure 4.18 for the mathematical set-up. Now first take the derivatives of x and y with respect to θ:

$$\frac{dx}{d\theta} = a\left(1 - \cos\theta\right); \frac{dy}{d\theta} = a\sin\theta,$$

so that

$$\left(\frac{dx}{d\theta}\right)^2 + \left(\frac{dy}{d\theta}\right)^2 = a^2[(1 - 2\cos\theta + \cos^2\theta) + \sin^2\theta] = 2a^2(1 - \cos\theta).$$

In a stable configuration, a pendulum bob at the end of a cord is kept in balance by the centripetal and gravitational forces, that is,

$$\frac{mv^2}{2} = mgy \Rightarrow v = \sqrt{2gy}.$$

A body's instantaneous velocity is given by $v = ds/dt$, where $ds^2 = dx^2 + dy^2$. Therefore,

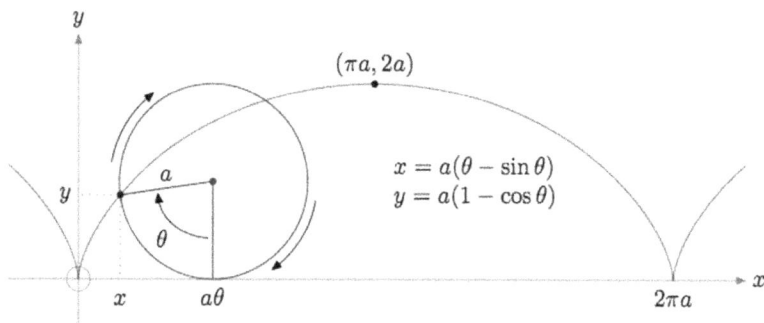

Figure 4.18. Mathematical set-up of a cycloidal path. (Credit: Thruston at the TeX/LaTeX Stack Exchange; licensed under CC BY-SA 3.0.)

$$dt = \frac{\sqrt{dx^2 + dy^2}}{\sqrt{2gy}} = \frac{a\sqrt{2(1 - \cos\theta)}\,d\theta}{\sqrt{2ga(1 - \cos\theta)}} = \sqrt{\frac{a}{g}}\,d\theta.$$

We can now determine the time required for the body to cover the full distance from the top of the cycloid to the bottom, that is,

$$t_{tot} = \int_0^\pi dt = \sqrt{\frac{a}{g}}\,\pi.$$

We can also easily determine the total time spent on the curve for any intermediate angle, θ_0, by replacing the argument on the right-hand side of the equation for the body's velocity to read $v = \sqrt{2g(y - y_0)}$, so that

$$t_{tot\,(arb.)} = \int_0^\pi \sqrt{\frac{2a^2(1 - \cos\theta)}{2ag(\cos\theta_0 - \cos\theta)}}\,d\theta = \sqrt{\frac{a}{g}}\int_0^\pi \sqrt{\frac{1 - \cos\theta}{\cos\theta_0 - \cos\theta}}\,d\theta.$$

Now use the half-angle equations [136],

$$\sin\frac{\theta}{2} = \sqrt{\frac{1 - \cos\theta}{2}}, \quad \cos\frac{\theta}{2} = \sqrt{\frac{1 + \cos\theta}{2}}, \quad \text{and } \cos\theta = 2\cos^2\frac{\theta}{2} - 1.$$

For the time spent by a body to travel from an arbitrary point on the cycloid to its bottom, we can now write

$$t_{tot\,(arb.)} = \sqrt{\frac{a}{g}}\int_{\theta_0}^\pi \frac{\sin\frac{\theta}{2}\,d\theta}{\sqrt{\cos^2\frac{\theta_0}{2} - \cos^2\frac{\theta}{2}}}.$$

By substitution of

$$u = \frac{\cos\frac{\theta}{2}}{\cos\frac{\theta_0}{2}} \quad \text{and } du = -\frac{\sin\frac{\theta}{2}}{2\cos\frac{\theta_0}{2}}$$

we eventually derive

$$t_{tot\,(arb.)} = -2\sqrt{\frac{a}{g}}\int_1^0 \frac{du}{\sqrt{1 - u^2}} = 2\sqrt{\frac{a}{g}}[\sin^{-1}u]_0^1 = \pi\sqrt{\frac{a}{g}} = \text{constant}.$$

It thus follows that the travel time to reach the bottom of the cycloid from any point on the curve is the same. In other words, we have shown that the cycloid is indeed a tautochrone. Using modern calculus, this derivation was relatively straightforward. However, Huygens did not have access to modern methods; calculus in its modern form would not be developed for another few decades at the time that Huygens obtained his proof of the cycloid's isochronicity.

In fact, Huygens solved a differential equation, but he used a mechanical, analytical approach. Huygens was a geometrist above all. Unlike many of his other investigations, however, Huygens' discovery of the cycloid's isochronicity did not follow from the natural progression of well-defined arguments; instead, he had to resort to a range

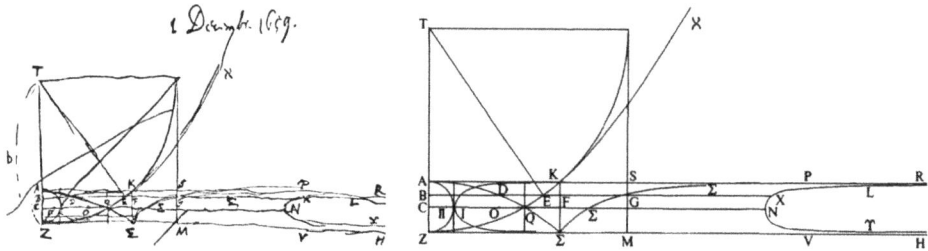

Figure 4.19. (*Left*) Huygens' basic diagram supporting the pendulum's isochronicity (Huygens C 1659 *Oeuvres Complètes de Christiaan Huygens,* **XVI**, p 392). (*Right*) Modern representation (Yoder J C 1988 *Unrolling Time* p 51; reproduced with permission.)

of manipulations to reduce the problem to manageable proportions. He started his investigation from a simple premise. In essence, he was interested in calculating the relationship between the period of a pendulum exhibiting very small oscillations and the time of free fall through the full extent of the pendulum [137].

His exploration of the cycloid's isochronicity started in earnest on 1 December 1659, as indicated in his basic diagram, which I have reproduced as the left-hand sketch of figure 4.19. The right-hand drawing is a modern representation, which we will use to follow Huygens' derivation.

The diagram represents both the physical configuration of the pendulum (including the path of its bob) and the mathematical configuration of the bob's motion. What presumably started from a simple diagram quickly became rather complex because of the need to add auxiliary lines to support his analysis. The key insight Huygens gained from his careful exploration of the diagram's physical and mathematical properties is that the simple pendulum's approximate isochronicity becomes exact isochronicity for cycloidal pendulums [138]. He also found that the force acting on the bob of a cycloidal pendulum is directly proportional to the angle subtended from equilibrium.

The sketches in figure 4.19 show a simple pendulum displaced from its position of rest, combined with the bob's circular path. Drawing on Galileo's laws of motion, Huygens realized that the speed of the bob along its path can be represented by a parabola of velocities, **ADΣ**, which is directly related to a body in free fall along the vertical axis, **AZ** (refer to the right-hand panel for clarity). Even this simple, first step in Huygens' reasoning shows the complexity of thought he was clearly capable of. After all, the diagram combines a physical pendulum configuration overlaid with a graph of the bob's motion, which represent two distinct frameworks without commonality.

We will now explore the steps Huygens took to arrive at his proof of the cycloid's isochronicity by drawing heavily on Yoder's eloquent explanation in her 1988 monograph *Unrolling Time* [139]. We start by considering a body falling from **A** to **Z** with uniform motion at a fixed speed equal to that attained at **Z** by a free-falling body under the influence of gravity along the trajectory **AZ**. In essence, this approach reduces the vertical gravitational fall to uniform motion using the terminal speed variant of the 'mean speed theorem' (we will deal with the factor of 2 introduced here along the way):

A uniformly accelerated body starting from rest travels the same distance as a body with uniform speed whose speed is half the final velocity of the accelerated body.

Huygens compared this body's motion with the pendulum bob's motion caused by gravity from an arbitrary point **K** on the circular trajectory **ZEK** (assumed to describe a small arc). Now compare the time of fall through the infinitesimal (*particula*) at **E** on the pendulum bob's arc with that through *particula* **B** on the vertical trajectory. Geometrical considerations show that the ratio of the two times is equivalent to **TE/BE**. From general principles originally derived by Galileo, it follows that the speed owing to gravity of a body at an arbitrary point on a curve (such as **E**) must be the same as that of a body in free fall from the same initial height **A** to the same vertical height **B**. The speed attained over the interval **AB** can be represented by the parabola **ADΣ**, as we saw above, with its base **ZΣ** equal to the length **AK**. The speed at an arbitrary point **E** is thus represented by the length **BD**; **ZΣ = AK** is the speed reached at **Z** by a free fall under the influence of gravity from **A**.

Although these manipulations seem rather awkward from our modern perspective, they allowed Huygens to proceed geometrically without the need to adopt a value for the gravitational constant, which he had not yet determined accurately at that point. We can now use the definition that the travel time is the ratio of distance covered to the speed with which the body is moving, so that the ratio of the times through the *particulas* **E** and **B** becomes **TE/BE × ZΣ/BD**.

At this point, Huygens introduced a new curve, **LXN**, which took this latter scaling into account by choosing **X** carefully:

$$\frac{\mathbf{BX}}{\mathbf{BF}} = \frac{\mathbf{BG} \times \mathbf{BF}}{\mathbf{BE} \times \mathbf{BD}},$$

where **BF = ZΣ**, that is, the terminal velocity, and **BG = TE**. This now leads to a new proportionality: the ratio of the time through the *particula* **E** to that moving uniformly through **B** is **BX/BF**. The next step involves a summation over all *particulas*, so that the ratio of the time through **KZ** to that moving uniformly through **AZ** becomes

$$\frac{\text{all } \mathbf{BX}}{\text{all } \mathbf{BF}} = \frac{\text{infinite space } \mathbf{APRXNHVZA}}{\mathbf{AZ} \times \mathbf{ZΣ}},$$

and if we now apply the mean speed theorem properly, we need to add the missing factor of 2 to the denominator of this ratio, which is, in fact, the time of gravitational fall through **KZ** with respect to that through **AZ**. The 'infinite space' referred to here is an unbounded region that may not actually be infinite; the denominator is the area of the rectangle defined by its sides.

Huygens struggles through the next steps of his derivation, eventually resorting to approximating the infinitesimal arc of the circular swing **ZEK** with the parabola **ZEא** – which is, in fact, the inverted image of the original parabola he drew, **ADΣ**.

He then draws a companion curve to a parabola (see figure 4.20), scaled by the mean proportion **BD/BK = BK/BL**. Because **BK** is constant, **BL** is inversely

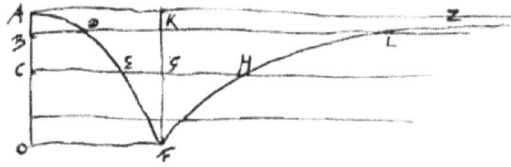

Figure 4.20. Continuation of Huygens' derivation of the isochronicity of the cycloid. (Huygens C 1659 *Oeuvres Complètes de Christiaan Huygens* **XVI** p 398.)

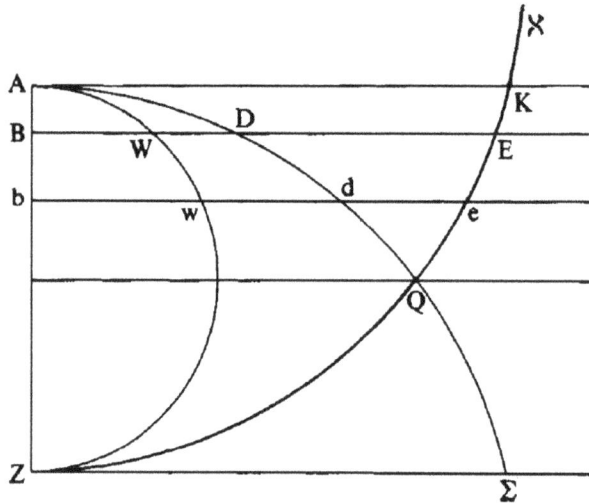

Figure 4.21. Introduction of the second parabola, adjusted from the original. (Yoder J C 1988 *Unrolling Time* p 56; reproduced with permission.)

proportional to **BD**, and since **BD** is equivalent to the speed at the *particula* **B** of a body falling through **AB** (and time is inversely proportional to speed), **BL** represents the time of fall through **B**.

Adding the times **BL** through all *particulas* on **AO** gives the 'infinite space' **OFHLZA**. The latter, Huygens states by reference to the mean speed theorem, is double the rectangle **AO** × **OF**. After all, the mean speed theorem implies that the time it takes to move through the trajectory **AO** under gravitational free fall is equal to twice the time it would take the same body moving at uniform speed of twice the terminal speed under gravitational free fall to cover that distance [140].

Now returning to an exploration of the curve **LXN**, both **BF** and **BG** are constant, so that the definition of **BX** depends only on its relationship to $(\mathbf{BE} \times \mathbf{BD})^{-1}$. **D** is located on the parabola **ADE** and **E** is found on the circle **ZEK**. Thus, given two points **X** and x on the curve **XNY**, they would be related by $\mathbf{BX}/bx = (be \times bd)/(\mathbf{BE} \times \mathbf{BD})$. However, if for small angles the circle's arc **ZEK** is equivalent to the parabola **ZEK** (see figure 4.21), $\mathbf{BX}/bx = bw/\mathbf{BW}$; **W** and w are located on a circle with diameter **AZ** [141].

Huygens then proceeds diligently to work through numerous proportionalities and an additional 'infinite space,' eventually concluding that the time of fall of the pendulum's bob from any arbitrary point **K** on the circular path of the simple pendulum to **Z** (see figure 4.19) is constant. (I will refrain from introducing the entire proof in detail, but refer the reader to Yoder's *Unrolling Time* instead. The important notion I want to get across is that, for small angles, Huygens approximated the circle by a parabola.) Therefore, through pure persistence and a rather convoluted approach Huygens proved that the circular trajectory through **K** is isochronous, provided that we only consider small arcs, as already noted by Galileo.

In retrospect, Huygens realized that his answer would be exact if **K** and **E** of the circular path **ZEK** (refer to figure 4.19) were both actually points on the parabola **ZEℵ**. However, the premise of his derivation, where the infinitesimal triangle at **E** is replaced by the triangle **TEB** requires **TE** to be perpendicular to the pendulum's trajectory. This does not apply if **E** is located on a parabola—and thus the parabola is not isochronous. This setback can be overcome by defining a new curve—which I have reproduced in figure 4.22—for which **DB/CB = CE/CF**. Here, **B** is located on the new curve, **DB** is perpendicular to the curve at **B**, **F** is a point on the parabola, and **CE** is fixed. After this tortuous and belaboured effort, Huygens' important discovery of the cycloid's isochronicity is a study in understatement:

'However, this I found to fit the cycloid, by the known method of drawing its tangent' [142].

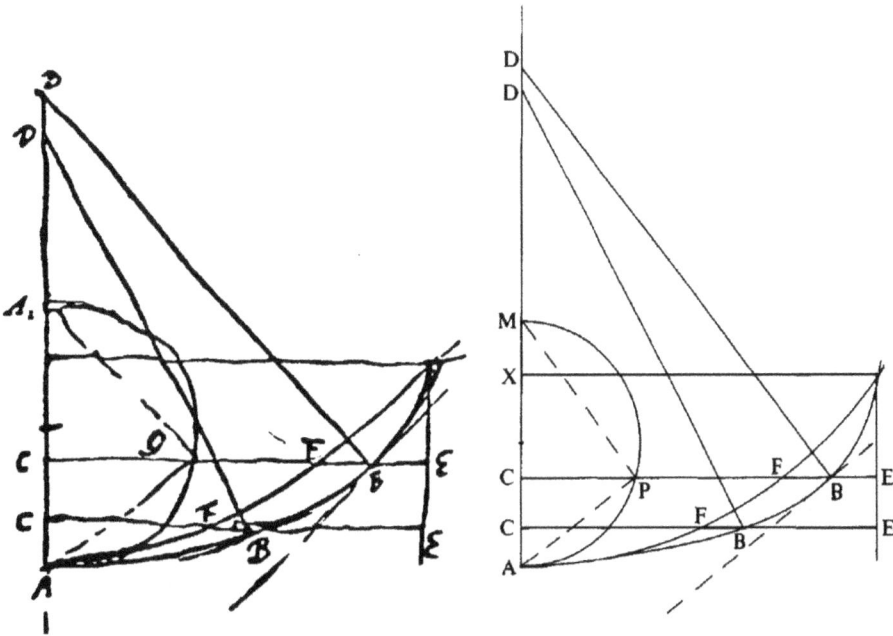

Figure 4.22. (*Left*) Huygens' discovery that the isochronous curve he has been looking for is a cycloid. (Huygen, C 1659 *Oeuvres Complètes de Christiaan Huygens* XVI p 400.) (*Right*) Modern representation with additional labelling. (Yoder J C 1988 *Unrolling Time* p 60; reproduced with permission.)

4.2.2 Huygens' theorems on the 'centrifugal force'

In *Horologium Oscillatorium*, Huygens derived the law of 'centrifugal force' for uniform circular motion: in the 17th Century, circular motion was considered more 'natural' than straight-line motion. Bertoloni Meli summarized the prevailing ideas and their subsequent evolution effectively and concisely thus:

> 'For Huygens and Newton, centrifugal force was the result of a curvilinear motion of a body; hence it was located in Nature, in the object of investigation. According to a more recent formulation of classical mechanics, centrifugal force depends on the choice of how phenomena can be conveniently represented. Hence it is not located in Nature, but is the result of a choice by the observer. In the first case, a mathematical formulation mirrors centrifugal force; in the second, it creates it' [143].

In fact, the main reason behind Huygens' exploration of the properties of uniform circular motion was to derive the correct value for the constant of gravitational acceleration. He first considered centrifugal motion on a plane, ignoring the weight of the body moving around in circular motion. Adding weight to the body, he generalized his approach to an object moving in a circular orbit on an inclined plane, balancing centrifugal motion and gravitational attraction. He further generalized this configuration by moving on from an inclined plane to the tangent of a paraboloid in which the body's orbital period is constant, irrespective of its altitude. In turn, this led him to the conclusion that …

> '[i]f its rotations were counted, in this way an exact measure of time would be had, by means of a more accurate pendulum' [144].

Huygens had initially sent his theorems on the centrifugal motion to the Royal Society in 1669, disguised as anagrams [145], requesting Oldenburg to …

> 'oblige [him] to safely retain them [his theorems] among the Royal Society's records … in order to avoid [future] disputes,'

presumably to be able to claim unequivocal priority of invention—although, as we will see below, Hooke and his colleagues were already making significant progress in this same field at that time. Although he never published any proofs of his theorems during his lifetime, notes pertaining to their validity appear in the Appendices [146] to his seminal work, *De Vi Centrifuga*, dated throughout the period 1659–1668, although it was not published until 1703. Its editors suggest [147] that Huygens started this work on 21 October 1659, while they add that the development of most of his theorems was most likely already in an advanced state by the end of that year.

In 1674 Huygens returned to his earlier studies of centrifugal motion. At the present time, the historical records—and in particular the timeline and ordering of Huygens' thoughts from that period—are somewhat confusing. It appears that the order in which Huygens' propositions were presented in his posthumously published work is not that in which the scientist arrived at his conclusions [148]. In the published version of 1703, which was compiled by the Dutch/Frisian mathematician Fullenius and his natural philosopher colleague De Volder for Huygens' *Opuscula Posthuma*, the editors reproduced only a subset of Huygens' theorems (and even offered their own proof of two propositions), following his listed order in *Horologium Oscillatorium*. The final ordering differs from Huygens' original in the sense that the propositions which were originally included at the beginning of the list appear in *De Vi Centrifuga* (as well as in the edited version in French of the *Oeuvres Complètes de Christiaan Huygens*) towards the end instead [149]. Yoder suggests that Huygens' intention to eventually publish his manuscript on centrifugal motion was likely overtaken by the rapid progress contemporary physicists—and, in particular, Newton—made in this field.

The date of 21 October 1659 associated with the start of Huygens' development of the theory related to centrifugal motion coincides with his efforts to repeat Mersenne's earlier experiments to determine the constant of gravitational acceleration, g. Contemporary measurements of g by Mersenne, Riccioli, and also by Huygens himself differed significantly from one another, which drove Huygens to seek an alternative way to determine the gravitational constant. On the same day, Huygens experimented with centrifugal motion and drafted his initial collection of propositions. Both of these research directions, that is, the determination of g and his investigations of centrifugal motion, were clearly linked, given his comments that '*[w]eight is the* conatus *to descend*' and that his explorations were '*concerning an accurate measure by means of the oscillations of a clock*' [150].

The fundamental concept Huygens developed in this context is the *conatus* of a body, that is, both its tendency to motion and the cause of the tension in a cord attached to a suspended or swinging body. The *conatus* is measured by the motion that results when it is freed from restraints, for instance, when the cord is cut. Huygens' most important insight was that the *conatus* for bodies suspended on cords or positioned on inclined planes is directly proportional to the forces expected to act upon them on the basis of the theory of statistics. In fact, the underlying, fundamental implication of this notion is Huygens' conclusion that the force responsible for a body's centrifugal motion is reciprocal to the gravitational force, a conclusion he derived based on his studies of horizontal circular motion.

A second important insight was that a body's motion following the severing of its restraint must be considered only very briefly after that instant of release, given that a body on a curved plane has the same *conatus* as one on the corresponding tangent plane, yet the motions of both bodies are only identical just after their release. Indeed, once the restraining cord has been severed, a body will continue to move uniformly along its orbit's tangent plane. For an observer in the rotating frame defined by the original circular orbit, the released body will continue to move as if it were in free fall, with the distance between a body continuing along the original circular orbit and its freely moving counterpart increasing in increments roughly proportional to the odd numbers 1, 3, 5, ..., as suggested by Riccioli.

Huygens calculated—for a given radius, r, or for the equivalent length of a cord— the velocity, v, with which a body needs to move in a horizontal circular orbit to result in the same tension in the cord as if the body were instead suspended from it. The prevailing condition which must be met is that the distance covered in equal time intervals in free fall and in release from circular motion must be the same—this is equivalent to saying that the *conatus* must be the same. Based on geometrical arguments applied to the law of falling bodies, Huygens worked out that the centrifugal *conatus* must be proportional to v^2/r.

Huygens' theorems on the 'centrifugal force' are, hence, equivalent to Newton's later centripetal-force derivation, that is, $F = mv^2/r$, where m is the object's mass. Note that this is the only place in Huygens' work where force appears. The majority of his vast body of work is instead based on the concept of conservation of energy. Huygens' preoccupation with exploring tensions in cords in *De Vi Centrifuga* eventually led to his investigation of the conical pendulum (which we discussed in chapter 4.1.2), in essence a body restrained by a cord that is balanced by gravity and the centripetal force.

Let us now consider Huygens' thought processes as reflected in *De Vi Centrifuga* in more detail. The following summary is heavily based on the eloquent exposition of this same aspect in Yoder's 1988 monograph *Unrolling Time* [151], to which I refer for additional details. Paraphrased in modern language, the first proposition we encounter in Huygens' earliest draft of *De Vi Centrifuga* declares that if a body is constrained by a cord to move uniformly and horizontally along a circular path with diameter d, and ignoring gravity, the tension on the cord exerted by the centripetal force will be equal and opposite to that caused by the body's weight—if it were suspended freely—and that the latter would trace its circular path in the same time as a body in free fall would cover the distance $\pi^2 d$ [152].

Mathematically, Huygens proceeded by superimposing a parabola onto a circle, where the former shape represents the horizontally projected path of a freely falling body and the latter is equivalent to the orbit of the body in the absence of gravity. He was already well versed in the mathematics required for this comparison, given that in his 1654 treatise *De Circuli Magnitudine Inventa* (*On Finding the Magnitude of the Circle*) he had worked out that any circle could be approximated with a parabola whose *latus rectum*—that is, the line segment parallel to the horizontal axis passing through the parabola's focus and with both end points on the curve—was set equal to the diameter of the circle [153].

The left-hand sketch of figure 4.23 shows Huygens's original sketch in support of his first proposition in *De Vi Centrifuga*. On the right-hand side, I have reproduced Yoder's version of this sketch, which includes the additional labelling and line segments we need to follow Huygens' rather elegant derivation of his proposition; I paraphrase Yoder's summary of Huygens' thinking about this problem below.

By definition, the *latus rectum* of the parabola **BF** (**CK** in the original sketch) is the length **BG**, that is, the diameter of the circle. Second, from the definition of the *latus rectum* of a parabola, it follows that $\mathbf{CB}^2 = \mathbf{CF} \times \mathbf{BG}$. If the distance **BF** is sufficiently small, **F** is considered to lie on both the circle and the parabola. Assuming uniform horizontal motion, we can calculate the time it takes for a body moving around the circle **BFG** to reach **F** from its starting points at **B**. That same body, in free fall and released from **B** at the same time at the body moving

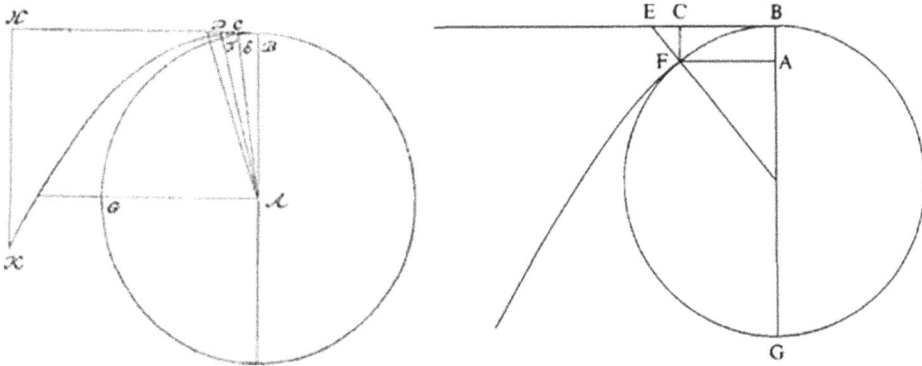

Figure 4.23. (*Left*) Huygens' sketch of the parabola that most closely matches the corresponding circle. (Huygens C 1659 *De Vi Centrifuga*, in: *Oeuvres Complètes de Christiaan Huygens* **XVI** p 259.) (*Right*) Updated sketch, supporting Huygens' basic equation, $EB^2 = EF \times BG$. (Yoder J C 1988 *Unrolling Time* p 21; reproduced with permission.)

around the circle, would travel a distance from **B** to a point just inside the length **BE**. Here, **E** is the intersection of the circle's extended radius that passes through point **F** and the tangent to both curves at **B**. Therefore, the length **EF** represents the separation of the tangential ('free falling') and circular paths. In other words, with respect to point **F**, **EF** is the distance through which the body moves away from **F** with accelerated motion owing to the centripetal force in the same time interval. For infinitesimally small distances **BF**, **EF** = **CF** and **EB** = **CB**. Since for the parabola we have $CB^2 = CF \times BG$, by definition, $EB^2 = EF \times BG$ [154].

Thus, Huygens derived a quantitative expression for the centripetal force. If the body is released at **B**, it would move in accelerated fashion (because of the centripetal force) through a distance $EF = BF^2/BG$ in the same time as a body moving uniformly in circular motion through arc **BF** = **BE** for sufficiently small intervals. For the same small intervals of length and time, the gravitational force acting on the body released from **B** in a state of free fall and the centripetal force exerted by the cord on the body in uniform circular motion are reciprocal and balance each other. Indeed, Huygens showed both forces (tensions) to be equivalent, which allowed him to extend his reasoning. In the same time interval a body traces a full circular path with diameter d, that is, a distance πd. Under uniform motion, it can thus move through a distance $(\pi d)^2/d = \pi^2 d$ because of either centrifugal acceleration or gravity.

Huygens continued his derivations of the laws of motion of bodies moving in uniform circular motion on 23 October 1659, which eventually resulted in his novel insight that, for constant velocity, the tension (*attractionis vis*) on the cord keeping the body in its orbit is inversely proportional to the orbit's radius [155]: indeed, $F = mv^2/r$. He next moved away from purely horizontal planes, thus adding a third dimension in the form of an inclined plane (and the body's weight, which he had ignored up to that point). The mechanical principle of the inclined plane is shown in Huygens's sketch reproduced in figure 4.24, provided in support of his Lemma I.

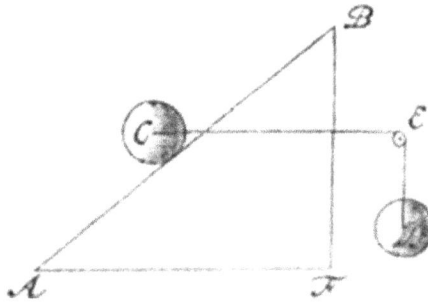

Figure 4.24. Huygens' mechanical principle of the inclined plane. (Huygens C 1659 *De Vi Centrifuga*, in: *Oeuvres Complètes de Christiaan Huygens* **XVI** p 281.)

This figure shows a balance between the horizontal force on body **C** exerted by a weight **D** and the gravitational force pulling down weight **C**, that is, we have a balanced configuration if (weight **D**)/(weight **C**) = **BF/AF** = 1. Combining his earlier insights into uniform circular motion with his derivations of the forces acting upon bodies on inclined planes would allow him to develop the theory underlying the conical pendulum [156], as we saw in chapter 4.1.2.

In parallel to Huygens' efforts, his counterparts across the English Channel pursued their own investigations into the *conatus* of the pendulum. During a reading at the Royal Society on 23 May 1666, which resulted in a paper titled *Of the Inflection of a Direct Motion into a Curve by a Supervening Attractive Principle*, Hooke stated ...

'*for in a circular pendulum the Degrees of Conatus at severall Distances from the perpendicular are in the same proportion with the sines as those arches of Distance as is evident by the figure.*'

The *conatus* Hooke referred to in this passage was the '*conatus* to descend' along the tangent of the pendulum bob's path **FH** or **CE** in figure 4.25. This is equivalent to the component of the gravitational force required to restore the pendulum to its lowest-energy (equilibrium) position. Hooke initially did not directly address the centripetal force, equivalent to the '*conatus* to the centre' in the plane of motion, which he considered in the context of planetary motion in the solar system. First, he simplified the problem to a perpendicular pendulum suspension at rest, adopting ...

'*a common principle of mechanicks that the conatus of a body descending in an inclining plaine to that of one descending perpendicular is in reciprocall proportion to the length of those plaines along which the body falls*' [157].

With reference to figure 4.25, Hooke then showed geometrically that the ratio of the *conati* along **CE** to that along **FH** was equivalent to the lengths **CD** to **FG**, commenting that ...

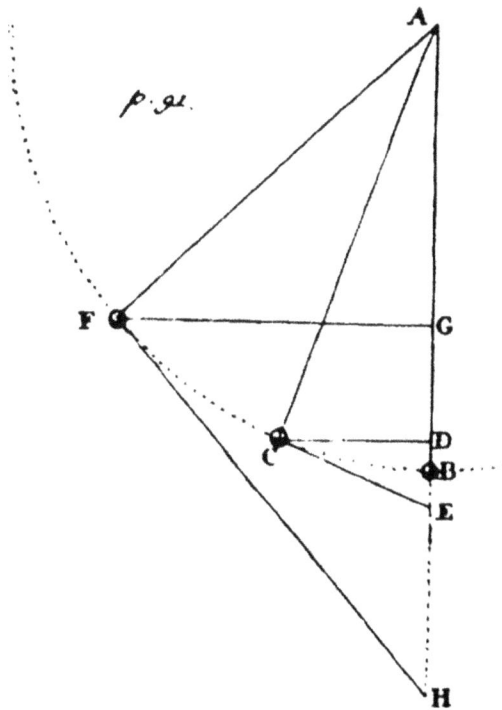

Figure 4.25. Hooke's diagram from his lecture of 23 May 1666. (Birch Th 1756 *History of the Royal Society* vol **2** p 91; redrawn by Patterson L D 1952 *Osiris* **10** p 290; Fig. 1; reproduced with permission.)

'*consequently the* conatus *of returning to the centre in ~~the same~~ a pendulum is greater and greater, according as it is farther and farther removed from the* centre' [116].

This is significant for our understanding of the equation of motion of the pendulum, and it is also equivalent to Huygens' seventh theorem on the 'centrifugal force,'

On the curved surface of a parabolic conoid, which has its axis oriented perpendicularly, all revolutions of a body traversing circumferences parallel to the horizon, be they large or small, are completed in equal times, which times are each equal to two oscillations of the pendulum of which the length is one-half the latus rectum *of the generating parabola* [158].

Hooke was just as interested in the effects of gravity as Huygens, and his research on the subject spanned more than 20 years, with particular focus on planetary motions. In 1674, in his treatise titled *Attempt to Prove the Motion of the Earth*, he put forward a theory of planetary motion based on the principle of inertia and a balance between an outward force responsible for centrifugal motion and an inwards-oriented gravitational force exerted by the Sun. A few years later, on 6 January 1679, he wrote to Newton about his ...

'supposition ... that the Attraction always is in a duplicate proportion to the Distance from the Centre Reciprocall, and Consequently that the Velocity will be in a subduplicate proportion to the Attraction and Consequently as Kepler Supposes Reciprocall to the Distance' [159].

While his suggestion about the velocity dependence was incorrect, Hooke had thus independently reached the same conclusion that this gravitational attraction would vary inversely as the square of the distance from the Sun, which Huygens had arrived at almost two decades earlier. This did not stop Hooke from claiming priority of invention, however. He was highly competitive and—apparently—not an easy person to get along with. When Newton presented the first book of his *Principia* to the Royal Society in 1686, Halley's contemporary report reveals that Hooke claimed that Newton had heard about the *'notion [of] the rule of the decrease of Gravity, being reciprocally as the squares of the distances from the Centre'* from him [160]. However, it is now abundantly clear that ...

'by the late 1660s, [the notion of an] inverse proportion between gravity and the square of distance was rather common and had been advanced by a number of different people for different reasons,' [161]

including by Huygens and Newton himself [162], the latter in the context of planetary motions. Nevertheless, Newton accepted and acknowledged that Hooke was among the scholars—including Wren and Halley [163]—who had independently derived the inverse square law pertaining to the solar system. Newton also told Halley that his interaction with Hooke had inspired his latent interest in astronomy,

'yet am I not beholden to him for any light into that business but only for the diversion he gave me from my other studies to think on these things & for his dogmaticalness in writing as if he had found the motion in the Ellipsis, which inclined me to try it' [164].

4.2.3 Technical improvements to the escapement mechanism

Despite the quarrels and controversies Hooke was involved in throughout much of his active career, he *did* contribute significantly to the development of accurate timepieces, thus securing his place in history. Around 1655, during the early days of the development of pendulum clocks as increasingly accurate timepieces, Hooke had been encouraged to read Riccioli's *Almagestum Novum* (1651). This led to the first of a long string of discoveries and significant original contributions to science. In fragments from his unfinished autobiography dating from around 1697, Hooke wrote that ...

'in the Year 1656 or 57, I contriv'd a way to continue the motion of the Pendulum, so much commended by Ricciolus in his Almagestum, *which Dr. Ward had recommended to me to peruse'* [165].

In turn, Hooke's efforts to improve the pendulum directly led to his invention of a predecessor of the 'anchor' or 'recoil' escapement, which he first demonstrated to the Fellowship of the Royal Society at their meetings in January 1668 and May 1669 [166]. The records of the Royal Society's meeting of 6 May 1669 [167] state that ...

'Mr. Hooke produced a new kind of pendulum of his own invention, having a great weight appendant to it, and moved with a very small force; viz. by such a contrivance, that a pendulum of about fourteen feet long, so as a single vibration of it is made in two seconds, with an excursion of half an inch or less, having a weight of three pounds hanging on it, and moved by the sole force of a pocket-watch, with four wheels, shall go fourteen months and cause very equal vibrations.

He shewed [showed] *two several contrivances for it; one was with a pin upon the balance of a pocket-watch, making a bifurcated needle to vibrate at one end, and on the other end the pendulum: Another was with a thread fastened on one end to the balance of the watch, and on the other end to the pendulum, and so moving it to and fro.'*

The Royal Society's secretary Oldenburg subsequently engaged Huygens in correspondence about Hooke's novel escapement in a series of letters exchanged between May and July 1669 [168–172]. Although Huygens initially expressed his doubts about the accuracy one could achieve with such a new timepiece, given the very small swing amplitudes involved, Oldenburg tried to assure him of its viability and, in fact, of the improved precision of clocks equipped with such escapements—commonly characterized by errors of some 10 seconds over the course of a day—compared with those featuring verge escapements.

The commonly used anchor escapement as we know it today appeared in print for the first time on 20 January 1675 in *Manuscript E* of the *Codices Hugeniani* (also known as the *Codices Hugeniorum*), a collection of nine bound books (A–I) and many additional documents left by Huygens in his will to Leiden University [173]. This date precedes Thomas Tompion's (the 'Father of British Horology;' 1639–1713) application in England of the anchor escapement, in 1676, under the direction of Sir Jonas Moore [174], English mathematician, surveyor, Ordnance officer, and patron of astronomy. Therefore, Huygens insisted that he had satisfied the priority of invention, which was however strongly contested by Hooke, who asserted that he had invented the anchor escapement to secure isochronism of pendulum clocks on land, using a rigid pendulum with a massive bob that traced a small arc with high regularity. This configuration hence no longer required the use of chops for isochronous operation. On balance of the surviving historical evidence, it is indeed likely that Hooke was justified in claiming to have invented the anchor escapement, around 1657 [175,176]. The London clockmaker William Clement (1633–1704), famous for making the 1671 turret clock of King's College, Cambridge, is credited with popularizing the anchor escapement, from approximately 1680. He is also sometimes credited with inventing the anchor escapement. The historical evidence pertaining to the priority of invention is not entirely clear-cut. Clement's possible claim to the invention of the anchor escapement dates from 1670, when the old tower clock of St. Giles Church in Cambridge ground to a halt. Clement was contracted to make a modern, pendulum-based replacement. Clement made a name as maker of high-quality 'long-case' clocks, which he equipped with anchor escapements. In his treatise *Horological Disquisitions* (1694), the London craftsman and clockmaker John Smith (1647/8–1727?) wrote:

> *'From Holland, the fame of this Invention* [the pendulum] *soon past over into England where several eminent and ingenious Workmen applyed themselves to rectify some Defects which as yet was found therein; among which that eminent and well-known Artist Mr. William Clement, had at last the good Fortune to give it the finishing Stroke, he being indeed the real Contriver of that curious kind of long Pendulum, which is at this day so universally in use among us'* [177].

Many of his long pendulums were also equipped with the long, 1¼-second pendulum. This 'Royal Pendulum' length enabled the novel inclusion of both minute and second hands on the clock face, in addition to the common hour hand.

Figure 4.26 is a reproduction of Huygens' first sketches of his anchor escapement; the scribbled names of 'Libnitz' (Leibniz) and Römer (probably referring to Ole

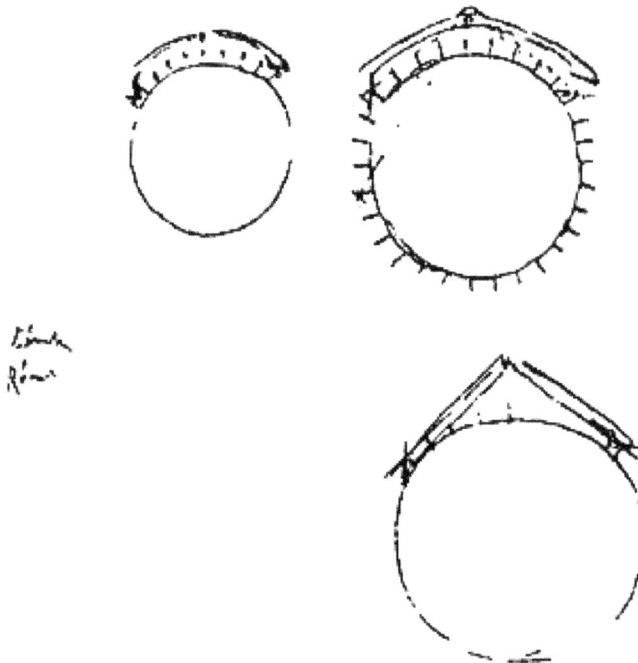

Figure 4.26. Huygens' first sketches of the anchor escapement, 20 January 1675. Note the scribbled names of 'Libnitz' and Römer. (Huygens C 1675 *Oeuvres Complètes de Christiaan Huygens* **XVIII** p 605.)

Rømer) in the left-hand margin may imply that the Dutch scholar had heard about the novel anchor escapement from these two contemporary scientists. The earliest known anchor clock that is still operational was constructed in 1670, probably by London clockmaker Joseph Knibb (1640–1711). It is located in the tower of Wadham College, Oxford.

Anchor escapements (see figure 4.27) were used to reduce pendulum swing amplitudes to 4–5°. Although verge escapements can also be designed to allow very small swing amplitudes, for instance by employing long pallet arms ('limbs') and forcing a large distance between the horizontal escape wheel and the pivot arbour for the pallets and the crutch, the introduction of anchor escapements triggered a drive towards smaller swing amplitudes in general. The newly fashionable small swing amplitudes also significantly reduced friction at the pivot point (just above the vertical escape wheel), as well as air resistance. In turn, this meant that pendulums equipped with anchor escapements needed less power to keep swinging and, hence, suffer from less wear.

Anchor escapements also allowed use of heavier pendulum bobs for a given driving force, thus providing the pendulum with a larger Q *factor* and higher accuracy. In general, however, application of a smaller driving weight means less friction in the bearings of the gear train, between the gear teeth, and between the brass escape wheel teeth and the metal pallet surfaces. For optimal operation, the escape-wheel teeth have to be tall, pointed, and tapered (to maximize strength). The radial length of each tooth, that is, the distance from the centre of the escape wheel to the tip of the tooth, on the one hand and the angle between pairs of teeth on the other has to be identical for regular, accurate operation with minimal friction. In addition, the silk cords commonly used for pendulum rods were soon replaced with a brass suspension spring—composed of a short strip of flat metal—that simply flexed around the shorter arc of the cycloidal chops, thus decreasing the friction even more.

The anchor escapement works similarly to the verge escapement we discussed in Chapter 3, but the verge rod is replaced by an anchor-shaped device: see component h in figure 4.27. The two pallets at the ends of the anchor's main lever are key to operation of the pendulum clock. While the anchor escapement swings back and forth on its pivot, the pallets at the ends of the escapement lever alternately catch and release escape-wheel teeth on either side. Compared with the verge escapement, the pallets of the anchor escapement are located at much greater distances from the axis of rotation around the pivot point, which in turn allows the use of much smaller angles of rotation to obtain the same arc.

Initially, Huygens and other contemporary scholars favoured long (and, hence, necessarily slow) pendulums for stability, specifically because a smaller swing allowed use of a much longer and heavier pendulum. Huygens initially proposed to include a reduction gear between the escapement and the pendulum crutch to reduce the arc of the swing, but the increased friction associated with such a configuration rendered this suggestion impractical. Eventually, more successful pendulum-clock operation came from short, fast-moving pendulums. Nevertheless, a key advantage of using long pendulums is that such a configuration would reduce wear in the escapement. On the other hand, a heavier pendulum

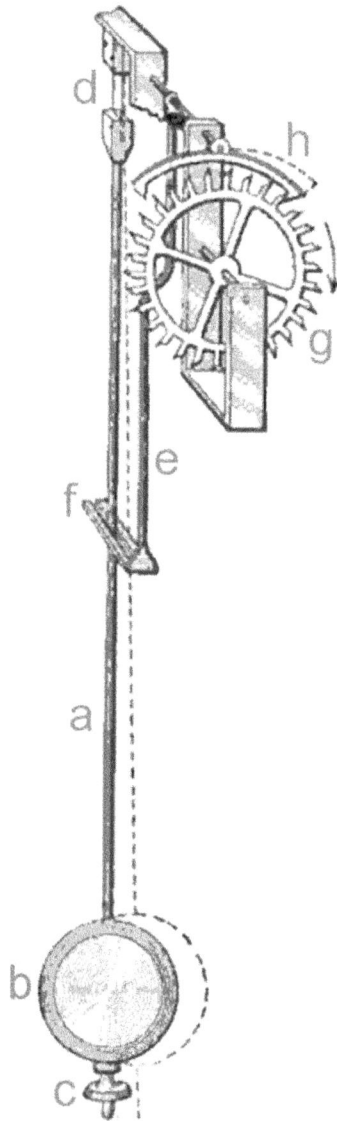

Figure 4.27. Pendulum and anchor escapement. (a) Pendulum rod; (b) Bob; (c) Rate adjustment nut; (d) Suspension spring; (e) Crutch; (f) Fork; (g) Escape wheel; (h) Anchor. (Coleman S E 1906 *The Elements of Physics* (Boston, MA: D C Heath & Co.) p 109, Fig. 87; this modified version obtained from Wikimedia Commons, licensed under the Creative Commons Attribution-ShareAlike 3.0 License.)

would, in principle, generate more friction. However, heavy pendulums have more angular momentum, and so their motion is not significantly affected by interactions with the escape wheel. Ultimately, the motion of long and heavy pendulums most closely resembles simple harmonic oscillation—although one of the main disadvantages of escapements, 'recoil,' still has to be dealt with. Recoil is the backward motion of the escape wheel—the temporary reversal of the entire gear train's motion at the end of the swing—during part of the clock's cycle, which interferes with the pendulum's free motion. It causes additional and excessive wear in and to the gear train and the gear teeth, while it also leads to inherent operational inaccuracies and is, therefore, detrimental to good timekeeping.

To overcome the disadvantages of recoil and excessive friction, the English mathematician and astronomer Richard Towneley (1629–1707) invented an improved version of the anchor escapement around 1675, now commonly known as the 'deadbeat' or 'Graham' escapement: see figure 4.28. It was first used by Tompion, mentor to the English clockmaker George Graham (1673–1751) after whom this type of escapement is often erroneously named, in a clock constructed for Moore and in two precision clocks he built for Greenwich Observatory in 1676 [178], as attested in correspondence between Towneley and the first Astronomer Royal, John Flamsteed [179]. Flamsteed used his new precision pendulum clocks to determine that the Earth rotates at constant speed (see chapter 6.2.2), an application suggested by Huygens in *Horologium*. We will look into Flamsteed's contributions in detail in chapter 6.

The metal pallets with which deadbeat escapements are equipped have two distinct faces. The shapes of the outside faces, commonly referred to as the 'locking' or 'dead' faces, are curved, concentric with the axis of rotation, labelled *b* in figure 4.28. The slanted inside faces are known as the deadbeat anchor's 'impulse' faces [180]. When, during the pendulum's oscillation cycle, the escape wheel is released and one of its teeth is snugly aligned ('locked') with one of the dead faces, the escapement's force is neutralized, that is, it is not transferred to the pendulum crutch and there is no recoil, so that the pendulum can swing freely for most of its outward

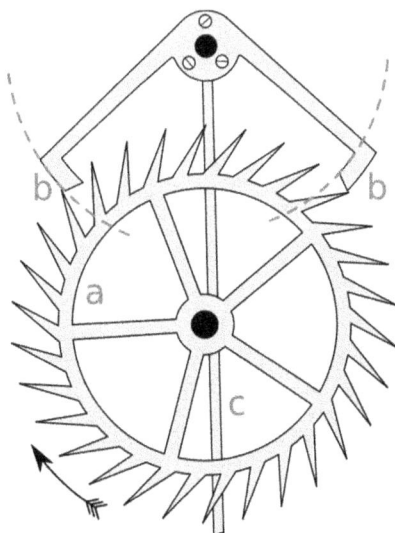

Figure 4.28. Deadbeat anchor escapement. (a) Escape wheel; (b) pallets, showing concentric locking faces; (c) pendulum crutch. (Britten F J 1896 *Watch and Clock Maker's Handbook* 9th edn (London: E F & N Spon), p 108; this modified version obtained from Wikimedia Commons, licensed under the Creative Commons Attribution-ShareAlike 3.0 License.)

swing and return. When the pendulum swing approaches the bottom, the locked tooth slides onto the pallet's slanted impulse face. In turn, this causes the escapement wheel to turn and transfer momentum to the pendulum. This latter sliding motion is associated with a degree of friction, but to a significantly smaller extent than that experienced by the original anchor escapement, because the deadbeat escapement's design has eliminated the recoil force by virtue of the snug match between the pallet's dead faces and the escapement wheel's teeth.

The improved performance of the deadbeat escapement (3 to 5 seconds per day rate errors) compared with the regular, recoil anchor escapement is predominantly owing to improved isochronicity of the former. The deadbeat escapement is not affected by changes in the pendulum clock's drive force [181] caused by, for instance, the weakening force of the gradually unwinding mainspring or irregularities in the friction affecting the gear train. In the absence of recoil, an increased drive force prompts a pendulum to swing faster while attaining a larger swing amplitude. Although these changes leave the oscillation period unchanged, the larger amplitude implies that the circular error may cause a small change in period. Although this was initially considered a reason to doubt the deadbeat escapement's improved accuracy, carefully designed devices can, in fact, be made more accurate than their non-deadbeat counterparts [182], particularly by detaching the drive force from the escapement's locking and impulse functions—an innovation introduced in the 18th Century [183] and, hence, beyond the scope of our narrative.

Additional improvements to the marine clock design were to come also in Huygens' time, however. A renewed focus on spring-driven pendulums and the novel idea of using triangular suspension of the pendulum bob opened up new design possibilities to arrive at a viable marine timepiece that could be used for longitude determinations to within practically useful limits. A number of promising developments triggered in the mid-1670s eventually led to the next set of long-range sea trials in the early to mid-1680s.

References and notes

[1] 1657-05-18: Mylon, Claude – Huygens, Christiaan; *Oeuvres Complètes de Christiaan Huygens* **II** No. 388 (p 29)

[2] Gunther R T 1923–1944 *Early Science in Oxford* (Oxford: Oxford University Press) **6** 10; **8** 146

[3] Robertson J D 1931 *Evolution of Clockwork* (London: Cassell & Co. Ltd) p 172

[4] Huygens C 1658 *Horologium* (The Hague) p 1

[5] Bonelli M L 1960 Di un Orologio di 'Gio. filip Trefler' di Augusta *Physis* **II** 242ff

[6] Bedini S A 1961 Agent for the Archduke, Another Chapter in the Story of Johann Philipp Treffler, Clockmaker of Augsburg *Physis* **III** 137–58

[7] Plomp R 1979 *Spring-driven Dutch pendulum clocks 1657–1700* (Schiedam: Interbook Int. BV) 15–6

[8] *Ibid.*

[9] Inventory number 3557

[10] Bedini S 1983 *Giuseppe Campani, Discorso 1660* (Milan: Polifilo edition) LX

[11] Piggott K 2011 *A Royal 'Haagseklok,' Appendix Three, Open-Research, Memo-Treffler: Johann Philipp Treffler's 1657/8 Pendulum Timepiece (DØcopy)* www.antique-horology. org/piggott/rh/memoranda/memotrefler.pdf [accessed 14 July 2016]

[12] *Ibid.*

[13] Whitestone S and Sabrier J C 2008 (December) The Identification and Attribution of Christiaan Huygens' First Pendulum Clock *Antiquarian Horology* **31/2** www.antique-horology.org/Thuret/Default.htm [accessed 14 July 2016]

[14] Hooke R 1666 On the Inflection of a Direct Motion into a Curve by a Supervening Attractive Principle *The Royal Society's Boyle Collection/Boyle Papers* **20** RB/1/20/34

[15] Hooke R 1666/1674 *Motion in a Curve. A Statement of Planetary Movements as a Mechanical Problem*; in: Gunther R T 1923–1944 *Early Science in Oxford* **VI** (*Hooke* Part I ref 3) pp 265–8

[16] Horrocks J 1637 *Philosophicall Exercises* (Royal Greenwich Obs.) 71v–73v

[17] Yoder J C 1988 *Unrolling Time* (Cambridge: Cambridge University Press) 31

[18] Gunther R T 1923–1944 *op. cit.* **6** 269

[19] Pugliese P 1989 Robert Hooke and the dynamics of motion in a circular path *Robert Hooke: New Studies* ed M Hunter and S Schaffer (Woodbridge: Boydell) pp 196–8

[20] *Ibid.* 276

[21] Derham W 1734 *The Artificial Clockmaker* (London: James Knapton)

[22] Ray J 1718 *Philosophical Letters* (London: William and John Innys) 28–9

[23] Gunther R T 1923–1944 *op. cit.* **6** 293

[24] Gunther R T 1923–1944 *op. cit.* **6** 294

[25] Huygens C 1663–1664 *Oeuvres Complètes de Christiaan Huygens* **XVII** 153: Note 2 and figures 61 and 62

[26] Patterson L D 1952 Pendulums of Wren and Hooke *Osiris* **10** 277–321

[27] 1667-12-04: Huygens, Christiaan – Huygens, Lodewijk; *Oeuvres Complètes de Christiaan Huygens* **VI** No 1614 (pp 167–8)

[28] Gunther R T 1923–1944 *op. cit.* **6** 10; however, for an opposing view, see Hall A R 1951 *Notes and Records of the Royal Society of London* **8** 16

[29] Smith J 1694 *Horological Disquisitions Concerning the Nature of Time* (London)

[30] Huygens C 1659 On Determination of the Period of a Simple Pendulum *Oeuvres Complètes de Christiaan Huygens* **XVI** 392

[31] Mahoney M S 2000 Huygens and the Pendulum: From Device to Mathematical Relation *The Growth of Mathematical Knowledge* ed H Breger and E Grosholz (Amsterdam: Kluwer) 17–39

[32] Huygens C 1659 *Oeuvres Complètes de Christiaan Huygens* **XVIII** 18–9

[33] Huygens C 1673 *Horologium Oscillatorium* Part II, Prop XXV

[34] 1673-10-18: Brouncker, William – Huygens, Christiaan; *Oeuvres Complètes de Christiaan Huygens* **VII** No 1962 (pp 344–5)

[35] Huygens C 1659 *De Vi Centrifuga*, in *Oeuvres Complètes de Christiaan Huygens* **XVI** 255–301; transl M S Mahoney www.princeton.edu/~hos/mike/texts/huygens/centriforce/ huyforce.htm [accessed 15 July 2016]

[36] *Ibid.*

[37] Yoder J C 1988 *op. cit.* 27

[38] *Ibid.* 28–9

[39] Crommelin C A 1947 Les horloges de Christiaan Huygens *Journal Suisse d'Horlogerie et de Bijouterie* **LXXII** 189–204

[40] *Ibid.*

[41] Huygens C 1659 *op. cit.* **XVI** 259

[42] Yoder J C 1988 *op. cit.* 31–2

[43] Crommelin C A 1947 *op. cit.*

[44] 1667-12-04: Huygens, Christiaan – Huygens, Lodewijk; *Oeuvres Complètes de Christiaan Huygens* **VI** No 1614 (pp 167–8)

[45] 1668-10-12: Huygens, Christiaan – Huygens, Constantijn; *Ibid.* **VI** No 1665 (pp 266–7)

[46] This derivation follows Crommelin C A 1947 *op. cit.*

[47] Capecchi D 2014 Attempts by Descartes and Roberval to evaluate the centre of oscillation of compound pendulums *Early Sci. Med* **19** 211–35

[48] Huygens C 1661 *Oeuvres Complètes de Christiaan Huygens* **XVI** 414 (Note 2)

[49] Huygens C 1673 *Horologium Oscillatorium* Part **IV-A**; transl Bruce I 2007, 2013 (most recent revision), www.17centurymaths.com/contents/huygens/horologiumpart4a.pdf [accessed 19 July 2016]; alternative translation by M S Mahoney 1977/1995 www.princeton.edu/~hos/mike/texts/huygens/centosc/huyosc.htm [accessed 19 July 2016]

[50] For a step-by-step guide, which I have followed here, see Mahoney MS 1985 *Diagrams and Dynamics Revisited* http://www.princeton.edu/~hos/Mahoney/articles/drawing/mpipaper.rev.v-1 [accessed 24 August 2017]

[51] Huygens C 1661 De centro oscillationis sive ad invenienda perpendicula simplicia isochrona propositis perpendiculis compositis *Oeuvres Complètes de Christiaan Huygens* **XVI** 415–27

[52] Bruce I 2013 *op. cit.*

[53] Lyons H 1944 The Founding of the Society: 1660–1670 *The Royal Society, 1660–1940* (Cambridge: Cambridge University Press) 23; record from the Society's *Record-book*

[54] There is some confusion about the actual date of Huygens' admission to the Fellowship of the Royal Society. Oldenburg indicates in his letter to Boyle of 2 July 1663 that Huygens was admitted on 27 June 1663, while Thomas Birch, in his *History of the Royal Society of London* (1756, vol **1**, p 263) implies that this happened on 22 June 1663: 'The council elected ... Monsieur Huygens and Monsieur de Sorbiere.'

[55] 1663-06-29: Huygens, Christiaan – Huygens, Constantijn Jr; *Oeuvres Complètes de Christiaan Huygens* **IV** No 1125 (pp 362–3)

[56] Mahoney M S 1980 *Christian Huygens: The Measurement of Time and of Longitude at Sea Studies on Christiaan Huygens: Invited Papers from the Symposium on the Life and Work of Christiaan Huygens* ed H J M Bos, M J S Rudwick, H A M Snelders and R P W Visser (Lisse: Swets & Zeidlinger) 234–70

[57] Patterson L D 1952 *op. cit.*

[58] Birch Th B 1757 *The History of the Royal Society of London* **IV** 480, 508

[59] Archives de l'Académie des Sciences 1666–1668 *Registres de l'Académie des Sciences* **II** *(Mathématiques)* 155

[60] Lavisse E and Rambaud A (ed) 1893–1901 *L'Histoire générale du IVe siècle à nos jours* **VI** (Paris: Armand Colin & Co) p 237

[61] Olmsted J W 1942–1943 The Scientific Expedition of John Richer to Cayenne (1672–73) *Isis* **34** 117–28

[62] 1669-05-06: Oldenburg, Henry – Huygens, Christiaan; *Oeuvres Complètes de Christiaan Huygens* **VI** No 1732 (pp 427–8)

[63] 1669-05-29: Huygens, Christiaan – Oldenburg, Henry; *Ibid.* **VI** No 1738 (pp 439–41)

[64] Olmsted J W 1942–1943 *op. cit.*

[65] *Ibid.*; Duhamel J B 1701 *Regiae scientiarium Academiae historia* (2nd edn: Paris) pp 104–12

[66] Guiffrey J (ed) 1881–1901 *Comptes des bâtiments du roi sous le règne de Louis XIV; Collection des documents inédits sur l'histoire de France* **I** col. 476 (dated 30 January 1670)

[67] Olmsted J W 1960 The Voyage of Jean Richer to Acadia in 1670: A Study in the Relations of Science and Navigation under Colbert *Proc. Am. Philos. Soc.* **104** 612–34

[68] *Ibid.*; annotations by O'Connor J J and Robertson E F 2012 *Jean Richer*, MacTutor History of Mathematics Archive; www-history.mcs.st-and.ac.uk/Biographies/Richer.html [accessed 19 July 2016]

[69] Dionne N E 1908 Inventaire chronologique des cartes, plans, atlas, rélatifs à la Nouvelle-France et à la Province de Québec, 1508–1908 *Proc. Trans. R. Soc. Canada* 3rd Ser vol **2** Part II p 8

[70] O'Connor J J and Robertson E F 2012 *op. cit.*

[71] 1670-01-22: Huygens, Christiaan – Oldenburg, Henry; *Oeuvres Complètes de Christiaan Huygens* **VII** No 1793 (pp 2–4)

[72] 1671-02-04: Huygens, Christiaan – du Hamel, Jean Baptiste; *Ibid.* **VII** No 1824 (pp 54–5); transl Olmsted J W 1960 *op. cit*

[73] 1672-02-13: Huygens, Christiaan – Oldenburg, Henry; *Ibid.* **VII** No 1866 (pp 140–3)

[74] Jones M 2008 Tycho Brahe (Tyge Ottesen Brahe) *Geographers vol* **27**: *Biobibliographic Studies* ed H Lorimer and C W J Withers 1–27

[75] 1690-05-11: de la Hire, Philippe – Huygens, Christiaan; *Oeuvres Complètes de Christiaan Huygens* **IX** No 2589 (pp 419–22)

[76] 1690-08-30: de la Hire, Philippe – Huygens, Christiaan; *Ibid.* **IX** No 2616 (pp 480–82)

[77] Cassini J D 1693 Les Hypothéses & les Tables des Satellites de Jupiter, reformées sur les nouvelles Observations *Mémoires de l'Académie Royale des Sciences Depuis 1666 jusqu'a 1699* vol **VIII** (Paris) 1729 315–505

[78] Konvitz J 1987 *Cartography in France, 1660–1848: Science, Engineering and Statecraft* chapter 1 (Chicago IL: University of Chicago Press) 5

[79] Halley E 1694 Monsieur Cassini his New and Exact Tables for the Eclipses of the First Satellite, reduced to the Julian Stile, and Meridian of London *Phil. Trans. R. Soc* **18** 237–56

[80] Richer J 1679 *Observations astronomiques et physiques faites en l'isle de Caienne* (Paris)

[81] Newton I 1687 *Philosophiæ Naturalis Principia Mathematica* Book III Prop 20

[82] Picard J 1682 *Mém. de l'Acad. (1666–1699)* **VII** 346–7, 435ff

[83] 1687-05-01: Huygens, Christiaan – de la Hire, Philippe; *Oeuvres Complètes de Christiaan Huygens* **IX** No 2455 (pp 130–3)

[84] 1661-12-30: Huygens, Christiaan – Moray, Robert; *Ibid.* **III** No 940 (pp 437–40)

[85] Huygens C 1661 *Ibid.* **IV** 67

[86] Huygens C 1673 *Horologium Oscillatorium* Part IV Prop XXII 141–142; see also Huygens C 1664 *Oeuvres Complètes de Christiaan Huygens* **XVIII** 331–3

[87] 1662-12-01: Huygens, Christiaan – Moray, Robert; *Oeuvres Complètes de Christiaan Huygens* **IV** No 1080 (pp 274–5)

[88] 1662-12-20: Huygens, Christiaan – Moray, Robert; *Ibid.* **IV** No 1083 (pp 280–1)

[89] 1662-12-14: Huygens, Christiaan – Huygens, Lodewijk; *Ibid.* **IV** No 1082 (pp 278)

[90] Huygens C 1664 *Ibid.* **XVII** 166

[91] Piggott K 2009 *A Royal 'Haagse Klok'* App 3 www.antique-horology.org/Oosterwijck/apx3.htm [accessed 22 August 2016]

[92] Leopold J H 1997 The Longitude Timekeepers of Christiaan Huygens *The Quest for Longitude* ed W J H Andrews (Cambridge, MA: Harvard University Press) 104 (Note 21)

[93] 1662-06-09: Huygens, Christiaan – Moray, Robert; *Oeuvres Complètes de Christiaan Huygens* **IV** No 1022 (pp 148–52)

[94] 1662-11-09: Huygens, Christiaan – Huygens, Lodewijk; *Ibid.* **IV** No 1073 (pp 256–7)

[95] Gunther R T 1923–1944 *Early Science in Oxford* (Oxford: Oxford University Press) **8**: 148

[96] Hooke R 1726 *Philosophical Experiments* (London: W. Derham) 4–6

[97] Astle T (ed) 1767 *Familiar Letters which passed between Abraham Hill Esq., … and several eminent and ingenious persons of the last century* (London: Johnston) 102–8

[98] 1662-12-14: Huygens, Christiaan – Huygens, Lodewijk; *op. cit.*

[99] 1664-09-23: Moray, Robert – Huygens, Christiaan; *Ibid.* **V** No 1256 (pp 115–7)

[100] 1664-01-09: Huygens, Christiaan – Moray, Robert; *Ibid.* **V** No 1200 (pp 6–7)

[101] Yoder J C 1988 *op. cit.* 152

[102] 1664-11: Huygens, Christiaan – Staten Generaal; *Oeuvres Complètes de Christiaan Huygens* **V** No 1278 (pp 152–4)

[103] Huygens C 1673 *op. cit.* 16–17; *Oeuvres Complètes de Christiaan Huygens* **XVIII** 115–7

[104] Transl M S Mahoney 1980 *op. cit.*

[105] *Ibid.*

[106] The first voyage took place from 29 April to 4 September 1663: *Oeuvres Complètes de Christiaan Huygens* **IV** No 1174 pp 446–451 (appendix to correspondence between Holmes and Moray); Huygens' response is contained in correspondence between 29 October and 9 December 1663: *Ibid.* **IV** 426–74

[107] Jardine L 2008 *Going Dutch: How the English Plundered Holland's Glory* (New York: Harper Press) chapter 10

[108] Patterson L D 1952 *op. cit*

[109] Holmes reported the 1664 Atlantic voyage to Moray; see 1665-01-23: Moray, Robert – Huygens, Christiaan, *Oeuvres Complètes de Christiaan Huygens* **V** No 1315 (pp 204–206); see also Holmes R 1665–1666 A Narrative concerning the success of Pendulum-Watches at Sea for the Longitudes *Phil. Trans. R. Soc.* **1** 13–5

[110] 1665-02-13: Moray, Robert – Huygens, Christiaan; *Oeuvres Complètes de Christiaan Huygens* **V** No 1329 (pp 233–8)

[111] Interestingly, from a letter of 6 January 1668 from Christiaan to Lodewijk Huygens (*Oeuvres Complètes de Christiaan Huygens* **VI** No 1618 pp 171–2), we learn that, according to Huygens' clockmaker Thuret, the Dutch Admiral Michiel Adriaanszoon de Ruyter had taken a pendulum clock on board his flagship on a voyage to Guinea and the Terra Nova (the newly discovered American continent), although it appears that he had no clue as to how to operate it and so no useful information was conveyed back.

[112] 1665-03-13: Moray, Robert – Huygens, Christiaan; *Ibid.* **V** No 1353 (pp 268–72)

[113] 1665-03-27: Moray, Robert – Huygens, Christiaan; *Ibid.* **V** No 1363 (pp 284–8)

[114] Huygens C 1665 *Journal des Sçavans* 23 February 1665 92–6

[115] Huygens C 1665 Kort Onderwijs aengaende het gebruyck der horologien tot het vinden der lenghten van Oost en West *Oeuvres Complètes de Christiaan Huygens* **XVII** 199–235; 1665-01-02: Huygens, Christiaan – Moray R; *Ibid.* **V** No 1301 (pp 185–189). Huygens engaged a printer on 19 February 1655 and sent the page proofs to Moray on 27 February 1665 so as to facilitate an English translation. The latter was published in the *Phil. Trans. R. Soc.* (1669: **4** 937–53). A French translation was postponed until more data had been acquired.

[116] Grimbergen K 2004 Huygens and the advancement of time measurements ed K Fletcher *Proc. Int. Conf. on Titan – From Discovery to Encounter* (Noordwijk: ESA Publ. Div.) 91–102

[117] 1664-16-12: Staten van Holland – Huygens, Christiaan; *Oeuvres Complètes de Christiaan Huygens* **V** No 1286 (pp 166–7)

[118] Huygens corresponded extensively with both Moray and Chapelain in the spring of 1665, see *Ibid.* **V**

[119] Mahoney M S 1980 *op. cit.*

[120] *Ibid.*

[121] Huygens C 1663–1664 Horloges marine [1663–1664] *Oeuvres Complètes de Christiaan Huygens* **XVII** 178–82

[122] 1668-05-11: Huygens, Christiaan – Huygens, Lodewijk; *Oeuvres Complètes de Christiaan Huygens* **VI** No 1639 (pp 216–8)

[123] 1669-09-04: Huygens, Christiaan – Oldenburg, Henry; *Ibid.* **V** No 1757 (pp 485–6)

[124] Huygens C 1669 *Ibid.* **VI** No 1766 (pp 501–503); App. to 1669-10-17: Huygens, Christiaan – Huygens, Lodewijk; *Ibid.* **VI** No 1765 (pp 499–500); also *Ibid.* **XVIII** 633–5

[125] Clément P (ed) 1861–1870 *Lettres, instructions et mémoires de Colbert* (Paris) **V** 294–5

[126] 1670-05: Huygens, Christiaan – Huygens, Lodewijk; *Oeuvres Complètes de Christiaan Huygens* **VII** No 1806 (p 27)

[127] 1670-02-10: Oldenburg, Henry – Huygens, Christiaan; *Ibid.* **VII** 5–6

[128] 1670-02-25: Vernon, Francis – Oldenburg, Henry; *Ibid.* **VII** 7–13

[129] Clark A 1898 *Aubrey's 'Brief Lives'* (Oxford: The Clarendon Press) vol **II** 58–9

[130] Evelyn J 1818 *The diary of John Evelyn Esquire, F.R.S.* (Scribner); diary entry of 28 August 1666

[131] Piggott K 1996 (reprinted 2011) *The pendulum – Æquations and tides* App. 7

[132] Bell E A 1947 *Christian Huygens and the Development of Science in the Seventeenth Century* (London: Edward Arnold & Co.) 59–60

[133] Turnbull H W 1939 James Gregory (1638–1675) *The James Gregory Tercentenary Memorial Volume* (London: Bell) 371ff

[134] Mahoney M S 1980 *op. cit.*

[135] Andriesse C D 1998 *Christian Huygens: Biographie* (Albin Michel)

[136] Weisstein E W *Tautochrone Problem* MathWorld (Champaign, IL: Wolfram)) http://mathworld.wolfram.com/TautochroneProblem.html [accessed 8 September 2016]

[137] Huygens C 1659 Travaux divers de statique et de dynamique de 1659 à 1666 (Deuxième partie. Dynamique. III) *Oeuvres Complètes de Christiaan Huygens* **XVI** 392

[138] Mahoney M S 2000 *Huygens and the Pendulum: From Device to Mathematical Relation The Growth of Mathematical Knowledge* ed H Breger and E Grosholz (Amsterdam: Kluwer) 17–39

[139] Yoder J C 1988 *op. cit.* 48–64

[140] Huygens C 1659 *op. cit.* **XVI** 398

[141] *Ibid.* 399

[142] *Ibid.* 397

[143] Bertoloni Meli D 1990 The Relativization of Centrifugal Force *Isis* **81** 23–43

[144] Huygens C 1659 De Vi Centrifuga *Oeuvres Complètes de Christiaan Huygens* **XVI** 308

[145] 1669-09-04: Huygens, Christiaan – Oldenburg, Henry; *Oeuvres Complètes de Christiaan Huygens* **VI** No 1758 (pp 487–490); App. to No 1757

[146] Huygens C 1703 De Vi Centrifuga App. I–VII *Oeuvres Complètes de Christiaan Huygens* **XVI** 302–28

[147] Huygens C 1703 *Ibid.* **XVI** 254 (Note 1)

[148] Yoder J C 1991 Christiaan Huygens' Great Treasure *Tractrix* **3** 1–13

[149] *Ibid.*

[150] Huygens C 1703 *op. cit.* **XVI** 303

[151] Yoder J C 1988 *op. cit.* 19–26

[152] Huygens C. 1703 *op. cit.* **XVI** 303

[153] Huygens C. 1659 Travaux de mathématiques pures 1652–1656 *Oeuvres Complètes de Christiaan Huygens* **XII** 165

[154] For an alternative method, see Huygens C 1703 *op. cit.* **XVI** 275–7

[155] *Ibid.* **XVI** 306

[156] Yoder J C 1988 *op. cit.*

[157] Hooke R 1666 *Royal Society Hooke MS 41*, photostat.; referred to by Patterson 1952 *op. cit.*

[158] Huygens C 1703 *op. cit.*; transl M S Mahoney www.princeton.edu/~hos/mike/texts/huygens/centriforce/huyforce.htm [accessed 31 August 2016]

[159] Turnbull H W (ed) 1960 *Correspondence of Isaac Newton, Vol. 2 (1676–1687)* (Cambridge: Cambridge University Press) 309 (document #239)

[160] *Ibid.*; pp 297–314 for the Hooke–Newton correspondence (November 1679–January 1679/80); pp 431–48 for the 1686 correspondence about Hooke's priority claim

[161] Gal O 2002 *Meanest Foundations and Nobler Superstructures: Hooke, Newton and the 'Compounding of the Celestiall Motions of the Planetts'* (Berlin: Springer) 9

[162] Whiteside D T 1991 The pre-history of the '*Principia*' from 1664 to 1686 *Notes and Records of the Royal Society of London* **45**(11–61) 13–20

[163] Newton I 1729 *The Mathematical Principles of Natural Philosophy*, Book 1, Scholium to Prop. 4, p 66 (transl A Motte)

[164] Turnbull H W 1960 *op. cit.*

[165] Purrington R D 2009 The First Professional Scientist: Robert Hooke and the Royal Society of London *Science Networks: Historical Studies* **39** (Basel/Boston/Berlin: Birkhäuser) 81

[166] Gunther R T 1923–1944 *op. cit.* vol. **6** pp 324–325, 351, and 353–356; Huygens C 1675 *Ibid.* **XVIII** 605–6

[167] Birch Th B 1756 *op. cit.* vol **II** p 361

[168] 1669-05-20: Oldenburg, Henry – Huygens, Christiaan; *Ibid.* **VI** No 1735 (pp 433–4)

[169] 1669-05-29: Huygens, Christiaan – Oldenburg, Henry; *Ibid.* **VI** No 1738 (pp 439–41)

[170] 1669-06-10: Oldenburg, Henry – Huygens, Christiaan; *Ibid.* **VI** No 1742 (pp 443–6)

[171] 1669-06-26: Huygens, Christiaan – Oldenburg, Henry; *Ibid.* **VI** No 1744 (pp 459–61)

[172] 1669-07-15: Oldenburg, Henry – Huygens, Christiaan; *Ibid.* **VI** No 1751 (pp 474–6)

[173] Yoder J G 2004 The Huygens manuscripts *Proc. Titan – From Discovery to Encounter* ed K. Fletcher (Noordwijk: ESA Publ. Div.) ESA SP-1278 43–54

[174] Robertson J D 1931 *op. cit.* 133

[175] Headrick M 2002 Origin and Evolution of the Anchor Clock Escapement *Control Systems Mag.* (IEEE) **22**(2) www.webcitation.org/5ko3yENRB [accessed 25 August 2016]

[176] Reid Th 1832 *Treatise on Clock and Watch-making, Theoretical and Practical* (Philadelphia, PA: Carey & Lea) 184

[177] Smith J 1694 *Horological Disquisitions etc.* (London: Richard Cumberland) 2–3; full text available from the Early English Books Online Text Creation Partnership http://quod.lib. umich.edu/e/eebo/A60473.0001.001 [accessed 28 August 2016]

[178] Betts J 1998 Regulators *Instruments of Science: An Historical Encyclopedia* ed R Bud and D J Warner (London: Taylor & Francis) p 121

[179] Flamsteed J., Forbes E. and Murdin L (ed) 1995 *The Correspondence of John Flamsteed, First Astronomer Royal* 1 (CRC Press); 1675-09-22: Flamsteed – Towneley pp 374 (No 229) 375 (Annotation 11)

[180] Headrick M 2002 *op. cit.*

[181] Glasgow D 1885 *Watch and Clock Making* (London: Cassel & Co.) p 293

[182] Rawlings A L 1993 *The Science of Clocks and Watches* 3rd edn (Upton: The British Horological Institute) 108

[183] Stoimenov M, Popkonstantinović B, Miladinović L and Petrović D 2012 Evolution of Clock Escapement Mechanisms *FME Trans.* (Faculty of Mechanical Engineering) **40** 17–23

Chapter 5

The long road to a practical marine timepiece

5.1 Spring-driven clock developments

Jean Richer's voyage to Cayenne in the early 1670s had forced Huygens to face the fact that his beloved pendulum clock design had both practical and more fundamental disadvantages for longitude determination at sea: the one-second pendulum could not be adopted for universal timekeeping. He hence started to explore alternative approaches to regulate his marine timepieces, with spring-driven clocks quickly taking the lead, given their promising potential. The period of a spring-driven timepiece depends only on the intrinsic properties of the spring itself:

$$P = 2\pi\sqrt{\frac{I}{b}},$$

where I is the moment of inertia of the balance wheel, a quantitative measure—in units of $kg\,m^2$—of the wheel's resistance to acceleration, corresponding to the torque required for the balance wheel to reach a given angular acceleration about its axis of rotation. The spring constant, b, quantifies the spring's stiffness or its 'elastic moment' (see chapter 3.1.2) in physical units of $N \cdot m\,rad^{-1}$. Coil springs tend to have reliably invariable spring constants over a large angular range.

   The idea to develop a spring-driven clock was not new by any means. During one of his Cutlerian lectures at the Royal Society's Gresham College, delivered in 1664, Hooke had already publicly suggested the use of a spring instead of a pendulum to regulate sea-going clocks so as to stabilize their operation at sea. As early as August 1665, Robert Moray informed Huygens of Hooke's ...

'altogether new invention, or rather twenty of them, for measuring time exactly as your pendulum clocks, as well on sea as on land, for, according to him, they cannot in any way be disturbed by changes in position, or even the air. It is, in a word, to apply to the balance, instead of a pendulum, a spring, which can be done in a hundred different ways; and he even went so far as to tell us that he has

undertaken to prove that one can so adjust the oscillations that small and large will be isochronous. It would take too long to describe these in detail, and he claims to be publishing the whole thing in a little while' [1].

Huygens responded [2, 3] that he had been made aware of a similar invention during his visit to Paris in 1660. This idea had been proposed jointly by Artus Gouffier (1627–1696), duke of Roannes, and Blaise Pascal, whose design had subsequently been converted into a practical model by the Parisian clockmaker Giles Martinot (1658–1726). Huygens was, however, concerned that a pitching and rolling ship would affect the spring-balance wheel's movements by introducing small irregularities, which he suspected would be difficult to eliminate, while he also expressed concerns about possible temperature effects on the spring's oscillations. It turns out that Huygens was ahead of his time regarding this latter aspect, as we will see below.

Despite his misgivings, however, within the following decade, Huygens seems to have made a U-turn as to the practical use of spring-driven marine timepieces. According to the editors of Huygens' papers [4], some time after July 1673, the scientist appears to have noticed a hitherto unrecognized property of an object that is moving along a cycloidal path: the effective motive force acting on a body—that is, the force inducing the body to move, which Huygens called the *incitation parfaite décroissante* or *incitation*—at any point along a cycloid is proportional to the arc length from the vertex to that particular location [5]. He supported his reasoning by introducing figure 5.1, stating that the ratio of the motive, gravitational forces acting upon points **B** and **A** along the cycloid **AC** is equivalent to the ratio of the forces along the planes **EC** and **PC**, that is, as **PC** to **OC** or, in other words, as the arc length **BC** to the arc length **AC**.

From his analysis of vibrating strings and of balls rolling along curved surfaces, making the crucial step to considering springs occurred naturally. Huygens proceeded by drawing springs in the form of both straightened leaves and helix-shaped coils, fastened at one end: see figure 5.2 [6]. He realized that the *incitation* exerted by a spring is in direct proportion to the force required to hold it at rest in its straightened

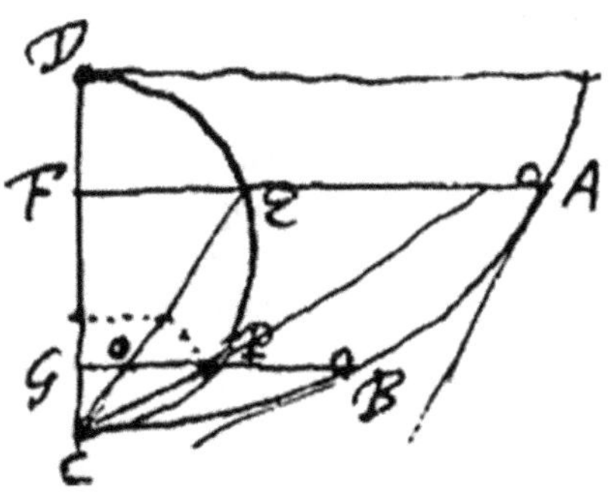

Figure 5.1. Sketch in support of Huygens' proof that the effective motive force acting on a body at any point along a cycloid is proportional to the arc length from the vertex to that particular point. (Huygens C 1673 *Oeuvres Complètes de Christiaan Huygens* XVIII p 489.)

[Fig. 16.]

Figure 5.2. Huygens' sketch of two springs, one in the shape of a stretched leaf, the other as a helix-shaped coil. (Huygens C 1673 *Oeuvres Complètes de Christiaan Huygens* XVIII p 496.)

or coiled configuration. In other words, the *incitation* (force) of a spring is proportional to the angle by which the system is forced out of equilibrium. This property is more generally known as Hooke's law (1678): *ut tension sic vis* ('As the extension, so the force') [7]; for cycloidal motion, the resulting oscillation is isochronous. Huygens had thus arrived at an important realization: springs could be adopted as ideal isochronous oscillators.

Although Hooke had first stated his law in 1660, he did not announce the solution and the general law of elasticity until his lecture '*Of spring*,' given and subsequently published in 1678 [8], several years after Huygens derived the equivalent proportionality. It is, therefore, unknown whether Huygens was influenced by Hooke or if he derived the proportionality on his own. Huygens and Leibniz had clearly discussed the provenance of Huygens' results in person some time in the period 1672–1676 when both scientists spent time in Paris, as evidenced by Leibniz' comment from 1691 noting that

> '*Mr. Newton did not address the laws governing springs; it appears to me that I heard you say [some time ago] that you had examined them, and that you had demonstrated the isochronism of [their] vibrations*' [9].

Huygens himself insisted in his response to Leibniz, on 26 March 1691 [10], that his result was well-known, '*as experience teaches constantly*,' but without providing any further proof.

Soon after Huygens' theoretical breakthrough, he managed to turn theory into practice, apparently with the aid of his Parisian clockmaker Thuret. His first sketch of a watch balance regulated by a coiled spring, accompanied by the word 'εὕρηκα' (*eureka*), dates from 20 January 1675 [11]: see figure 5.3. Huygens' sketches are accompanied by a number of notes. At the top, we find a general description of the scene, '*Watch balance regulated by a spring*.' The two notes on the left of the drawing read:

> '*The spring should be held in the air in the drum and be riveted at the side and at the spindle (arbour)*,'

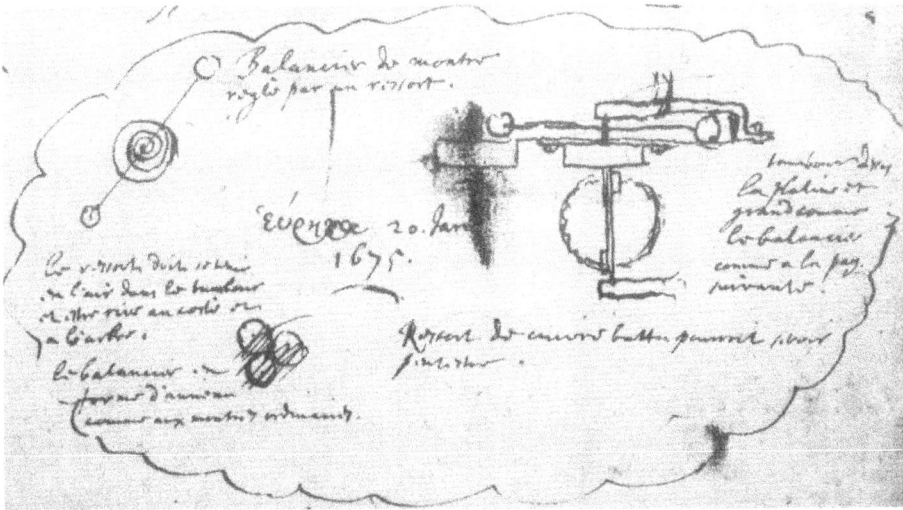

Figure 5.3. Huygens' drawings of a spring balance and escapement, 20 January 1675. (*Oeuvres Complètes de Christiaan Huygens* VII p 408; Adversaria, Manuscript E; Leiden University Library, HUG 9, f.18r.)

and

'The balance in the shape of a ring, as in ordinary watches,'

while the note on the right reads, with reference to the sketch at the top of figure 5.4,

'The drum above the plate and as large as the balance, as on the following page.'

Below the sketch on the right, Huygens seems to have mused about the best material one could use for his new invention,

'[a] spring of beaten copper could serve perhaps.'

5.1.1 The Huygens–Thuret controversy

Huygens was keen to convert his sketch into a working model, so on the morning of 21 January 1675 he first showed his invention to his friend Pierre Perrault (c. 1608/11–1680), the Receiver General of Finances for Paris-turned-hydrologist, and then repeatedly but unsuccessfully tried to find Thuret in his workshop. During his third attempt, around noon on 22 January, he managed to convey his enthusiasm to the clockmaker, in confidence, whose response was equally enthusiastic. Given the nature of the notes accompanying Huygens' first drawings, it is likely that they are the result of that initial discussion between Huygens and Thuret [12], when both men must have bounced ideas off each other to arrive at a practically viable design for their first spring-driven, balance-wheel operated watch. The essence of Huygens' novel design was a spring that was affixed *at both ends*, but which could move freely

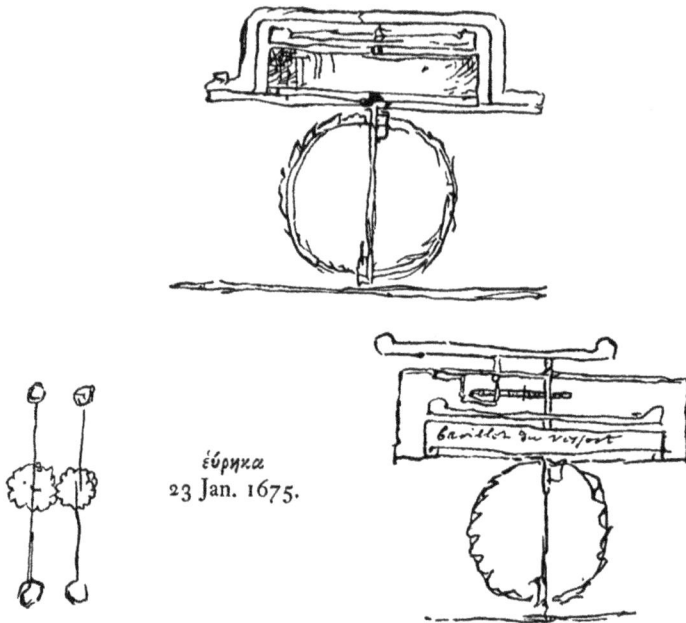

Figure 5.4. Various design improvements, including an improved escapement mechanism (*top*), suggested in the days following the initial *eureka* moment of 20 January 1675 (Huygens C 1675 *Oeuvres Complètes de Christiaan Huygens* **VII** p 409.)

in between. He must have insisted on retaining this design feature, which had not been suggested previously in relation to spring-driven clocks, given that both the French patent he would be awarded later that year and the description of his spring-driven timepiece, which he published subsequently in the *Journal des Sçavans*, includes this property. By reference to the earlier pendulum design, the balance wheel was equivalent to the freely swinging, adjustable weight, while the spring regulated the period of oscillation and could be adjusted separately by varying the coil's tension.

Manuscript E of the *Codices Hugeniani* contains a detailed account—at least from Huygens' perspective—of the controversy that subsequently developed between both men, from 21 January until 25 February 1675 [13]. Following their initial meeting, Thuret set out to construct a model watch according to Huygens' design, working non-stop until its completion around three o'clock that afternoon. Having completed his initial assignment, he subsequently continued to work on a somewhat different and improved version of a spring-driven watch of his own design. The next day, 23 January, he proudly demonstrated the improved design to Huygens, whom he had called away from a meeting, requesting that Huygens keep these developments secret. Further improvements were implemented during the next two days: see figure 5.4.

As we have seen before, Huygens was not only a formidable scientist, he also had a keen business acumen. Thuret's new spring-driven watch design (see figure 5.5) obviously lent itself very well for a patent application. Specifically for that reason,

Figure 5.5. (*Left*) Isaac Thuret's first spring-driven timepiece, signed *Thuret A Paris*, 1675–1680. (*Right*) The clock's escapement mechanism, signed *THVRET*. (Plomp R 1999 *Annals of Science* 56 pp 379–394.)

Huygens approached Colbert on 31 January 1675, showing the latter his new invention. Colbert indeed promised the scientist a French patent, a promise Huygens communicated to Thuret that same afternoon. The clockmaker, in turn, indicated his keen interest to be included as co-inventor. To his disappointment, however, Huygens refused him that honour, but he sweetened his message by pointing out that Thuret would profit most from the invention and that his practical skills would always be recognized.

Initially, Huygens agreed to Thuret's request to delay his patent application, at least until the latter had completed a model of their new watch for presentation to the French King. However, in the meantime (writes Huygens) Thuret seems to have gone to great lengths to publicly announce the significance of his own contributions to the new design. This prompted Huygens to abandon his earlier agreement, speeding up the patent application instead.

The men did not meet again until 4 February 1675, but the severity of their newly acrimonious relationship only revealed itself on 8 February, when Huygens discovered '*that Thuret had shown my invention to Mr. Colbert eight days before me*,' that is, already on 23 January 1675. A day later, on 9 February, Thuret initially denied Huygens' accusation, but he later confessed. Huygens added a note to Thuret's admission of his breach of their mutually agreed secrecy that he had decided to subsequently collaborate with the Parisian clockmaker Antoine Gaudron (1640–1714) instead. There are no surviving sources that can independently verify the extent of Thuret's improvements to Huygens' original design [14], but it is clear that some of the members in their mutual social circles—including Madame Colbert and Charles Honoré d'Albert de Luynes (1646–1712), Duke of Chevreuse and married to a daughter of Colbert—valued Thuret's contributions more highly than Huygens did. Indeed, Thuret appears to have designed an effective escapement which improved Huygens' original design. Nevertheless, Huygens could not be swayed.

Huygens repeatedly pointed out that Thuret apparently did not fully understand his design at first:

'In explaining it to him, he said (as yet barely understanding it), "I find that so beautiful that I still can't believe it is so"' [15].

Apparently, Thuret expressed concern about the practical feasibility of Huygens' spring-driven design, worrying that lateral vibrations might affect the regularity of the resulting oscillations. Huygens later said that he responded ...

'that what he said of the trouble with these vibrations was something contrived to make it appear that he knew something about the application of the spring, but that this itself showed that he had known nothing about it, because, if he had thought about attaching the spring by its two ends, he would have also easily seen that these vibrations were of no concern, occurring only when one knocked or beat against the clock and even then not undercutting the effect of the spring' [16],

although Mahoney has pointed out that Thuret may actually have been struggling with a similar design of his own around the same time, but without seeing the practical solution that Huygens had realized [17].

An attempt was made to broker peace between the antagonists. During a long discussion on 25 February 1675, Thuret announced that the watch he had been working on was close to completion, and that he was prepared to present it to Huygens shortly afterwards. Huygens was, however, only prepared to accept the watch in return for payment, thus showing that he considered the invention his own and that Thuret had merely lent his skills to facilitate its practical implementation. Thuret was given no choice but to eventually accept this denigrating proposition. In fact, he was made to sign a letter on 10 September 1675 in which he gave up any intellectual claims to the spring-driven watch invention [18]. It is likely that Thuret agreed to signing away his rights driven by commercial reasons: on 15 February 1675 Huygens had acquired the exclusive right [19] *'to have made watches and clocks of a new invention'* in France for a period of 20 years, which consisted of ...

'a spring shaped as a spiral which regulates the rotations of a free balance, larger and heavier than ordinary specimens.'

Having signed the letter relinquishing all intellectual rights to Huygens, Thuret was free to proceed as Huygens' craftsman of choice for the commercialization of his new spring-driven watches [20].

The text of the patent included the explicit right to develop marine clocks driven by a balance spring: see Thuret's first model in figure 5.5 (right). Thuret's clock was indeed designed specifically for use on board a ship, as evidenced by the fact that the top of the case is provided with an iron ball enclosed by a brass cylinder to facilitate free movement, that is, using a Cardan suspension. It is likely that, specifically for applications to marine timepieces, Huygens included drawings of escapement mechanisms equipped with two balances, rotating in opposite directions so as to minimize the effects of movement.

In addition, he drew two versions of his initial spring-driven escapement design: one according to the usual configuration, where the escapement was directly coupled to the balance wheel (see figure 5.3), and a second which included an additional balance wheel between the balance and the escapement. The right-hand panel of figure 5.5 shows that Thuret's marine spring-balance design indeed includes an extra gear between the balance wheel and the verge. This construction implies that the balance wheel has to rotate by approximately 150 degrees before the next pallet is released. In turn, this results in a slowly moving balance wheel characterized by a period of one second. I will discuss this latter design in more detail below.

Huygens' day-to-day account of the events that occurred between himself and Thuret between 21 January and 25 February 1675 was apparently composed at a later date, and hence it must be seen as at least somewhat one-sided. Nevertheless, we have a series of surviving letters in Huygens' *Oeuvres Complètes*, exchanged with Henry Oldenburg at the Royal Society, which shed some independent light on the priority of invention. In his letter of 30 January 1675, Huygens stated,

'About Pendulums, I shall tell you ... that I have recently come upon a new invention pertaining to Clocks which I am having [someone] presently working on and which appears to be successful. I include the secret here, in anagram form; as you know, I have done [this] before with new discoveries and for the same reason' [21].

The anagram of Huygens' discovery reads as follows:

4	1	3	5	3	7	3	1	2	3	4	3	2	4	2
A	B	C	e	F	i	L	m	n	O	r	S	t	u	x

which, rearranged to its original Latin, becomes *Axis circuli mobilis affixus in centro volutae ferreae* (*The axis of a moving circle [the balance wheel] attached to the centre of an iron coil*). Oldenburg responded on 12 February 1675 [22], noting that their mutual friends were keen to see the new invention in practice and that he expected Huygens to oblige. By that time, the row between Huygens and Thuret had already become common knowledge, which prompted Huygens to respond defensively on 20 February 1675 [23]:

'In my last letter of 30 January, I sent you the secret of a new invention pertaining to clocks, of which you since may have been informed. You will already know what it is, because the secret has not been well preserved here owing to the bad faith of the watchmaker I had commissioned to do the work. Already the day after I had informed him of this invention, allowing him to make a model, he promptly made another and proceeded to showing it off without my knowledge to Mr. Colbert and to several other people, saying that he was its creator.'

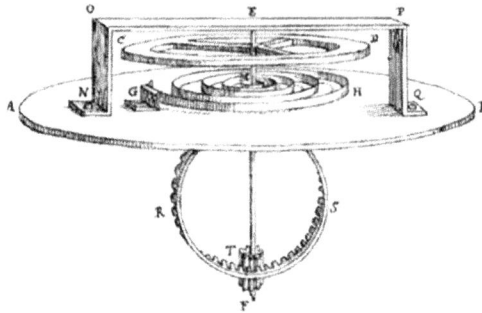

Figure 5.6. Huygens' sketch of his spring-driven watch design, showing the wound spring (**GH**) and the oscillating balance wheel (**C–D**) affixed to the verge (**X–T**). (Huygens C 1675 *Oeuvres Complètes de Christiaan Huygens* **VII** p 425.)

He then proceeded to describe his invention in detail, with reference to a figure which was published in the 25 February 1675 issue of the *Journal des Sçavans* [24, 25] (see figure 5.6), saying ...

> '*that this invention is composed of a spring that has been turned into a spiral, attached in its centre to the spindle of a circular balance which turns on its pivots; and with its other end to a piece that is attached to the backplate of the timepiece. Once the balance is put in motion, this spring's coils alternately turn backwards and forwards, and so it maintains, with the little help it gets from the gears of the timepiece, the movement of the balance. And this ensures that, even if its turns are smaller or larger, the periods going backwards and forwards are always equal to each other. If such timepieces are made small, they will be very precise pocket watches and as larger clocks they can be useful everywhere, particularly to find the longitude, both at sea and on land, since their perform-ance is equal to that of Pendulum Clocks, and no vehicle's movement will be able to stop them.*'

Huygens clearly felt that he needed to give his perspective of the acrimonious situation with Thuret which had developed, continuing ...

> '*The dishonesty of the craftsman, about which I have spoken to you, caused me a lot of trouble and frustration. But eventually, after I had explained and convinced Mr. Colbert of his bad behaviour, he has done me justice, and he has given me the King's exclusive right [Privilège du Roy]*[1] *for this invention; upon which my plagiarist, having seen that he had become embroiled in a very serious matter, not knowing what to do, requesting everyone he knew to urge me to forgive his mistake and offer him work as before, promising to testify to everyone that he is in no way part of the claimed invention.*'

---

[1] Note that at this time, Huygens already received a regular pension from the French King, Louis XIV.

The Huygens–Thuret conflict about the extent of Thuret's contribution to the invention of the spring-driven clock is often thought to have ended their cooperation, despite Huygens' note to Colbert that Thuret *'had been for a long time one of my friends'* [26]. However, a careful study of Huygens' correspondence seems to suggest the contrary [27]. For instance, in a letter to his brother Constantijn Jr. dated 9 August 1675, he writes ...

'What they said about my plagiarist is true ... but I have deferred to the King's advice, where Mr. Colbert has promised to send me a decision that would be equivalent to actually having registered the Privilège du Roy. I will wait and see what the effect of this development will be, and I am determined to obtain a confession from this rogue [ce coquin] and get the satisfaction I desire, or else I will leave everything in this country behind ...' [28].

His father seems to have attempted to defuse Huygens' anger; in the letter's margin, he added an admonition, *'ne saevi, magne sacerdos,'* implying that adversarial conduct is not good for business. Indeed, despite his hurt feelings, Huygens mentions Thuret's valued craftsmanship specifically in a letter to Oldenburg dated 21 November 1675, where he discusses a watch made for the Duke of York:

'It is made by Thuret, who makes up to now the best [watches] and with great demand. He is the one who has treated me so badly after I had confided him that invention. But, having retracted in the end by a letter which others had obliged him to write to me, and having come to me to ask pardon, I have no difficulties anymore to employ him' [29].

This sentiment is confirmed by considering the designs of Huygens' successive longitude clocks, without either remontoire or fusee. These closely resembled Thuret's initial design, except for the *pirouette* (so known because of the slow, large movements of the balance wheel) that was originally included, a combination of the large gear (**ROS**; see figure 5.6) and the pinion (**T**) between the balance wheel (**C–D**) and the escapement (not shown).

Rupert T Gould, the historian of horology, explained the efficacy of this latter device as follows:

'Huygens's method of applying the spring was original, for he geared up the balance, so that in stead of describing, like Hooke's, an arc of 120° or so, it revolved several turns at each beat. The device is termed a 'pirouette.' It is theoretically objectionable on account of the friction in the gearing, and never came into general use. He found himself, however, baffled by the effect of temperature on the strength of the spring' [30].

Despite the additional friction, Huygens had included the pirouette in an attempt to allow the balance wheel to make wider swings between beats, and therefore attain a lower oscillation frequency and fewer pulses on the verge. He expected this configuration to result in a more detached and therefore more accurate escapement. In pendulum clocks, this was achieved by application of long pendulums; in his new spring-driven watch design, he attempted to mimic that situation by including a slowly moving balance rotating over a wide angle. To limit the angle associated with the rotation of the verge, an additional gear was required. However, Huygens eventually (or perhaps immediately) abandoned the idea for practical reasons, despite being convinced that his *pirouette* configuration was the more precise design.

5.1.2 Claims and counterclaims

Always the opportunistic businessman, Huygens concluded his letter of 20 February 1675 to Oldenburg by suggesting that he would be interested in licensing his invention in England as well, offering a share of the proceeds to either Oldenburg himself or to the Royal Society. As we will see below, this letter subsequently became part of a second controversy, this time revolving around claims of priority between Huygens and Hooke. Although the development of spring-driven time-pieces was not the prerogative of either man—they had been in use since the 16th Century in chamber clocks, instead of falling weights (see also our discussion about the spring-driven fusee in chapter 4)—this did not stop them from engaging in a heated dispute about the priority of their clocks' operation by means of a balance wheel.

Huygens' novel invention was the implementation of such a balance wheel, which revolved back and forth isochronously, thus enabling replacement of the pendulum as regulator of his innovative timepieces. In response to a rival claim, most likely by the French abbot and physicist-cum-inventor Jean de Hautefeuille (1647–1724) regarding the invention of the spring balance, Huygens wrote:

'It will be seen how much they differ, since, besides a totally different form and application of the spring, I employ a balance that turns on its pivots, and my invention consists of the combination of these two things' [31].

Upon Huygens' announcement of his new invention to the *Académie Royale des Sciences* in 1674, De Hautefeuille rather opportunistically challenged Huygens' claim of the priority of invention both legally and intellectually, claiming that he himself had proposed replacing the pendulum by a thin strip of steel for timekeeping purposes [32, 33]. He included a detailed technical description as well as figure 5.7 in support of his legal claim. His design consisted of a balance wheel **ABCD** (see the right-hand sketch, labeled *Figure II*), with teeth underneath, a verge escapement **EF** equipped with pallets **G** and **H**, and a pinion wheel **IK**, which was meant to perpendicularly raise and lower a straight rack with teeth on one edge (a *cremaillere*). The latter was attached by one end to a spring **AB** (shown in the left-hand sketch, *Figure premiere*) and on the other end to a weight, **C**.

De Hautefeuille argued that a clock constructed in this manner could continue to oscillate without interruptions. He also admitted, however, that he had not been able

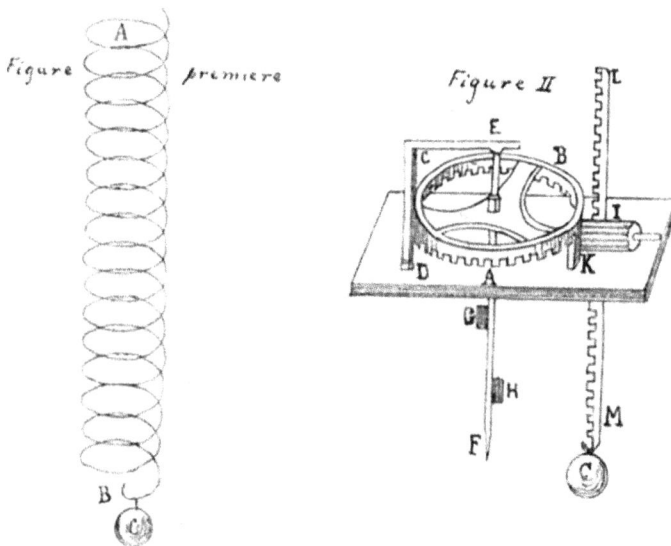

Figure 5.7. Sketch provided by Jean de Hautefeuille in support of his legal challenge to Huygens' priority of invention of the spring-driven, balance-wheel operated clock. (Huygens C 1675–1676, *Oeuvres Complètes de Christiaan Huygens* VII p 449.)

to make his design work in practice, but that this was merely a matter of involving a craftsman to pursue its construction—De Hautefeuille considered this aspect a mere '*accident*.' His legal and intellectual claims rested on the fact that he had pointed out that a vibrating spring could be used as a practical regulator, forcing a clock to beat regularly. Huygens summarily dismissed De Hautefeuille's claims, pointing out that the latter had certainly not been the first to suggest such a mechanism. The main problem in getting it to work related to the conversion of a spring's regular motion into the uniform movement of a practically viable escapement, and for that to work significantly more design effort was required, and so was advanced physical insight.

I already alluded to the more serious nature of Huygens' dispute with Hooke regarding the priority of invention. While the chronological history of their quarrel is well known and has been extensively recorded, the technical aspects of the dispute are less clear—yet they deserve careful clarification in order to place the dispute in its proper context. Springs have been used for two different purposes in watch designs, which is where the technical confusion arises. On the one hand, they have been used as part of the escapements in marine timepieces. It is clear from the surviving historical evidence that Hooke had been working on constant-force spring-controlled escapements since at least 1660.[34] Since this application counteracts irregularities in a clock's gear train, when implemented correctly it will allow clocks to run at very high accuracy. On the other hand, Hooke wondered about the '*use of springs instead of gravity for making a body vibrate in any posture.*' The resulting use of spring-balance applications to enforce isochronous movement was a second innovation dating from the mid-17th Century. It replaced the force of gravity affecting a swinging weight by the spring's elasticity ('*artificiall gravity*' in Hooke's own words), which would allow one to carry it around or take such clocks on journeys to geographically distant locations, thus facilitating continued accurate timekeeping on pitching and rolling ships during any voyage. The essence of the dispute between Huygens and Hooke was, therefore, not whether Hooke had applied springs to force isochronous motion of his timepieces, but whether he had used *spiral* springs, in particular, before Huygens, who used such springs routinely because of their efficiency [35].

However, Huygens had initially simply implemented this type of balance spring only because the balance in his first watch rotated more than 1½ turns; from around 1675, he also used spiral springs in the more conventional watches he ordered from Thuret. Hooke initially applied straight springs instead. It is often asserted that Huygens' employed his spiral spring because it could be easily adjusted to make his watches isochronous. Indeed, this is how isochronicity is achieved in modern watches equipped with detached 'lever' escapements, invented around 1755—well after Huygens' death—by the British clockmaker Thomas Mudge (1715–1794). However, watches in Huygens' time usually used verge escapements, which destroy the isochronicity of any kind of balance spring...

On 28 January 1675, at a meeting of the Royal Society, Oldenburg had read out Huygens' first letter to him in which the Dutch scientist wrote of his new invention [36]. Based on his diary entries, Hooke does not seem to have realized the importance of this missive for his own career [37]. The meeting's record includes the anagram communicated by Huygens, but not its solution. On 18 February 1675, Oldenburg read Huygens' second letter [38], now including the solution to the anagram. Oldenburg appears to have been a staunch supporter of Hooke's discoveries and inventions until at least the winter of 1674–1675 [39], but this development irreversibly changed the relationship between both men, who until that time had been on good terms [40]. Hooke had apparently been informed of the second letter's contents by Robert Boyle the day before, and he immediately challenged Huygens' claim of priority by reference to the Royal Society's journals and an appeal to its historian, Thomas Spratt: in Hooke's *Diary* we find as entry for this date,

> '*Zulichem's* [Huygens'] *spring watch spoken of by his letter [to Oldenburg].*
> *I shewd when it was printed in Dr. Spratt's book.*'

In addition, at the Royal Society's meeting of 13 January 1664, Brouncker went on record by stating that ...

> '*Mr Hooke had discovered to himself, Dr Robert Moray, and Dr Wilkins, an*
> *invention, which might prove very beneficial to England, and to the world,*'

which has been interpreted [41] as a reference to Hooke's spring-balance watch. The Royal Society's *Journal Book* included a reference from 29 August 1666 to Hooke's demonstration of his spring-driven watch from 15 March 1664,

'Mr Hooke produced ... a new piece of watch-work of his contrivance, serving to measure time exactly both by sea and land; of which he was ordered to bring in the description' [42].

Nevertheless, the Fellowship of the Royal Society remained unconvinced of the practical use of Hooke's watch. On behalf of the Royal Society, Oldenburg wrote that ...

'Mons. Huygens, notwithstanding [Hooke's claim], should be thanked for his communication and informed what had been done here; and what were the causes of its want of success' [43].

He subsequently wrote to Huygens that the Royal Society's Fellowship would ...

'suspend their judgement until they can have the advantage of the figure and a more ample description, principally in the face of Mr Hooke's having invented, some years ago, a similar thing, as he believes, which however did not then succeed entirely according to his hopes' [44].

Oldenburg had, perhaps on purpose and out of apparent jealousy [45], omitted any record of Hooke's progress on the spring-driven watch, while Spratt's *History of the Royal Society* only included a generic description of Hooke's 'invention,' referring to ...

'several new kinds of Pendulum watch for the Pocket, wherein the motion is regulated, by Springs, or Weights, or Loadstones, or Flies moving very exactly regularly' [46].

An independent reference to Hooke's spring watches is found in the journal of Count Lorenzo Magalotti (1637–1712), the Italian philosopher, author, diplomat, and poet, describing a demonstration of Hooke's watch at a meeting of the Royal Society in February 1668, ...

'a pocket-watch with a new pendulum invention. You might call it with a bridle, the time being regulated by a little spring of tempered wire which at one end is attached to the balance-wheel, and at the other to the body of the watch. This works in such a way that if the movements of the balance-wheel are unequal, and if some irregularity of the toothed movement tends to increase the inequality, the wire keeps it in check, obliging it always to make the same journey.'

There is no record of Hooke's demonstration in the Royal Society's *Register Book*, which was maintained by Oldenburg at the time. In fact, in the Royal Society's *Hooke Folio* there is clear evidence of Oldenburg's suppression of any reference to Hooke's demonstration. The *Hooke Folio* contains the original minutes

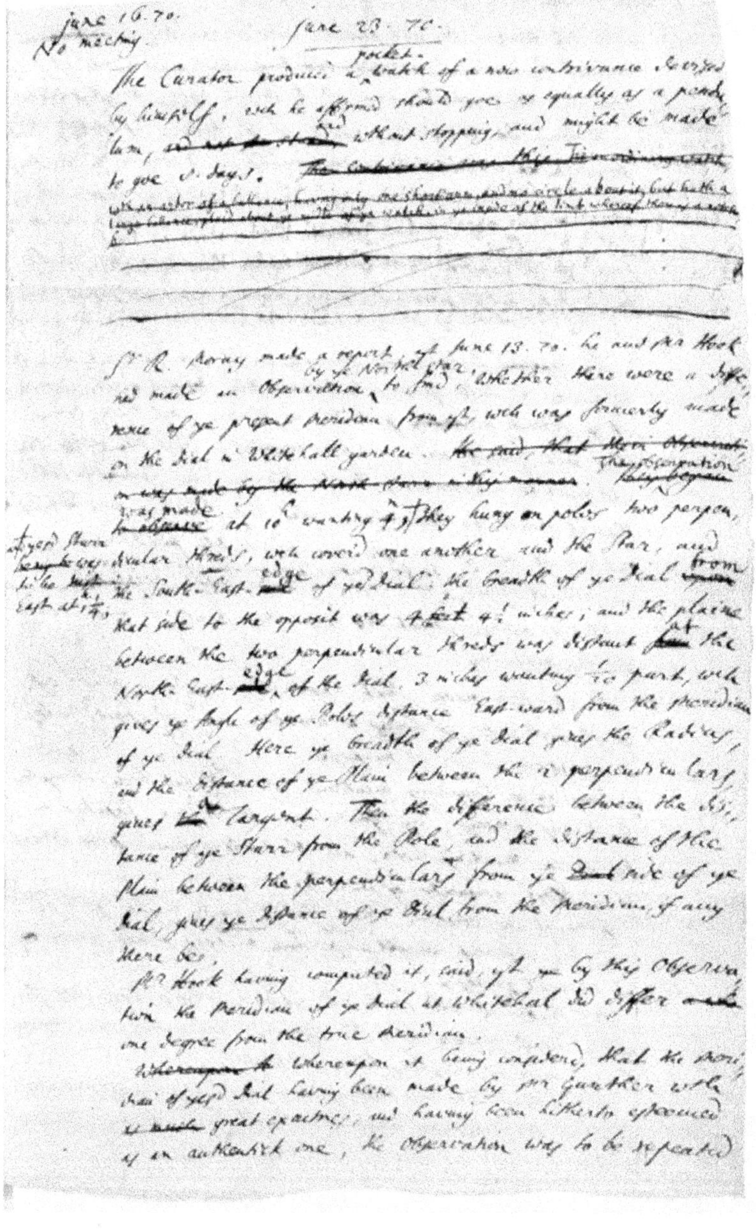

Figure 5.8. *Hooke Folio* (2006), p 83, providing evidence that any reference to Hooke's demonstration of 23 June 1670 has been crossed out. (© The Royal Society.)

of the Royal Society's meeting of 23 June 1670, which *do* contain a reference to Hooke's invention. However, these notes have been crossed out (see figure 5.8), so that they were not included in the *Journal Book* when Oldenburg's minutes were transcribed. The *Hooke Folio* was rediscovered only in 2006 [47, 48].

More precise descriptions of Hooke's spring-driven watch escapement are found in the correspondence of Moray, the Royal Society's first President, with Huygens, either directly or through Oldenburg. Specifically, in a letter dated 30 September 1665, Moray instructed Oldenburg to ask Huygens about his use of a spring for isochronicity:

'... *if he doth not apply a Spring to the Arbour of the Balance, and that will give him occasion to say somewhat to you; if it be that, you may tell him what Hooke has done in that matter, and what he intends more.*'

Hooke strongly disagreed with the Royal Society's assessment that his invention had been unsuccessful and attempted to convince others that he was entitled to the invention's priority. Indignantly, he wrote in his diary that '*[t]he Society inclined to favour Zulichems* [Huygens]' [49]. He proceeded to show a spring-driven watch of his own design to the mathematician Sir Charles Scarborough (1615–1694), the King's physician, on 19 February 1675, and subsequently proceeded to search the Royal Society's *Register Book* for evidence of his own priority, grumbling that '*Zulichems spring [was] not worth a farthing.*' He uncovered a reference to his own spring-driven watch from 1666, and in late February/early March 1675 he turned to Thomas Tompion to convert his design into a practical timepiece, simultaneously engaging in lengthy discussions about spring-driven watches with Christopher Wren [50].

As we have seen, meanwhile Huygens had offered Oldenburg the English patent for his own spring-driven watch design, provided that the latter could secure the King's approval. Hooke learned about this development on 6 March 1675 from his friend Sir Jonas Moore (1617–1679), the English mathematician, surveyor, astronomer, and Ordnance Officer. Hooke was understandably greatly dismayed by this news, since he would be prevented from continuing his own efforts to develop a spring-driven marine timepiece if Huygens and Oldenburg were successful. He angrily accused Oldenburg of abusing his position as Secretary of the Royal Society for his own profit, announcing that Huygens had learnt the secrets of his spring-driven design from him (Oldenburg).

Hooke's *Diary* entry for 6 March 1675 states that he was ...

'*At Sir J. Mores [Moore's]. he told me of Oldenburg's treachery his defeating the Society and getting a patent for Spring Watches for himself*' [51].

He was clearly incensed by these developments, as evidenced by a subsequent comment in his *Diary*, calling '*Oldenburg a raskall for not registering things brought into the Society ...*', and noting on 10 June 1675 that he had '*... reproved Oldenburg for not Registering Experiments,*' while '*Brouncker took his* [Oldenburg's] *part*' [52].

The quarrel between Oldenburg and Hooke was played out in the pages of the learned societies. Following Oldenburg's reading of Huygens' letters, he published the Dutch scientist's claim to the invention in the March 1675 issue of the

Philosophical Transactions of the Royal Society, of which he was the editor. Hooke responded by outlining his own version of the situation as a postscript to an article in *Helioscopes* of October 1675, where he published a summary of his Cutlerian lecture (*A Description of Helioscopes*), explaining that in 1660 he took his invention to '*several Persons of Honour*'—Boyle, Moray, and Brouncker (with whom he fell foul over the ensuing 'Oldenburg–Brouncker conspiracy'). Hooke added,

> '*[o]f these things the Publisher of the* Transactions *was not ignorant, and I doubt not but Mr Hugens hath had an account, at least he might have read so much of it in the* History of the Royal Society *as was enough to have given him notice of it*' [53].

Meanwhile, Huygens became increasingly concerned about these developments, particularly in the face of his potential loss of income if Hooke's discoveries turned out to be viable. On 1 July 1675, he wrote to Oldenburg,

> '*I do not know how you put up with the ill-founded boastings of this man, and whether you have not considered whether, if he had had so useful and important an invention, he would have failed to avail himself of it and put it into effect*' [54].

While Hooke was correct in claiming that he had publicly discussed spring-driven watches in the 1660s, his descriptions of such devices had been very general and deliberately vague. That cautious attitude now cost him dearly in his dispute with Huygens, since it was judged insufficiently detailed to claim the priority of invention, and his secrecy meant that there were no witnesses who could be rallied in support. The public quarrel between Oldenburg and Hooke went back and forth, with attacks and counterattacks appearing in the *Philosophical Transactions* and *Lampas*[2]. Hooke's strongly worded piece concluded,

> '*To his [Oldenburg] upbraiding me with his having published some things of mine; I answer, he hath so, but not so much with mine as with his own desire, and if he send me what I think worth publishing I will do as much for him, and repay him in his own coyn. Lastly, Whereas he makes use of We and Us ambiguously, it is desired he would explain whether he means the Royal Society, or the Pluralities of himself. If the former, it is not so, as I can prove by many Witnesses; if the later, I neither know what he is acquainted with, or what has been imparted or explained to him*' [55].

'*Speque metuque Procul hinc procul ito. Ho*'
('Go far away from both hope and fear.')

---

[2] Hooke responded to Oldenburg's paper in the *Philosophical Transactions* by adding yet another postscript to the summary of his next Cutlerian lecture in *Lampas*; although the journal was officially published in 1677, it was already available in print by October 1676.

Oldenburg had the final word in the matter. Hooke was publicly disowned on 2 November 1676 when the Council of the Royal Society declared its support of '*Mr. Oldenburg's integrity and faithfulness to the Royal Society*,' commending Oldenburg's trustworthiness and honesty '*in the management of the intelligence of the Royal Society.*' This enabled Oldenburg to publish rebuttals in the September and November 1676 issues of the *Philosophical Transactions*, declaring that ...

'*the publisher of this tract* [Philosophical Transactions] *intends to take another opportunity of Justifying himself against the Aspersions and Calumnies of an immoral Postscript put to a Book called* Lampas, *published by Robert Hooke: Till which time, 'tis hoped, the Candid Reader will suspend his Judgement*' [56].

However, it was probably not Oldenburg who had spilled the beans following Hooke's lectures at both Gresham College and to the Royal Society's Fellowship. Indeed, as we saw at the start of this chapter, it is likely that Hooke's friend Sir Robert Moray had informed Huygens and the foreign members of the Royal Society of Hooke's progress in 1664, specifically to support Hooke's priority of invention. Nevertheless, Hooke's friend and biographer, the English naturalist Richard Waller (d. 1715), writing in 1705, clearly believed that Oldenburg had betrayed Hooke's secret to Huygens:

'*It is probable that their intimacy procur'd what he knew; and it is evident that Huygen's discovery of this was first publish'd in the Journal des Sçavans, and from thence in the Philos. Transact. for March 25th 1675, about ten Years after that Letter of Sir Robert Morays, and fifteen after Hooke's first discovery of it.*'

Huygens had offered the English patent to Oldenburg or the Royal Society, since he could not apply for one himself on account of his foreign citizenship. In fact, Oldenburg was not eligible for the same reason, but neither Oldenburg nor Hooke realized this at the time. Hooke developed a distrust of Oldenburg and the Royal Society's impartiality. His *Diary* entry for 3 September 1675 indicates that he '*Writ against Oldenburg*,' probably in the form of the postscript to his article in *Helioscopes*, on 25 September of that year followed by some satisfaction that '*Sir Chr Wren read my papers ... against Oldenburg and approved.*' In November 1676, we read in his *Diary*,

'*Saw the Lying Dog Oldenburg's transactions. Resolved to quit all employments and to seek my health,*'

although he apparently did not follow through, because in October 1676, he once again '*Resolved to Leave the Royal Society,*' complaining of '*Great intrigues of* [the Royal Society's] *Councell*' and the '*Grubendolian Councell,*' where he used an anagram of Oldenburg's name commonly used in Royal Society circles at the time. He had clearly become an angry and bitter man, additionally calling the Royal Society's secretary William Croune (Croone; 1633–1684) '*a Dog.*' Hooke won a moral victory of sorts, however:

'*May 20, 1677. Oldenburg fled at my sight.*'

Hooke's biographer Waller may well have hit on the correct explanation as to why Hooke allowed this dispute to develop, despite claiming to have invented a spring-driven timepiece some 17 years earlier:

'*It must be confess'd that very many of his Inventions were never brought to the perfection they were capable of, not put into practice till some other Person … cultivated the Invention, which, Hooke found, it put him upon the finishing that which otherwise possibly might have lain 'till this time in its first Defects: whether this mistake from the multiplicity of his Business which did not allow him a sufficient time, or from the fertility of his Invention which hurry'd him on, in the quest of new Entertainments, neglecting his former Discoveries when he was once satisfied of the feazableness and certainty of them, I know not*' [57].

Turning away from the learned society for a moment, between March and October 1675, Hooke and Tompion worked frantically towards the development of a reliable spring-driven watch, trying a '*perpendicular Spiral spring*' and a '*Double spiral spring,*' without relying on the Royal Society to act on their behalf in securing a royal patent. King Charles II, meanwhile, clearly did not want to get involved in the dispute. From Hooke's *Diary* we know that the King granted him a number of audiences in April 1675, giving him a chance to provide updates on his practical progress.

'*April 7. The King most graciously pleas'd with it and commended it far beyond Zulichem's. He promised me a patent*' [58].

On the same day, the King refused Oldenburg his sought-after patent, although on 10 April 1675, the King informed Hooke that '*unless we made hast with the watch he would grant the patent*' to Oldenburg. On 17 May, Hooke sent his prototype spring-driven watch to the King, who received it '*very kindly*' and subsequently '*affirm'd it very good*' at an in-person meeting with Hooke, facilitated by Sir Jonas Moore. To establish priority, Hooke opportunistically had it engraved: '*Robert Hooke inven. 1658. T. Tompion fecit, 1675*' (*Robert Hooke, invented 1658. T. Tompion, constructed 1675*). Hooke and Tompion constructed a second watch for the King, which was completed in late August of that year and which the King agreed '*did very well.*' He is also said to have constructed watches for James II, the Duke of York (1633–1701), Prince Rupert of the Rhine (1619–1682), and many other notables [59]. Nevertheless, Hooke did not get a royal patent. Neither did Oldenburg in the end.

It seems clear, nevertheless, that both Hooke and Huygens contributed significantly to the development of a spring-driven watch, although the priority of invention remains hidden in the mists of history. Huygens is generally credited with the invention, however, since he produced the first accurate and reliable spring-driven, balance-wheel-operated timepiece in collaboration with Thuret.

5.1.3 Slow progress

Huygens believed the spring-driven watch design to represent the most promising approach to address the problem of determining one's longitude at sea. He considered mechanical problems as the most important reason why, at that point in time, the spring balance could not (yet) compete with the pendulum in terms of accuracy, but he most likely considered these problems primarily a challenge for clockmakers. Indeed, despite misgivings by Newton, who was convinced that the timing accuracy had to be solved by scientists [60], the problem was eventually laid to rest by clockmakers.

Huygens must have been well aware of the limitations of his early spring-driven watches, which allowed him to keep the time to an accuracy of 4 to 5 minutes per day. In a letter to Oldenburg of 11 July 1675 [61], he admitted that the accuracy of his new watches was lower than that of clocks with a long pendulum. In addition, he seems to have realized that his balance spring construction was rather sensitive to movement. As we have seen, Huygens tried to remedy this latter sensitivity by introducing a second balance wheel, rotating in the opposite sense, but this required too much force and was subject to too much friction in the gear train. He concluded his letter to Oldenburg by referring to a simpler solution he was working on, most likely the idea to double the number of revolutions from 120 to 240 per minute [62].

The bitter priority disputes with Thuret, Hooke, and De Hautefeuille took much of Huygens' attention, preventing him from making sustained progress on the development of a viable and accurate marine timepiece from the initial pocket-watch design for a number of years. Nevertheless, he had realized soon after his initial design efforts that a larger-size spring-balance-driven clock would be more stable on choppy seas, an idea he communicated about freely by 1679. Indeed, the French *privilège* of 15 February 1675 specifically refers to a new portable watch construction that is '*plus grand et plus pesant qu'aux ouvrages ordinaires*' (*larger and heavier than ordinary specimens*) [63], while the resolution of 25 September 1675 of the Staten van Hollant en Westfrieslant (the Provincial States of Holland) to award an octroy (patent) includes the phrase '*in grooter formaet gemaect sijnde*' (*made larger*) [64]— although the *octroy* itself does not mention the larger size [65]. In addition, in his *Memoire concernant l'Académie Royale des Sciences* [66] from 1679, which he sent to Paul Pellison (1624–1693), official historian of the French King and author of the *Histoire du Roy* (*History of the King*) [67], Huygens stated that ...

'because ~~the~~ pendulum clocks necessarily suffer from the ship['s] movements, there is more hope to succeed with balances with a spiral spring, but constructed in a large size because the accuracy is believed to depend on the dimension, and it would be very worthwhile to do this experiment' [68].

Indeed, his experiments seem to have borne this out, given that in October 1682 he wrote to Jean Gallois (1632–1707), the French scholar and abbot, that ...

'after the tests I have recently made, I venture to promise clocks just as accurate as our long pendulums by means of my invention of the spiral spring which can easily resist the largest movements of the sea' [69].

Meanwhile, Hooke's efforts were met with considerably less enthusiasm. Since approximately 1659, he had been working on improving his designs for smaller-size, *portable* timepieces. However, his contemporaries in England saw little merit in pursuing these developments:

'All I could obtain was a Catalogue of Difficulties, first in the doing of it, secondly in the bringing of it into publick use, thirdly in making advantage of it. Difficulties were proposed from the alteration of Climates, Airs, heats and colds, temperature of Springs, the nature of Vibrations, the wearing of Materials, the motion of the Ship, and divers others. Next, it would be difficult to bring it to use, for Sea-men knew their way already to any Port, and Men would not be at the unnecessary charge of the Apparatus, and Observations of the Time could not be made well at Sea, and they would nowhere be of use but in the East and West India voyages, which were so perfectly understood that every common Sea-man almost knew how to pilot a ship tither. And as for making benefits, all People lost by such undertakings; much has been talked about the Praemium for the Longitude, but there has never been any such thing, no King or State would ever give a farthing for it, and the like; all which I let pass...' [70].

By 1682, Huygens' efforts to perfect his marine clock design had become sufficiently promising to attract the attention of and formal encouragement from the Directors of the Dutch East India Company, the 'VOC,' as a potentially viable approach to determining longitude at sea. As a result, the VOC issued a number of resolutions in support of taking Huygens' clocks on a voyage to the Cape of Good Hope. The first resolution of support was issued on 31 December 1682 [71]. It specifically authorized the Mayor of Amsterdam, Johannes Hudde (1628–1704), governor of the VOC and a mathematician by training, to ...

'correspond with Mr. Huygens and one van Ceulen about the invention and construction of very accurate timepieces, which would not deviate from each other by more than a second per day [24 hours], in which way the East and West could be found ...'

The Van Ceulen referred to in the VOC resolution was the Dutch watchmaker Johannes van Ceulen (1629–1695), who lived across the street from Huygens in The Hague. That same month, Huygens asked Gallois in confidence to arrange permission for an extension of three to four months of his leave of absence from the *Académie Royale des Sciences* so as to supervise sea trials in the Dutch Republic [72].

Meanwhile, Huygens had also taken note of Oldenburg's warning of 1675 that temperature and humidity differences might affect the performance of his metal springs:

'I do not doubt, sir, that you have considered the effects of the temperature of the air, especially those of heat, on springs of that type ... It appears rather difficult to prevent them, or to remedy them, and to ensure that the periods of oscillation are always equal to each other in duration. But, I say, your intelligence has certainly foreseen [this] and found a means to protect the device against this inconvenience' [73].

Like Huygens in 1665, Oldenburg also appears to have been ahead of his time, since the effects of temperature changes on metal were not yet understood in the late-17th Century, beyond the fact that blacksmiths had long been using the expansion properties of iron to fit bands around wagon wheels. By May 1675, Huygens had performed initial experiments to ascertain the effects of temperature differences on the springs, in response to objections raised by Father Chérubin d'Orléans (1613–1697), a monk at the *Convent d'Orléans à Angers*. He reported in a letter to the French scholar Henri Justel (1620–1693) that ...

'I have not found that heating the balance spring in a flame produces any slower vibrations than when it is cold' [74].

However, Huygens was wrong. His subsequent experiments with springs and spring-regulated clocks had made him conclude by 1683 that springs are indeed very sensitive to changes in both temperature and humidity, in the sense that cold conditions accelerate the oscillations of a clock spring [75], since temperature variations affect a spring's elasticity. Huygens was baffled by these differences, which had also been observed in the context of the thermal expansion and contraction of pendulum rods—so that the position of the centre of gravity and therefore the accuracy of pendulum clocks was seasonal. However, this change in length of the pendulum rod with temperature was barely acknowledged in the 17th Century. In fact, Huygens even refuted [76] Godfried Wendelinus' assertion that the oscillations of a pendulum were faster in winter than in summer, and he never accepted that the length of the pendulum was temperature dependent. It took another century for these insights to become commonplace.

Modern physics teaches us that the effect of increasing the temperature on a balance spring is that the modulus of elasticity—also known as *Young's modulus*—is reduced (while the diameter of the balance is increased, but this latter effect is negligible compared with the change in Young's modulus). Solid bodies deform when force is exerted onto them. Young's modulus is a measure of the stiffness of the body, that is, it quantifies the extent to which a body returns to its original shape after the force is removed. As we saw before, the oscillation period of a spring balance is given by

$$P = 2\pi \sqrt{\frac{I}{b}},$$

where I is its moment of inertia and b is the spring's elastic moment. When the temperature increases, thermal expansion results in an increased diameter of the spring balance. This, in turn, increases its rotational inertia and, in response, the watch will run more slowly. At the same time, a temperature increase causes expansion of the spring in terms of its thickness, width, and length. The increases in the latter two dimensions have opposite effects on the spring's strength and effectively cancel each other. The increase in thickness results in a slightly increased strength, but this effect is rendered unimportant by the reduced elasticity of the metal. As a direct consequence, the spring produces less force for a given angle of rotation. In practical terms, the reduction in the elasticity of the material for steel or brass springs given by Young's modulus results in a temperature dependence of a spring balance's oscillation period of order 10 seconds per degree Celcius per day. (Changes in the temperature will also have an effect on the viscosity of the lubricating oils applied, but this effect is negligible compared to that owing to the change in the spring's elasticity.)

Huygens had experimentally found this temperature dependence by the mid-1680s, but he was unable to compensate for these effects in his practical spring watches. It is unclear whether Hooke had independently observed this temperature effect. Temperature compensation was not realized during the period covered by our narrative. It took until 1714 before an approximate solution to this problem had been achieved. During this time, George Graham made a number of valiant attempts at constructing an isochronous and, therefore, temperature-insensitive pendulum. He cleverly used the varying expansion rates of different metals, where the temperature-induced expansion in one metal is counterbalanced by the expansion of the other, so that the actual length of the pendulum remains unchanged. He was, however, ultimately unsuccessful in this pursuit. Instead, Graham designed mercury-compensated pendulums, where mercury held in a glass cylinder was meant to work similarly to a mercury thermometer: when the ambient temperature—and therefore the pendulum length—increases, the mercury contained in the glass cylinder expands as well (and vice versa), so that the pendulum's centre of gravity remains unchanged. Full compensation was only achieved in 1753 by the English clockmaker John Harrison (1693–1776). In essence, developments and improvements in metallurgy led to improved springs; 'regulators' made of bimetallic strips were used to compensate for changes in both the balance wheel's moment of inertia and the effective length of the springs, which in turn helped to compensate for temperature changes.

In addition, high humidity might cause the metal components in 17th Century watches to oxidize (that is, rust). William Palmer (d. 1737), the watchmaker, suggested to keep watches for use at sea close to a fire [77] so as to maintain them at a constant temperature. A similar suggestion had at that time already been put to the British Parliament by the contemporary London watchmaker Stephen Plank in an early effort to secure its Longitude Prize [78]. He proposed, in fact, to keep the watches in a brass box on top of a stove to avoid changes in the ambient temperature. However, in high temperatures, the watches' lubricant will dry more quickly, which in turn would increase their wear and tear through friction, and hence affect their stable operation.

5.2 Return to the marine pendulum design

Given the state of science in the late 17th Century, however, by late 1683 Huygens lost interest in the use of spring-driven marine timepieces to determine longitude at sea, because he could not compensate adequately for these temperature effects. Of course, history tells us in retrospect that he was on the right track, but that he lacked the technical knowhow... From an historical perspective, abandoning his *pirouette* had been a simple and straightforward first step. Freeing the balance from temperature effects and the escapement from coupling took more time and effort; both would be realized in the 18th Century. Neither the earlier design features of the remontoire nor the fusee had to be reintroduced, so that—in hindsight—the Huygens–Thuret spring-driven clock represented the start of a truly promising development. Notably, the fact that the Huygens–Thuret clock did not include a fusee demonstrates that Huygens was convinced that his balance spring would, on its own, retain isochronicity—pretty much like the cycloidal chops in his pendulum clocks were meant for the same purpose. This choice is, however, somewhat puzzling, given that he was also aware of issues with the varying driving moment the absence of a fusee introduced.

5.2.1 The tricord pendulum

Nevertheless, by the end of 1683 Huygens decided to start exploring other forms of isochronous oscillators, which would have '*the effect of the spring without the spring*,' that is, he wanted to retain the type of motion facilitated by a spiral spring-driven escapement but without its temperature dependence. In his original cycloidal-pendulum design for use at sea, dating from around 1672, Huygens had explored whether mounting a pendulum with two wires, attached to a spring-driven clock on the front and back, would increase its stability on a rocking ship: see figure 5.9. This early triangular marine pendulum's suspension was Cardanic, that is, it initially used a ball-and-socket mount, later replaced by a gimbal mount, so as to minimize the effects of a ship's rocking movement in the plane of the pendulum's motion. In addition, the box located below the pendulum bob contained a number of heavy lead pieces, also aimed at improving the clock's stability at sea.

He returned to this line of inquiry in 1683, which eventually led to his creation of an intricate three-dimensional pendulum regulator, the *pendulum cylindricum trichordon* or tricord pendulum, whose *eureka* moment occurred on 4 December 1683 [79]. His initial sketches of the tricord pendulum are reproduced in figure 5.10.

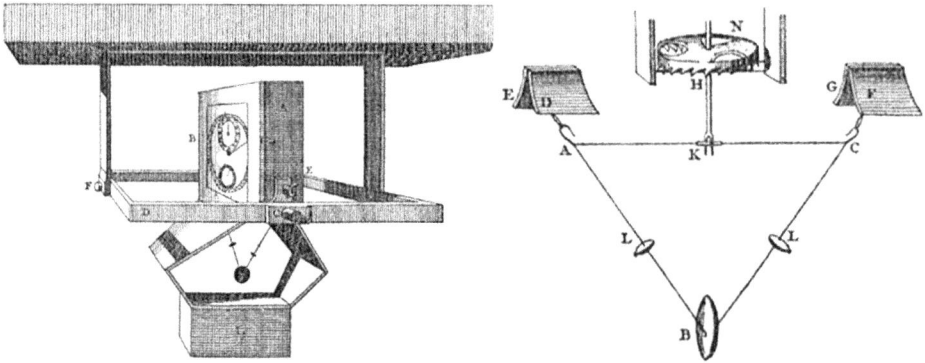

Figure 5.9. (*Left*) Huygens' marine cycloidal pendulum clock with double suspension, 1672–1673. (*Right*) Detailed suspension. (Huygens C 1673 *Horologium Oscillatorium*, Part I.)

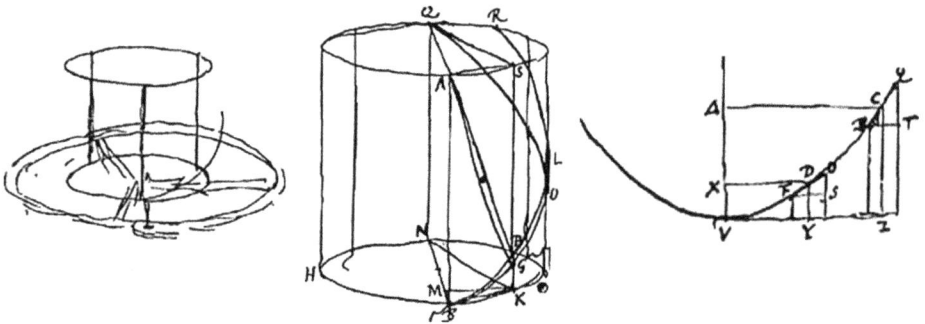

Figure 5.10. Huygens' initial sketches of his tricord pendulum, 4 December 1683. (Huygens C *Oeuvres Complètes de Christiaan Huygens*, **XVIII**, p. 527.)

His new pendulum design consisted of three equal-length, parallel wires suspended from a fixed mounting and attached at equidistant points on the inside of a heavy ring or cylinder, either solid or hollow. The latter could rotate freely and also move up and down along the device's perpendicular axis. The underlying idea was ingenious: when the ring or cylinder is moved away from its equilibrium position through rotation about its axis, it will move slightly upwards and the wires will deviate from their vertical orientation. Upon releasing the cylinder, it will perform torsional oscillations about its equilibrium position without the need for a spring balance (and, hence, unwanted temperature effects).

From Huygens' *Oeuvres Complètes* we learn that the Dutch scholar set out to derive the tricord pendulum's isochronicity on 7 December 1683. His mathematical analysis of this three-dimensional contraption was less than straightforward. He started by considering each of the three wires as individual pendulums with bob weights equivalent to one third of that of the ring. He then referred to the displacement principle to show that the actual space-curve generated by the end points of the wires traced sinusoidal curves, similar to parabolas (see the middle drawing in figure 5.10), which approximated isochronous paths, particularly for small angles, although they were not perfect cycloids. One of his drawings—shown in figure 5.11 (top)—also included small curved chops at the wires' suspension points, presumably to force the pendulum wires to follow cycloidal paths. (Huygens eventually opted for the use of longer wires to approximate isochronous operation, instead of isochronal chops.) The mathematics behind the calculation of their curved shapes is particularly impressive given the device's three-dimensional design. He also

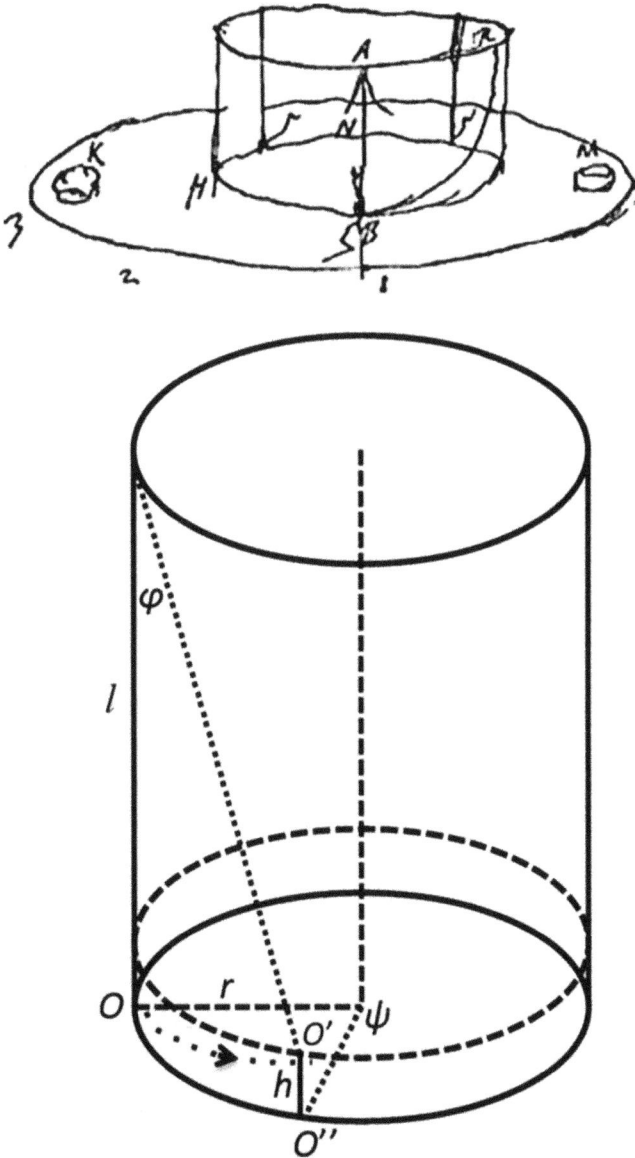

Figure 5.11. (*top*) Huygens' drawing of his tricord pendulum with cycloidal chops (*Oeuvres Complètes de Christiaan Huygens* **XVIII** p 531). (bottom) Geometry of the tricord pendulum (adapted from Crommelin 1938.)

derived that the tricord pendulum's period depends on both the square root of the length of the wires and the size of the ring, suggesting that calibration could be achieved by radially adjusting small weights on the ring, although it appears that Huygens did not derive the actual period of his practical tricord pendulum [80].

Let us now consider the mathematical background to the tricord pendulum in more detail, following Crommelin's excellent exposition [81]. Note, however, that while Crommelin developed the arguments based on modern calculus, Huygens did not have access to those tools. Instead, he proceeded on the basis of geometric considerations—and, in fact, he simplified the operation of his three-dimensional contraption to the basic principles of the circular pendulum in his final notes of 10 June 1684.

Our basic premise is that the device contains three wires of equal length, l, which are arranged in a circle of radius r and initially oriented vertically, so that both their suspension points and the points where they are attached to the swinging body define horizontal planes when the device is in equilibrium. The potential and kinetic energies, E_p and E_k, of a simple pendulum swinging over small angles θ from the vertical direction can be written as

$$E_p = \frac{1}{2}A\theta^2 \qquad \text{and} \qquad E_k = \frac{1}{2}B\left(\frac{d\theta}{dt}\right)^2,$$

where A and B are constants, so that the period of the pendulum becomes

$$P = 2\pi\sqrt{\frac{B}{A}}.$$

We now take for the coordinate θ the pendulum's rotation angle and refer to the pendulum's deviation from the vertical axis as φ. I refer the reader to the bottom panel of figure 5.11 for the relevant geometry. The bottom extremity of one of the pendulum's wires is at rest in a point O; upon being perturbed, it will be moved to O' following the curve OO'. The vertical displacement $O'O''$ is given by

$$h = l(1-\cos\varphi) = 2l\sin^2\frac{1}{2}\varphi \approx \frac{1}{2}l\varphi^2.$$

The final approximation only holds for small angles φ, expressed in radians. The sketch in figure 5.11 implies that $l\varphi = r\psi$ and for small angles we can assume that $OO' = OO''$, so that

$$h = \frac{1}{2}\frac{r^2}{l}\psi^2.$$

Therefore,

$$E_p = \frac{1}{2}\frac{mgr^2}{l}\psi^2; A = \frac{mgr^2}{l}$$

and

$$E = \frac{1}{2}I\left(\frac{d\psi}{dt}\right)^2; B = I,$$

where I is the pendulum's moment of inertia. This thus yields

$$P = 2\pi \sqrt{\frac{lI}{mgr^2}},$$

which applies to any body whose moment of inertia is known. For instance, for a thin ring,

$$I = mr^2, \qquad \text{so that } = 2\pi \sqrt{\frac{l}{g}},$$

while for a massive cylinder,

$$I = \frac{1}{2}mr^2, \qquad \text{so that } = 2\pi \sqrt{\frac{l}{2g}}.$$

5.2.2 Initial trials on the open sea

In view of the VOC's encouragement, on 17 December 1683 Huygens sent his design drawings and an approximate model of his new invention to his clockmaker Van Ceulen. Huygens seemed keen to test his novel tricord design in practice; writing to the Dutch mathematician Bernhard Fullenius, he stated that ...

> 'at the request of the [East India] Company, I have undertaken the construction of clocks to determine the longitude, possessing as constant a regularity as those with the three-foot pendulum, but such as should not be disturbed by the motion of the sea. I found the task to be more difficult than I had initially thought, although it is not completed yet there is little doubt that it will succeed' [82].

That final sentence turned out to be overly optimistic, as we shall see shortly. Van Ceulen soon produced two of the novel clocks, which—following some further adjustments in early 1684—enabled Huygens to make a convincing case to the VOC in July 1684 for financial and practical support. At their meeting on 27 July 1684, the governors of the VOC approved a financial contribution to Van Ceulen so as to 'complete the work to perfection,' recording that ...

> 'The Lord Mayor Hudde has stated that he is authorized by the resolutions of 31 December 1682 and 28 February 1684 to ... spend one to two thousand guilders on [the endeavour] ... despite not having achieved the aim ... proposing whether the assembled [governors] could approve payment of two hundred silver ducats to the aforementioned van Ceulen, a man of little wealth, on account of the work so far completed' [83].

The editors of Huygens' Oeuvres Complètes comment that Van Ceulen's financial circumstances were not dire, but he was likely 'a man of little wealth' compared with the rich businessmen on the VOC's governing council [84].

On 13 August 1685, Huygens travelled to Amsterdam with his two new marine clocks, hoping to suspend and regulate them on board a VOC galleon or 'flatboat.' To his regret, however, *'for the wind was perfect for a trial'* [85], he did not manage to test them in practice, since Hudde had been absent, prompting Huygens to return home to The Hague after a week of waiting in vain. Hudde, upon his return to Amsterdam, promised Huygens to send a VOC galleon to the port of Scheveningen for the latter's sea trials [86]. However, as we learn from a letter of 6 September 1685 [87] which he sent to his brother Constantijn Jr, Huygens preferred to undertake his initial sea trials during a 2–3 day voyage on the relatively calm Zuyderzee, the former inland sea in the centre of the present-day Netherlands (see figure 5.12).

Huygens' next journey to Amsterdam, on 9 September 1685, turned out to be more fruitful than his earlier trip in mid-August. The VOC's governing board had

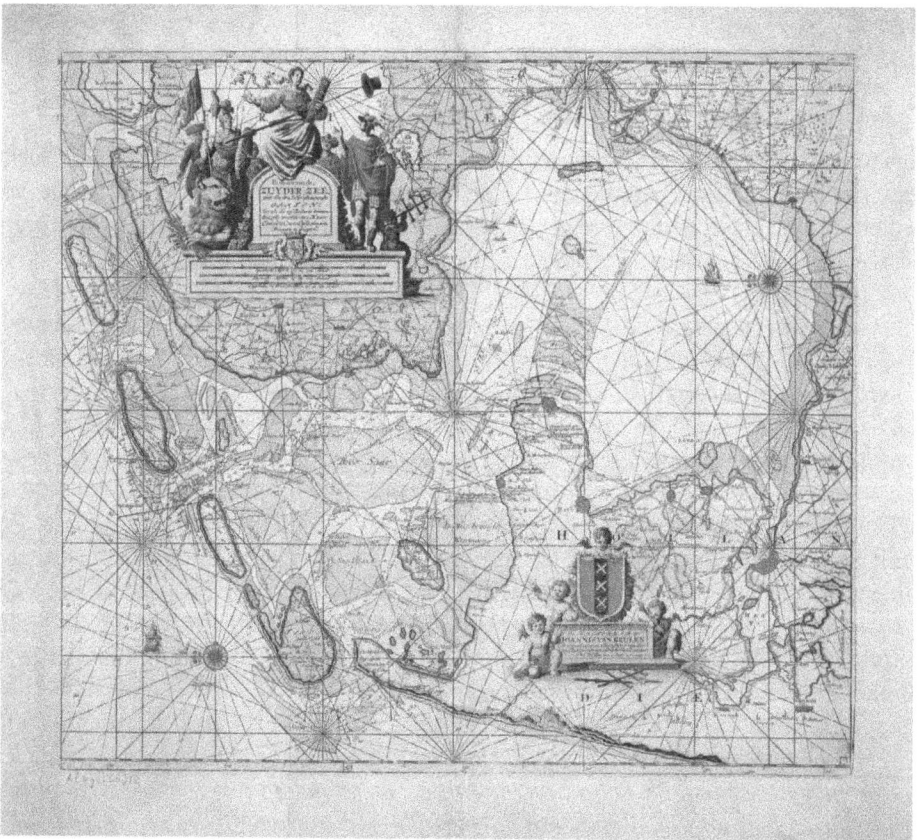

Figure 5.12. Map of the Zuyderzee by Johannes van Keulen (no relation to the watchmaker Johannes van Ceulen), ca. 1680. Amsterdam is shown as the half-round red area at the bottom right, Texel is the teardrop-shaped island at the bottom left and Enkhuizen Harbour is found at the tip of the promontory extending from the Amsterdam coat of arms at the bottom. North is to the left, East at the top. (Stichting Provinciale Atlas Noord Holland; reproduced with permission, Noord-Hollands Archief; inventory number NL-HlmNHA_560_001459_G.)

assigned Huygens a ship and captain, Barent Fockes, at their meeting of 30 August 1685 [88]; they also authorized Hudde to pay Van Ceulen a second tranche of 200 ducats ('*ducatons*') for his work and compensate the blacksmith he had solicited with a fee of 70 guilders ('*gulden*'). Hudde was charged with retaining oversight, so as to make sure that the project would lead to 'enlightenment' of the state of maritime navigation. Huygens met with a very courteous and amenable Barent Fockes, who had however been given instructions to sail to the VOC's main anchorage at the northern Dutch island of Texel rather than merely onto the Zuyderzee [89]. This would involve a 7–8 day voyage instead of the 2–3 days Huygens had anticipated, but the governors of the VOC insisted on a trial on the open sea. After all, Huygens wrote to his father, initial tests on open water would need to be successful, because '*otherwise it would be useless to continue further afield.*' Clearly somewhat worried, he continued that ...

> '*I have found the ship quite small, although it has made the voyage to the East Indies, ..., so that if [on this ship] the clocks can maintain reasonable time, we do not have to worry about their performance when suffering from storms on a large ship to the [East] Indies.*' [90].

On 11 September they embarked for their voyage from the port of Amsterdam, the only test ever undertaken by Huygens himself[3]. However, because of a severe storm [91], the ship's captain was forced to abandon the journey and seek refuge at Enkhuizen Harbour, fearing damage to his sails. They eventually completed their voyage to Texel a few days later, after the storm had subsided. Despite the rough conditions, one of the clocks continued to run smoothly [92]; the second clock ran intermittently. Huygens fell ill on board the ship [93], presumably because of a lack of sleep and physical exertion during the voyage; he returned home from Texel in early October 1685 to recover from his ailment. Nevertheless, he was significantly strengthened in his conviction that his clocks were seaworthy [94]. He therefore informed Hudde on 26 October 1685 that his main envoy, Thomas Helder (d. 1687), was ready and prepared to take his marine clocks on an endurance voyage to the Cape of Good Hope [95, 96]. The VOC subsequently made provision for two of Huygens' clocks and two attendants to join its fleet on a voyage to the Cape.

5.2.3 The first long-range sea trial to the Cape of Good Hope

It is rather curious and perhaps hard to understand why Huygens reverted to the use of gravity-driven pendulums, since his invention of the spring-driven balance wheel had all but obviated the effects of gravity. After all, Huygens and his contemporaries had long established that gravity-driven marine clocks should be avoided on

---

[3] On 6 September 1685, the VOC governors request that Huygens be accompanied on his forthcoming sea trial by one Johannes de Graef or someone with similar experience and skills. It is unclear whether this directive was met.

pitching and rolling ships... Moreover, despite the initially promising practical results with the tricord pendulum, further tests were abandoned around 1685.

Huygens prepared detailed instructions for Helder, his assistant Johannes de Graaff (spelt 'de Graef' by the editors of Huygens' *Oeuvres Complètes*—this reflects the typical variants of spelling in use at the time), and the clockmaker Willem van der Dussen, collected in the *Instructie en onderwijs aangaande het gebruijk der Horologiën tot het vinden der Lengde van Oost en West* (*Instructions and education as regards the use of the clock to find the longitudes of East and West*) [97]. This new set of instructions was based on his earlier collection, the *Kort Onderwys* (1665), with additional guidance and rules regarding the mounting, regulating, and maintenance of the timepieces—the latter were a translation into Dutch of the instructions provided to De la Voye in 1668 [98]. It is likely that Huygens decided to add practical guidance because of his displeasure with Richer's lackluster compliance with his instructions during the earlier and largely unsuccessful sea trials.

In his lengthy set of instructions, comprising 34 elaborate articles, Huygens appears to have reverted to the use of the type of marine clocks he had been working on already around 1671, which were spring-driven cycloidal pendulums with triangular (that is, not *trichordial*) suspensions. He included figure 5.13 specifically to explain the clock's mounting and operation at sea:

'*III. **ABCD** is a metal brace, at the top in the shape of a cross, with **EF** oriented perpendicularly to **BC**. The largest frame is **GH**, rotating on its axes **A** and **D**, which pass through the bottom of the aforementioned brace. **IK** is a smaller frame, which rotates inside frame **GH** with its axis **R**, as well as an axis on the opposite side which cannot be seen here. Attached to this frame **IK** immediately below the axes **R** are the downwards-running metal supports **LM**, **NS**, which converge towards the bottom, and which are held together down there, with a screw protruding underneath, where a weight **T**, which has a large hole in its centre, is snugly joined, and which at the bottom with a small backplate and the nut **O** is affixed so that it does not move. This weight is about 40 pounds, the heavier the better. The timepiece is **P**, which must first be placed inside the frame **IK** once the latter has been fastened to the top cover. This is done with the screws **E** and **F**. And in such a way that the arm **EF** is oriented along the ship's length, and the entire cross **EFBC** is kept horizontal or thereabouts when the ship is at rest, so that the timepiece will not hit the arms **BA** and **CB** when the ship is slanted. But in particular one must pay attention that this cross is affixed snugly to the top cover, if necessary by adding some wedges or blocks on top of the ends at **B** and **C**. Because otherwise the timepiece will be subject to a small, but mostly invisible, motion due to the force of the pendulum, which will cause the clock to run faster and no longer maintain its regularity.*'

The first VOC-sanctioned long-range voyage was commanded by master Karsten (Carsten) de Gilde on board the East Indiaman *Huis te Zilverstein* (also called *Silversteyn*), a so-called 'spiegelretour' ship built for transportation of cargo, which was owned by the VOC's Amsterdam Chamber. The voyage, with 96 soldiers and

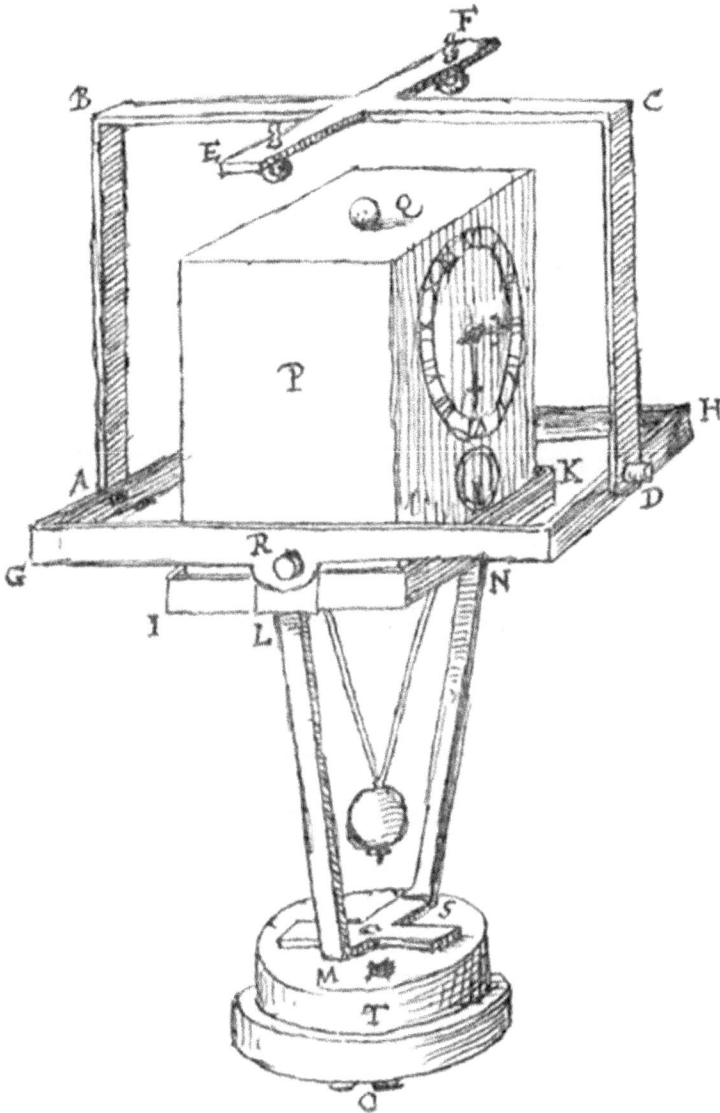

Figure 5.13. Huygens' explanatory drawing of the first marine timepieces taken by the VOC on a voyage to the Cape of Good Hope in 1686. (Huygens C 1686 *Oeuvres Complètes de Christiaan Huygens* **IX** p 56.)

138 passengers on board [99], commenced on 24 May 1686, arriving at the Cape on 26 September of that year (for an artist's impression, see figure 5.14). Unfortunately, much of the voyage was affected by rough seas, and Helder did not manage to obtain any useful measurements. Helder's journal refers to mishaps with the clocks on the outward leg of the voyage; Huygens later commented specifically that '*going they were not of service*' [100]. The ship continued on its voyage past the Cape to Batavia in the Dutch East Indies on 7 October 1686, leaving Helder and De Graaff ashore with their clocks.

Figure 5.14. Arrival of a VOC fleet, including the spiegelretour ship Noord-Nieuwland, at the Cape of Good Hope in Table Bay, 1762. (Oil on canvas, Anonymous, 18th Century. Iziko William Fehr Collection. Source: Wikimedia Commons; Creative Commons Public Domain Mark 1.0.).

Their return voyage started when the East Indian return fleet (for an artist's impression, see figure 5.15) set off from the Cape en route to the anchorage at Texel. On invitation of Master Marten Wildvang, they embarked on the East Indiaman *Wapen van Alcmaer* on 20 April 1687 [101], another 'spiegelretour' ship which had departed for the VOC's home anchorage from Masulipatnam (India) on 8 October 1686, but notes on the clocks' performance do not begin until 10 May 1687 [102]. It is likely that this delay was caused by a number of unfortunate events which happened on board the *Alcmaer*: its captain, the chief mariner, and several crew members succumbed to an unknown illness they had acquired off the coast of the Cape. Helder, who also held the rank of chief mariner, was made captain for the rest of the voyage back to Texel [103]. He must have considered this a turn of good fortune, thinking that it would provide him with more opportunities to acquire useful information about the clocks' operation and stability at sea. However, when the *Alcmaer* returned to Texel on 15 August 1687, Helder was no longer on board. He had died shortly after having left the Cape, in late April 1687, and many of his notes had disappeared; 14 others on board the ship also perished at sea. Huygens' second envoy, De Graaff, took over and managed to acquire enough measurements for Huygens to trace back the ship's course.

Significantly delayed, in part caused by obligations arising from his father's passing as well as his own poor health, Huygens eventually submitted his findings

Figure 5.15. The VOC return fleet setting off from Batavia, Dutch East Indies, 7 August 1648. (Oil on canvas, Anonymous, ca. 1674. Collection Stedelijk Museum Alkmaar, Netherlands; inventory number 20636. Reproduced with permission.).

regarding his clocks' accuracy—*Rapport aengaende de Lengdevindingh door mijne Horologiën op de Reys van de Caep de B. Esperance tot Texel A° 1687 (Report regarding longitude determination by my clocks on the voyage from the Cape of Good Hope to Texel in the year 1687)*—to Hudde on 24 April 1688 [104]. He requested that his entire report not be read at the VOC governors' meeting, but that it be assessed by competent examiners with up-to-date knowledge of maritime navigation. He conceded that there was a small problem with the longitudes determined by his clocks—indeed, the measurements seemed to imply that the *Alcmaer* had sailed right through Ireland and Scotland… Figure 5.16 shows the route according to the raw measurements from Huygens' clocks as well as that logged by the ship's navigators.

As we will see below, Huygens attributed the ship's apparent deviation from the navigators' route—implied by the raw measurements from his clocks—to the effects of the Earth's rotation: the 'spinning-off' effect on bodies, and hence their loss of weight, was greater at the Equator than towards the Poles. We will return to the scientific implications of these results in chapter 6. Meanwhile, we should keep in mind that Helder's voyage occurred well after Richer's sailing to Cayenne in 1672, when the latter had determined that the period of a pendulum-driven timepiece depends on one's geographic location. Nevertheless, as late as Helder's journey in 1687, Huygens remained ambivalent about the validity of Richer's conclusions. However, when he took into account this latitude dependence on the assumption

Figure 5.16. The route determined by the Alcmaer's navigators is indicated by the westernmost curve (green), while that based on raw timing measurements is represented by the easternmost curve (yellow). Upon correction of the latter for the effects of the Earth's rotation, Huygens arrived at the curve closest to but slightly to the east of the westernmost route (red). (*Oeuvres Complètes de Christiaan Huygens* **IX** p 273.).

that any variation in the gravitational force was fully proportional to the Earth's rotation, the *Alcmaer*'s route determined by his clocks became very close indeed to that reported in the ship's log. We will discuss the underlying assumptions made by Huygens in chapter 6; we will examine the corrections needed on the basis of measured changes in the length of the seconds pendulum in detail below. The updated route (see the red line in figure 5.16) showed that the *Alcmaer* had clearly not sailed right through Ireland and Scotland on its way to the VOC's anchorage at Texel, and that the ship's terminus after 117 days at sea was just 17 seconds of arc (approximately 19 km) East of the actual longitude of the VOC's anchorage—for the time an unprecedentedly accurate determination of a ship's position [105]. It amounted to a loss of a mere 68 seconds of clock time over the course of the voyage from the Cape to Texel. Huygens hence proudly proclaimed,

> '*I have found that the route of the vessel was much better marked on the map than it was without this correction; so much so that arriving at this port, there was not 5 or 6 leagues of error in the longitude thus adjusted. This presupposes that the aforementioned Cape had been well surveyed by the Jesuit Fathers when they passed by there on the way to Siam* [Thailand] *in the year 1685, and that it is located some 18 degrees more to the east than Paris, which I know moreover to be scarcely far from the truth*' [106].

This result hence finally convinced him of the reality of the latitude dependence of a pendulum's period, thus pitting Huygens against Newton for the rest of his life. We will return to the scientific disagreements between Huygens and Newton in chapter 6.

It is unclear whether anyone not directly involved with the VOC or in the assessment of Huygens' claims was aware of the significant discrepancies that had come to light when tracing back the route followed based on the clocks' raw measurements. There is no surviving evidence that Huygens commented in any correspondence on the *Alcmaer*'s apparent trajectory through Ireland and northern Scotland, although a memorandum by David Gregory (1627–1720), the inventor, from 11 November 1691 includes a passage which suggests that Huygens may have discussed the problem of the latitude dependence of the gravitational force during a journey to England in the summer of 1689:

> '*By observations of a ship from the Cape of Bonne Esperance* [the Cape of Good Hope] *to Texel on board which was a two of these Clocks, the course of the ship was on the coast of Ireland on the supposition the weight was the same in all parts of the earth or the pendulys vibration in equal times, but if the the* [sic] *other hypothesis of the less weight at the Equator be true the course will be (as it was) by the north of Scotland but both systems bring the ship to Texel*' [107].

The throw-away comment on the ship's course through Ireland could not have come from any of Huygens' written communications, since he referred to this impossibility in neither his report to Hudde, nor his *Discourse on the Cause of*

Gravity (1690). The source of this information may, indeed, have been Huygens himself, or perhaps Nicolas Fatio (Faccio) de Duillier (1664–1753), the Swiss mathematician, with whom Huygens had shared an early morning coach ride from Hampton Court to central London on 10 July 1689 [108]. The coach ride was also shared with Newton, but Gregory had—at that time—not yet started his correspondence with the latter [109].

In the passage from Huygens' *Addition* to his *Discourse on the Cause of Gravity* included above, he referred to an accurate determination of the Cape's longitude by the visiting French Jesuit missionary Guy Tachard (1651–1712). In 1685, Tachard joined an expedition sent to the East by the French King Louis XIV. The expedition, which was led by Alexandre, Chevalier de Chaumont (1640–1710), the first French ambassador to the King of Siam, also included the two ambassadors of the 1684 First Siamese Embassy to France, Father Bénigne Vachet of the *Société des Missions Étrangères de Paris*, and a mission of six Jesuit priests and scholars—the 'superior' Jean de Fontanay, Francois Gerbillon, Joachim Bouvet, Louis Le Comte, Claude de Visdelou, and Tachard—who would subsequently try to reach China to study its arts and scientific developments.

In addition to their missionary work, the main purpose of the Jesuits, referred to as 'royal mathematicians,' was to determine the longitudes and variations of the compass of the places they visited, correct existing maps and navigational instructions, collect scientific knowledge, and acquire interesting books for the King's library. Their instruments, provided by the *Académie des Sciences*, included a pendulum clock with a period of one second. The expedition arrived in Table Bay on 31 May 1685 and stayed until 7 June. De Fontaney and Tachard were well received and allowed to set up a temporary observatory in a pavilion in the garden of the VOC's compound.

Tachard's astronomical observations at the Cape, which started on the night of 2 June 1685, were mainly aimed at obtaining a more accurate determination of its longitude. He timed the beginning of a transit of Jupiter's brightest Galilean moon, Ganymede. To calibrate his clock with respect to the local solar time, he observed the Sun's altitude on the morning and afternoon of both 3 and 4 June 1685. Tachard compared the timing of the eclipses of Ganymede with the times predicted by Cassini's ephemeris tables [110]. The latter implied that on 4 June 1685, the satellite was due to reappear from Jupiter's shadow at $08^h25^m40^s$ local time in Paris; Tachard observed the event at $09^h37^m40^s$ local time at the Cape. This difference of 1 hour and 12 min implied that the Cape was located 18 degrees East of Paris[4]. Huygens combined this latter difference with the location of Texel, 3^h35^m East of Paris, determined by Giovanni Battista Riccioli and included in his *Geographiae* [111], which was based on observations of lunar eclipses, thus yielding a longitude difference between the Cape and the Texel anchorage of 14^h25^m.

---

[4] Tachard's result indicated that the Cape was three degrees further West in longitude than shown on most common sea charts, a difference Huygens also commented on. However, his method lacked precision and despite the care with which the observations were made, the final result still placed the Cape almost two degrees too far East.

Interestingly, in the exchange of letters related to the Jesuits' determination of the longitude of the Cape, once again we encounter Huygens' short temper and lack of diplomacy if challenged. In the letter to Hudde accompanying Huygens' report, he referred to another letter from the Dutch scholar and manuscript collector Isaäc Vossius (1618–1689) to Coenraad van Beuningen (1622–1693), VOC governor, where Vossius questioned the accuracy of the Jesuits' measurements, stating that …

'the clock of Mr. Christiaen Huijgens performs excellently, but if one were to calibrate it based on the Eclipses, it will indicate during the 24 hours of a day and a night no more than 22 hours' [112].

Huygens clearly bristled at the suggestion, countering that in Vossius' letter, …

'where he objects to the observations of the Jesuits at the Cape of Good Hope and in general against observations of the longitude based on the Satellites of Jupiter, but both without any reason, since he has little knowledge of Astronomy and of the relevant type of observations … Because one cannot fathom what the meaning is of these words' [113].

The close match of the end point of Huygens' corrected route with the independently known longitude of the Texel anchorage convinced him of the viability of his clocks as accurate marine timepieces. He summarized his conviction in the opening paragraph of his report to the VOC's governors:

'I can bring very good news concerning this invention, for I have found that by using the aforementioned clocks the longitudes between the Cape of Good Hope and Texel have on the whole been measured very well, and the total longitude between these two places [has been measured] so perfectly that it only deviates by 5 or 6 leagues, which I admit I have seen with exceptional satisfaction, it being certain proof of the possibility of this very-long-sought-after affair' [114].

Huygens' reference to an accuracy of his positional determination of 5 or 6 leagues corresponds to approximately 30 km or 27 minutes of arc at the latitude of Texel. This is somewhat uncertain, given that the league was not a universally defined distance measure at the time; it is approximately three modern nautical miles. If Huygens' claim of such a high accuracy was indeed correct, this implies that he would have met the accuracy requirements (better than 30 minutes of arc, or 55 km at the Equator) for the award of the full prize money of £20 000 for which provision was made in the British Longitude Act of 1714. It is, therefore, rather curious that Huygens' report to Hudde was apparently never widely circulated among contemporary scientists nor translated into French, Latin, or English. This may indicate that the strength of Huygens' claim was perhaps not as great as implied by his bold assertion that his clocks could be used to determine the longitude unequivocally. Hudde never scrutinized Huygens' report in detail, citing a lack of time [115]. However, on 14 May 1689 Huygens' report and the map of figure 5.16

were passed on to De Volder for formal review. The latter was generally supportive of Huygens' conclusions, but he cautioned the VOC governors that they should not place too much weight on the results of a single experiment [116]. His careful scrutiny of Huygens' analysis revealed that Huygens had made a mistake in his calculations pertaining to the ship's longitude on 8 June 1687, which also affected all subsequent calculations [117]. The corrected values resulted in the improved accuracy of the *Alcmaer*'s longitude with respect to that of the coast of Texel of 17 minutes of arc which is generally cited.

Let us now consider Huygens' claim and its associated uncertainties in some more detail to ascertain whether it actually holds up under close scrutiny. First, it is important to realize that De Graaff's measurements include significant gaps, particularly from 20 April to 10 May 1687, from 10 to 27 May 1687, and from 8 to 24 July 1687: while the ship's mariners' best estimates (or, in Huygens' own words, '*guesses*') based on dead reckoning were logged on every single day when the ship was at sea, De Graaff's measurements covered a grand total of only 21 days. Presumably this sparse temporal coverage was caused by the need for the Sun to be visible at both sunrise and sunset, as Huygens had instructed:

'*At the Rising and Setting of the Sun, when it is half above the Horizon, marke the time of the day, which the Watches, then shew; and though you have in the mean time sayl'd on, it is not considerable. Then reckon by the Watches, what time is elaps'd between them, and add the half thereof to the time of the Rising, and you shall have the time by the Watches, when the Sun was at South; to which is to be added the Aequation of the present day by the Table. And if this together makes 12 hours, then was the Ship at Noon under the same Meridian, where the Watches were set with the Sun. But is the summe be more than 12, then was she at Noon under a more Westerly Meridian; and if less, then under a more Easterly; and that by as many times 15 degrees, as the Summ exceeds or comes short hours of 12: as the calculation thereof hath been already delivered*' [118].

Despite the good agreement between the course recorded in the ship's log and Huygens' corrected trajectory, the latter was by necessity based on interpolation. In his report to the VOC, Huygens commented that …

'*the differences between the mariners and the corrected clocks are usually about 1 or 2 degrees, and always less than 3 degrees. And it should amaze no one that the mariners' reckoning would be 3 degrees off the true longitude on such a long voyage, because of the uncertainty in their guesses, from unknown currents and the ship's falling behind, as well as from its uncertain advancement.*'

This assessment is indeed correct for the northward leg from the Cape to the northern tip of Scotland. Only one measurement in the table shows a deviation in excess of Huygens' statement, 3 degrees and 24 min observed on 8 June 1687, but this position is not included on Huygens' map, because it is partially based on the less accurate clock 'B.' However, for the second part of the voyage, the southward leg from the north of Scotland to the anchorage at Texel, the difference in position on 5 August 1687 is 3 degrees 17 min (and the measurements of 1 August are also suspect), but Huygens seems to have ignored these discrepancies. In fact, he summarily dismissed the mariners' logs for this final leg, suggesting that the North Sea maps used by the ship's mariners were incorrect.

With hindsight, we can place these claims in their proper context. While Huygens' calculations placed Texel 14 degrees 25 min West of the Cape, the mariners' course located the island more than 15 degrees West of the Cape, while contemporary world maps consistently maintained a difference of 18 to 19 degrees. The discrepancy between Huygens' and the mariners' longitude determinations at the ship's final destination may, to some extent, be explained if we realize that the last measurement in the ship's logs, made on 15 August 1687, was obtained when Texel was in sight, but the *Alcmaer* was still three (German) nautical miles West of the island—a distance of 22.2 km or 20 minutes of arc, which Huygens apparently did not correct for in his tables. If he had done so, the total difference in longitude implied by his clocks' measurements would have been 37 minutes of arc, or 41.2 km at the latitude of Texel, corresponding to a cumulative loss of 2 minutes and 28 seconds over the course of the 117-day voyage from the Cape. Clearly, this would still have been an unprecedented accuracy for the time.

Nevertheless, I note that Huygens was apparently aware of the three-mile positional discrepancy between the *Alcmaer* and the coast of Texel, given his conclusion in the report submitted to Hudde:

> '*One sees here then how perfectly the clocks have measured the longitude between these two places, because on August 15, just before putting in at Texel, this longitude of the clocks had been 56 min. 34 s of time, which equals 14 degrees 8 and ½ minutes. Thus, the difference is only 16 and ½ minutes, which is about a quarter of a degree, which at the parallel of Texel equals only 2½ miles. Or if one adds to it the 3 miles to the west where the place of this observation was estimated, then the difference amounts to 5 and ½ miles, which one should consider small in light of such a long voyage.*'

Huygens would have been able to make a much stronger case for his claims if he could have obtained independent calibration data for his clocks' longitude measurements. Unfortunately, the *Alcmaer* did not call into any ports en route to the north of Scotland, so that we do not have intermediate data points on the northward leg. Nevertheless, by the end of the voyage the *Alcmaer* passed two landmarks for which independent longitude determinations were available, Fulo (Foula) Island, to the southwest of the Shetland Islands, on 29 July 1687 and the southern tip of the Shetland Islands on 1 August 1687. On these dates, the cumulative corrections to the clocks' measurements were relatively small [119]. However, instead of using the known longitudes of these landmarks as his calibration, Huygens concluded that the longitudes of both locations required independent verification. The discrepancies between the longitudes implied by his clocks and those indicated on the map of the Dutch cartographer Dirck Rembrantsz van Nierop (1610–1682) from the 1650s used

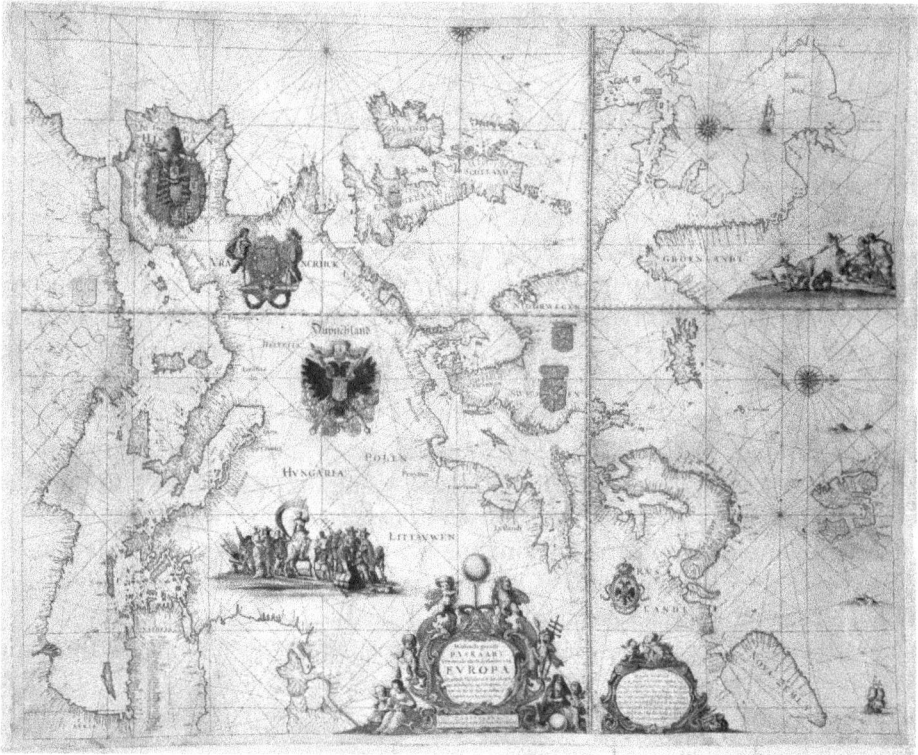

Figure 5.17. Map of Europe by Dirck Rembrantsz van Nierop and Pieter Goos, 1658–1660. (Source: Wikimedia Commons/Library of the University of Amsterdam; Creative Commons Public Domain Mark 1.0.)

by the *Alcmaer*'s mariners (see figure 5.17) gave rise to concern: the map gave the location of Fulo Island three degrees West of that implied by Huygens' clocks. The wider ramification of this discrepancy was that the entirety of northern Scotland and its outlying islands would have been rendered a few degrees too far to the West [120].

Before jumping to possibly unwarranted conclusions, we should first consider the quality of the map used by Huygens for reference. In his report on the accuracy of the determination of longitudes based on measurements made with his clocks on board the first VOC-sponsored voyage to the Cape, he stated that part of the accompanying map, that is, the area for latitudes north of 27°N, was derived from a *'chart of Europe with advancing degrees made by D. Rembrandts van Nierop'* [121]. He also stated that he had adjusted the location of Africa so as to maintain a longitude difference of 14 degrees 25 minutes between Texel and the Cape. In addition, he retained a difference in longitude between Tenerife and Texel of 22 degrees, placing Tenerife 36 degrees 25 minutes West of the Cape. This difference was 38 degrees on the map used by the *Alcmaer*'s mariners and 41 degrees on 'ordinary' maps; present-day maps place Tenerife 35 degrees 3 minutes West of the Cape.

However, careful reanalysis of Huygens' claims [122] suggests that the map included in the *Oeuvres Complètes* and reproduced in figure 5.16 differs from that accompanying Huygens' report to the VOC. Schliesser and Smith suggest that Huygens most likely took Rembrantsz' map of Europe and adjusted the location of the African continent so as to map the Cape and Paris onto Tachard's coordinates. They also point out that the map of figure 5.16 contains a glaring error, in the sense that the position indicated for 15 August 1687 is located to the West of the course estimated by the ship's mariners rather than to the East as implied by both the table and text of his report to the VOC. It is thus likely that the map included in the *Oeuvres Complètes* is a poorly rendered reproduction of the original map; the positional mistake most likely originated from an absent-mindedly made sign error, in the sense that Huygens may have swapped East and West of the reference route.

Although Huygens requested to see the actual map used by the *Alcmaer*'s mariners, he likely never received a copy [123, 124]. In his report to the VOC, he indicated that he tried to plot the ship's course on a variety of maps, before submitting his final version, that is, the trajectory shown in figure 5.16. The *Oeuvres Complètes* include a different map showing the mariners' originally logged course, the trajectory suggested by the clocks' raw data and the corrected course, which I have reproduced in figure 5.18 [125]. This map is characterized by a difference of 41 degrees 40 minutes between the longitudes of Tenerife and the Cape, while Texel's location is approximately 18 degrees 20 minutes West of the Cape (and 3 degrees 35 minutes East of Paris). The coastlines look rather different from those in figure 5.16, and the three trajectories differ more significantly from each other than those in figure 5.16. This latter aspect can most likely be explained by reference to the longitude zero-points adopted for the maps in figure 5.16 and figure 5.18. The former used the Cape as reference, while the latter adopted the longitude of Tenerife as its zero-point. The resulting differences between the longitudes of Tenerife and the Cape were 38 degrees for the map of figure 5.16 and 41 degrees for that of figure 5.18, which in turn led to the larger differences seen in the map of figure 5.18. It appears that the mistake pointed out above was not made when drawing this map [126].

From a scientific perspective, perhaps the more important question this first VOC-sponsored sea trial aimed at addressing was the length of the seconds pendulum and its variation throughout the voyage, from the anchorage at Texel to the Cape. Early efforts by Richer and the French *Académie* had provided intriguing evidence of a latitude-dependent effect, although Huygens was unwilling to fully accept these results until he had been able to obtain independent confirmation. As late as May 1687, Huygens was clearly confused by the conflicting reports that did the rounds [127] regarding the presence or absence of variations in the seconds pendulum length in '*different climates*' (see also chapter 4.1.4). This is evidenced in a letter he wrote to the French polymath Philippe de la Hire (1640–1718) of 1 May 1687 [128], where he also pointed out that he had sent two of his clocks to the Cape to resolve this issue once and for all. Indeed, one of the major conclusions of Huygens' 1688 report to the VOC governors was that the length of the seconds pendulum had to be shorter at the Equator than at more temperate

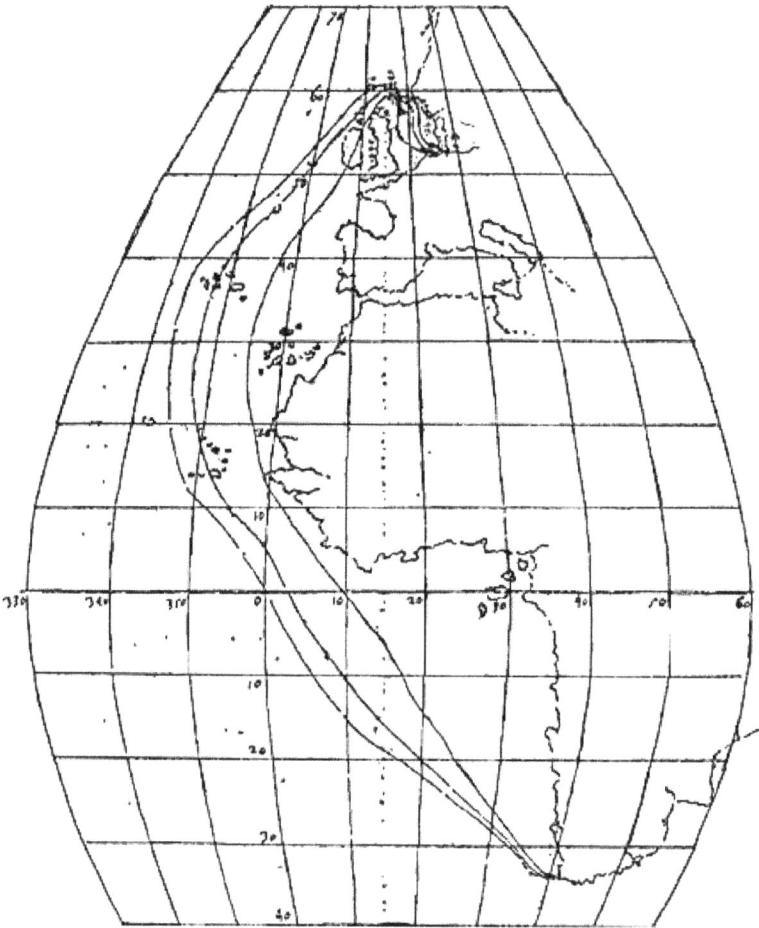

Figure 5.18. The *Alcmaer's* course as logged by the ship's mariners, the raw course indicated by Huygens' clocks and that following his corrections for the effects of the Earth's rotation, shown on an 'ordinary' map. (*Oeuvres Complètes de Christiaan Huygens* **XVIII** p 640.)

latitudes, and thus that the length of the seconds pendulum could not be used as a universal measure.

Schliesser and Smith [129] pointed out that at the time of Huygens' 1688 report, the length of the seconds pendulum at Paris had been established to within $^1/_{10}$ of a *ligne*, approximately 0.2 mm, so that we can be confident of the measurement accuracy that could be achieved at the time. Huygens was suspicious of the body of previous measurements collected by the time the first VOC-sponsored sea trial went underway:

> '*But we cannot entirely trust these first observations, the occurrence of which we do not see as conspicuous in any way—and we can trust still less, given what I believe, in those that are said to have been made in Guadeloupe, where the shortening of the Paris pendulum had been found to be two* lignes. *We must hope*

that in time we will be informed exactly of these different lengths, at the Equator as well as in other regions; and certainly it is something that well deserves being researched with care, even if it would only be to correct, according to this theory, the motions of the pendulum clocks, in order to make them serve as a measure of longitudes at sea' [130].

Given the opportunity of the 1686–1687 sea trial, Huygens expressly instructed Helder in detail how to measure the length of the seconds pendulum throughout the voyage, [131] emphasizing the need to maintain small swing angles of less than 2 to 3 thumbs, corresponding to 3–4½ degrees. Such small swing angles would limit any measurement uncertainties in the pendulum length to less than 0.085 *lignes* [132]. This was important, because a difference in the swing angle between 4 and 10 degrees could explain Richer's 1672 observations of differences in the pendulum length between Paris and Cayenne. Richer himself did not specify the angles he had allowed; he only commented in his report on the Cayenne expedition that they had been kept very small. Unfortunately, we do not have any surviving notes by Helder related to his having measured the pendulum length on his voyage to or from the Cape. It is therefore unclear whether any such measurements were in fact made. Huygens noted in his 1688 report that he had provided his envoy with clear instructions [133], but the editors of his *Oeuvres Complètes* suggest that these may never have been received [134]. As such, Huygens's corrected course of the return voyage from the Cape as indicated by his clocks was based on theoretical corrections related to the Earth's centripetal force, not on measurements of the actual variation in the seconds-pendulum length with latitude.

5.2.4 Second long-range trial to the Cape

Since this first test on board a VOC ship was felt to have left too many open questions as regards the precision of the resulting longitudes, De Volder had recommended that the VOC undertake a second sea trial. Once again, Johannes de Graaff was enlisted to take charge of the clocks, which this time Huygens had commissioned from the prominent Hague clockmaker Pieter Visbagh (Visbach; 1634?–1722), formerly apprentice of and subsequently successor to Coster. De Graaff had meanwhile been made VOC 'onderkoopman' (a rank just below that of merchant) [135]. Huygens and De Graaff engaged in extensive correspondence to ensure that the test would go well and that De Graaff was fully prepared for any unforeseen circumstances. The latter was joined on the voyage by his assistant Pieter van Laer [136] and the clockmaker Jillis (Gilles) Meijbosch [137]. The outward leg—on board the 'spiegelretour' ship *Brandenburg* commanded by Master Evert Verbrugge [138]—commenced on 29 December 1690 with 250 passengers on board, arriving at the Cape on 4 June 1691 [139]. The return ship, the flute *Spierdijk*, sailing under the command of Master Sieuwert de Jong, returned from the Cape on 26 June 1692, arriving at Veere harbour, south of Rotterdam, on 10 October 1692 [140].

Unfortunately, this second sea trial was anything but an unqualified success. First, only one of the two clocks taken on board the *Brandenburg* was operational during the outbound voyage. Yet no useful data could be obtained during the first leg from Texel to the Cape Verdian port of St Jago (São Tiago) because of inclement weather conditions at the anchorage that had prevented any initial calibration, despite repeated attempts to do so by De Graaff [141–144]. Huygens commented on this failure to secure a baseline for their observations in a detailed rebuttal of De Graaff's measurements:

> '*It follows directly from the opening paragraphs of the Journal which was kept on the outward voyage how, since the mounting of the clocks on the* Brandenburgh, *from 22 December 1690 until the 29th of the same month, everything possible has been done to calibrate the clocks' performance based on observations of the Sun, but all in vain owing to the continuous overcast weather conditions, as well as because of the distance between ship and shore. Therefore the clocks could not be put into practice until the island of S. Jago at Capo Verde*' [145].

The second clock's performance was affected by a broken spring throughout the voyage [146, 147]. Upon arrival at the Cape, De Graaff fell seriously ill for about three weeks. In addition, the need to obtain proper measurements and calibration of the clocks based on the Sun's motion, for which he needed at least three weeks, caused the envoys to delay their return voyage by a year [148, 149]. To add insult to injury, when he finally embarked on the return voyage, he did not install the clocks correctly on the *Spierdijk*, so that once again no useful data were obtained [150]. Huygens appeared to have been uncharacteristically forgiving: '*Sed errare humanum est,*' *but to err is human...*

De Graaff's measurements, combined with Huygens' proposed corrections for the Earth's rotation based on the *Alcmaer*'s voyage a few years earlier, resulted in a course that deviated significantly from that logged by the mariners[5]. However, in a letter [151] to De Volder dated 19 April 1693, Huygens conceded that De Graaff's measurements needed correction and that he himself had made mistakes: De Graaff had, apparently, misunderstood Huygens' instructions; he had added the corrections derived by Huygens when subtraction was required and vice versa... [152] In addition, the envoy had made an error of 24 seconds in his calculations. Although this latter error propagated throughout the measurements for the remainder of the voyage, the resulting deviation was fairly small, of order one-tenth of a degree [153]. Despite the array of errors affecting De Graaff's measurements, in his final report [154] Huygens significantly understated the rather disastrous results, reporting that '*the clocks have not proved such a success as we had hoped for.*'

He attempted to salvage some of their results, suggesting that the maps employed by the VOC's mariners were in need of revision. The longitude difference between

---

[5] De Graaff's report is no longer available, but we can glean much about its contents from Huygens' 1693 commentary.

Figure 5.19. (*Left*) *Africae accurata tabula,* Amsterdam: Nicolaes Visscher, 1666–1690(?). (Courtesy: Special Collections Department, Bryn Mawr College Library.) (*Right*) Globe of Joan Blaeu, Amsterdam, 1645–1648. (Courtesy: Amsterdam Museum; reproduced with permission.)

St Jago and the Cape resulting from the clocks' measurements was 48 degrees 14 minutes; this was in good agreement with '*the newest maps and globes*' of the Dutch mapmakers Nicolaes Visscher II [155, 156] (1649–1702) and Jan (Joan) Willemsz Blaeu(w) [157], but it differs significantly from the modern value, 42 degrees 3 minutes. Figure 5.19 displays a reproduction of Visscher's Africa map and a photograph of a Blaeu 'Large Globe,' with which Huygens had apparently compared his tabulated corrections to the mariners' estimates and De Graaff's measurements, '*My reckoning … accords very well with the map [of] the Large Globe of Blaeuw*' [158]. He interpreted the close match between the longitude determined based on De Graaff's measurements at the Cape after a voyage of two months with that given by Visscher's map as evidence of his clocks' reliability:

'*It is now noteworthy how significantly these 2 courses differ from each other, and if someone were to think that this might have been caused by the clock's malfunctioning, I would say that this has a low probability. Because with the longitude [difference] between S. Jago and the Cape as measured with the clock, it would be very unusual if it had run too quickly and then again too slowly, nevertheless yielding a final result that would have been the same as if it had been running regularly*' [159].

He attributed the discrepancy between the mariners' estimated longitude difference and that resulting from De Graaff's measurements as owing to the unaccounted-for effects of ocean currents in the mariners' measurements, reminiscent of what had happened during Holmes' second sea trial of the Huygens–Bruce clocks in 1664 (see chapter 4):

'One knows that there is a constant general current from East to West, and it transpires that the ship has been affected by a current that has slowed it down, since from 19 April until 9 May (during which days the most significant curvature in the course has occurred with respect to that indicated by the clock) the derived latitudes have been systematically smaller according to the Journal than those based on estimates' [160].

Huygens was thus confident that his clocks' performance had held up during the sea trial, announcing that ...

'I have thus shown that the clocks have either performed successfully, or were not enabled to do so' [161].

He confidently stated that his clocks had ...

'succeeded very accurately and precisely in the measurement of Longitude, namely, on the outbound voyage from the island St. Jago to the Cape of Good Hope' [162, 163].

However, in addition to issues related to the accuracy of the measurements, the trial also suffered from a number of basic flaws in the set-up of the experiments. For one thing, no attempt was made to check Huygens' corrections independently, either by obtaining measurements with a seconds pendulum or by calibrating the measurements using the longitudes of well-known landmarks. These could have been firmed up by land-based observations of the ephemerides of Jupiter's moons at the Cape while De Graaff was incapacitated by his illness, yet the latter does not appear to have made such measurements. Indeed, this was one of Huygens' main recommendations in his final report, which was however largely ignored:

'It would still be very helpful if one investigated the true longitude at some important places with regard to the Meridian of Texel or Amsterdam, by observing the satellites of Jupiter.'

In fact, Huygens specifically suggested that measurements taken on both legs of the voyage *at the same place* should be compared with the accurately known timing of Jupiter's satellites to provide the final proof. Unfortunately, and to Huygens' significant chagrin, his recommendation fell largely on deaf ears [164]. Nevertheless, Huygens' suggestion to use the satellites of Jupiter for calibration purposes brings the efforts to achieve an accurate determination of longitude at sea full circle: use of Jupiter's moons, whose ephemerides act as a very accurate natural clock, had been suggested from the outset [165]. Practical considerations called for more straightforward timing measurements, however.

Largely because of Huygens' insistence on the accuracy of his pendulum clocks, Huygens and De Volder, who had been called upon once more as independent assessor, engaged in a heated correspondence. The accuracy required for the trial to be deemed successful had been clearly communicated by the VOC. Initially, a VOC resolution of 31 December 1682 called for an accuracy of better than 1 second deviation per 24 hours, although this was subsequently relaxed to a performance requirement of better than 2 seconds per 24 hours in a new VOC resolution dated 28 April 1684. In a letter of 24 March 1693, Huygens commented [166] to De Volder on the accuracy of the observations taken during the second voyage; he concluded that a positional deviation of 10 minutes 51 seconds was incurred over a period of six days. Although this was within the VOC's requisite level of accuracy, this would leave Tenerife almost 5 degrees off from contemporary map measurements.

De Volder rightly pointed out that Huygens' approach to proving his clocks' accuracy was somewhat self-contradictory and, in fact, inconsistent. In his 1688 report on the first sea trial, Huygens' confidence had been boosted by the close correspondence between the longitude determination at the Cape based on his clocks and that reported by Tachard, yet in his discussion of the results of the second test he completely ignored Tachard's measurement and instead adopted Visscher's map as his benchmark [167]. The resulting difference was significant, implying a change in the Cape's longitude from 36 degrees 25 minutes East of Tenerife to 41 degrees 14 minutes East of the Canary island [168], a difference of 4 degrees 49 minutes. De Volder suggested that the longitudinal difference between Tenerife and the Cape was '*a bit more than 38 degrees which appears to accord with the maps of the [East India] Company*' [169].

Huygens wrote in his rebuttal that none of the contemporary maps in common use included a reliable location for St Jago, for which the ephemerides of Jupiter's moons would have to be observed locally [170]. He also commented that one could not have seen any landmarks along the way, pointing out that ...

'*[i]t is true that, following these corrections the calculated course now runs very much to the West of the course estimated by the mariners, deviating by up to $8\frac{2}{3}$ degrees, and only 2 degrees from the coast of Brazil. But it is indeed possible that the general current from East to West has carried the ship along, which the mariners could not have noticed themselves. It is said that one could have seen the Brazilian coast from 30 German miles, which I don't believe, except that it is also unknown with any certainty how close the coast drawn on the map is to its true Longitude*' [171, 172].

Nevertheless, and although he also pointed out that the resulting longitudinal difference between St Jago and the Cape of 48 degrees agreed well with that indicated by Visscher's map, Huygens eventually conceded that the jury was still out on the performance of his clocks:

'*I do not want to pretend that one could conclude from this or based on the previous trial of A°. 1687 that the perfection [of my method of] Longitude measurement has been demonstrated conclusively*' [173].

However, he was keen that De Volder put in a good word for him with the VOC governors, thus diffusing the bad impression De Graaff's report had made [174]. It has been suggested [175] that his concession regarding his clocks' accuracy was driven by his ongoing development of a new marine clock design, the *balancier marin parfait* ('perfect marine balance'), which he wanted to start promoting shortly. This suggestion is supported by his final comments in a letter to the VOC governors of 6 March 1693 [176], which he repeated in a note to De Volder of 19 April 1693 [177], where he hinted at forthcoming developments:

'*I have on this occasion invented something quite different and incomparably better, which I have in hand at the present moment, whereby any little difficulty in the use of this invention will once and for all be removed, of which in due course I hope to give Your Excellencies further particulars, ...*'

The largely unsuccessful experiments on board VOC vessels in the late 1680s and early 1690s had taught Huygens that even such a simple aspect as the mounting of the clocks could have a major impact on their performance. He thus concluded that ...

'*[o]ne of the great defects of the clock with a suspended pendulum on a ship is that the force of the pendulum causes a small movement in the whole clock and this [varies] more or less according to the freedom of the axles of the iron frame, from which arises the inequality of the hours*' [178].

Huygens had started work on a radically new type of clock around the time of the first sea trial of his tricord pendulums on the Zuyderzee in 1684 [179]. The need for access to a reliable clock to determine one's longitude at sea remained unabated. Clearly, the method of longitude determination still hinged on a reliable clock. Huygens kept his hope alive with the development and construction of his perfect marine balance in 1693, looking forward to the next set of sea trials with a significantly improved marine timepiece.

The breakthrough came from an unexpected angle. Huygens had been working on a solution to the problem posed by Florimond de Beaune (1601–1652), the French jurist and amateur mathematician. De Beaune had challenged Descartes in a letter from 1638 to find a solution to the differential equation

$$\frac{\mathrm{d}y}{\mathrm{d}x} = \frac{\alpha}{y - x},$$

which is now considered the first application of the so-called 'inverse tangent' method used for finding the properties of a curve from studying its tangent. Huygens published his analysis in *Acta Eruditorum* in October 1693 [180], having derived a new curve that would ensure isochronous operation of a clock without being adversely affected by the rocking and pitching motions of ships at sea. His basic premise underlying an isochronous pendulum was that its driving force must be proportional at any time to the extent the pendulum is out of equilibrium.

He first considered the arrangement of figure 5.20a, which he had invented on 13 January 1693 and continued to develop throughout January and February 1693 [181].

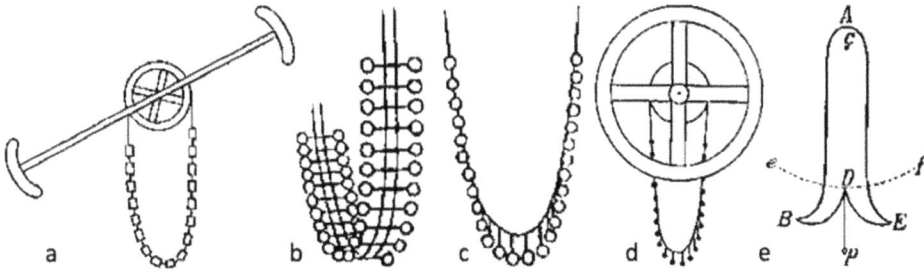

Figure 5.20. Huygens's development of the 'balancier marin parfait' (see the text for details.) (Huygens C 1693, *Oeuvres Complètes de Christiaan Huygens* **XVIII** pp 546–61.)

This followed early sketches from December 1692 and a subsequent mathematical analysis of the isochronicity of the contraption [182]. Huygens' new isochronous regulator consisted of a vertical balance wheel equipped with a chain that linked two systems of small, equal, and equidistant weights. (He also experimented with the inclusion of a float which was partly submerged in oil or mercury.) While swinging, each of these systems would rise and fall, and thus they would alternately increase and decrease the force exerted on the chain by the weights. As we saw in chapter 4, Huygens had originally invented such a system in 1659 to counterbalance the centrifugal motion of the conical pendulum, forcing it back into a paraboloidal envelope (for instance, see figure 4.10) [183]. The resulting torque acting on the wheel was directly proportional to its displacement from equilibrium, thus leading to isochronous oscillations.

Specifically, if m is the mass of one centimetre of the chain, the force exerted on the central component is

$$K = \frac{1}{4} hmg,$$

where h is the displacement height of the chain from equilibrium, while the arc subtended by this same displacement would be $s = \frac{1}{2}h$. Therefore,

$$\frac{K}{s} = \frac{1}{2} mg,$$

independent of h, and hence the oscillations are expected to be isochronous. The same exercise can be done based on the force moment, M, and the angle of rotation, φ, which similarly leads to a ratio of M/φ that is independent of h [184]. Following the same arguments as for the tricord pendulum, we can now work out the device's period.

Huygens expected the chain to oscillate slowly, even on choppy seas, so that the effects of the latter would be greatly reduced. Initial tests showed that the device did not operate satisfactorily, however: large oscillations were completed more slowly than expected. Huygens attributed this behaviour to a combination of increased air resistance associated with larger oscillations and the large number of weights that had to be kept going while their angles with respect to the vertical direction were changing constantly. The latter disadvantage could be remedied by means of a rosary-type arrangement, such as that shown in figure 5.20b or c, while the former problem prompted him to change the pendulum design to that shown in figure 5.20d.

Huygens was certainly concerned with reducing friction as much as possible. The editors of Huygens' *Oeuvres Complètes* suggest [185] that the English clockmaker Henry Sully (1680–1728), who is widely credited with inventing a means to achieve just that, may have benefited from ideas first put forward by Huygens. Sully implemented 'rollers' (anti-friction wheels or disc bearings, formally known as a 'frictional rest escapement') to reduce the mechanical friction of the pivots in his clocks (see figure 5.21 [186], but it is likely that he received advice from Newton, Leibniz, the Swiss mathematician and physicist Daniel Bernouilli (1700–1782), and George Graham, the English inventor and clockmaker, with all of whom he is known to have corresponded. In fact, Bernouilli noted that he had seen Sully's rollers, while Graham wrote in 1724 that he had seen a very similar system in an old pendulum clock [187]:

'Your way of reducing friction on the axes is very good. I have not seen anything like it in our Art, except just once, in more than twenty years: it was the top pivot of the balance of an old pendulum clock, which was contained between three turning wheels, placed there for this purpose. It appeared that the wheels had not been made by the manufacturer of the clock, and that another hand had added them.'

In his response, Sully is said to have written,

'With regard to my method to reduce friction on the axes, I ignored that what one would have used in the performance of our Art; but having seen a large wheel which served to turn a millstone suspended in nearly the same way, I felt that we could make good use of it in our clock making; I believe that I have applied it usefully in this machine.'

Indeed, Huygens already provided a sketch of a similar system in 1693, apparently developed independently, although it is not known whether he actually constructed a working model. It has also been suggested [188] that Huygens may have learnt the details of this escapement from either Hooke's duplex escapement design (which had two escape wheels but was first described in detail only around 1696) or from Bürgi's cross-beat escapement. The latter was invented by the Swiss clockmaker Jost Bürgi (1552–1632). This type of escapement is a variation of the verge escapement I discussed in chapter 4, which contain two foliots that rotate in opposite directions. In this configuration, the impulse is given by a pallet to one escape wheel, with locking by means of a second wheel.

In the 1859 *Proceedings of the Society of Antiquaries of Scotland* [189], Alexander Bryson (1816–1866), horologist and President of the Royal Scottish Society of Art in 1860–1861, extolled the virtues of the duplex escapement, in particular its ability to retain stability:

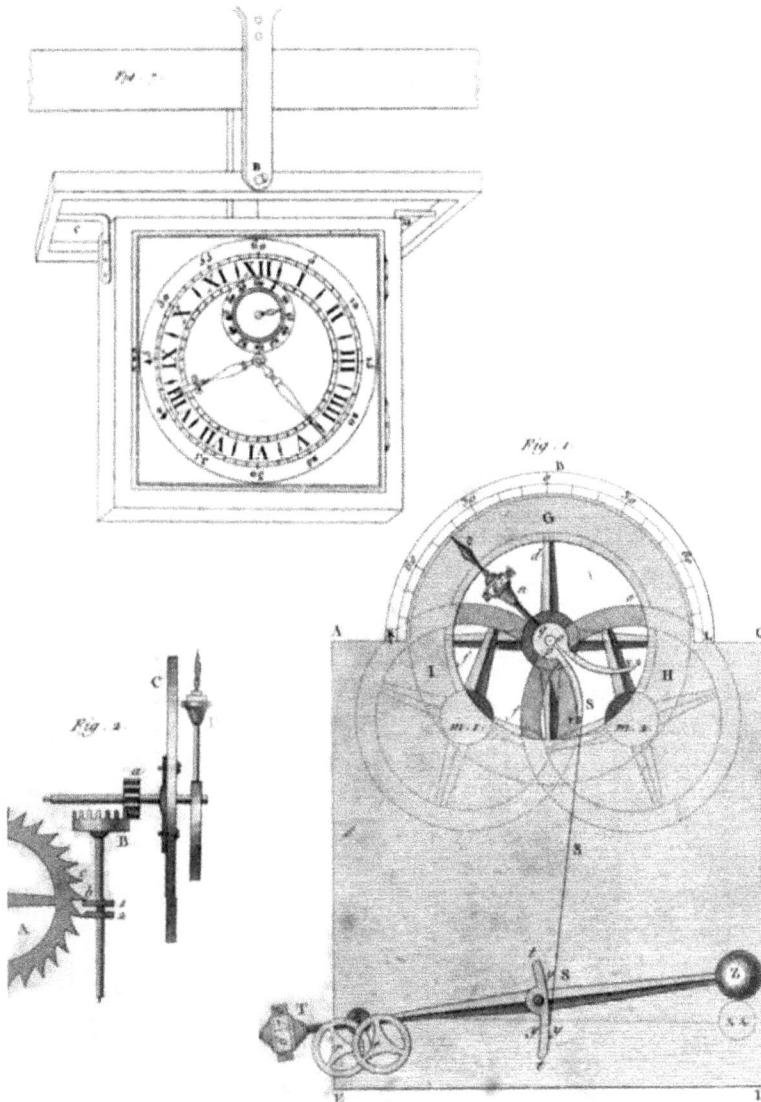

Figure 5.21. Henry Sully's 1724 marine clock design (*bottom right*), escapement (*bottom left*) and gimballed suspension (*top*). Note the anti-friction rollers in the bottom right-hand sketch. (Berthoud F 1802; reproduced from Wikimedia Commons under the CC-PD-Mark/PD-Art/PD-old-100 licenses.)

'[Hooke's] watch had the balance spring and double balance, with a duplex escapement, also Hooke's invention … Had Hooke foreseen that his duplex escapement would stand the test of upwards of two hundred years' experience, he would have closed with this agreement, and no doubt would have gained the prize of 100 000 florins offered by the States of Holland for ascertaining the longitude at sea. … After his disappointment Hooke concealed his invention of the duplex escapement for many years …'

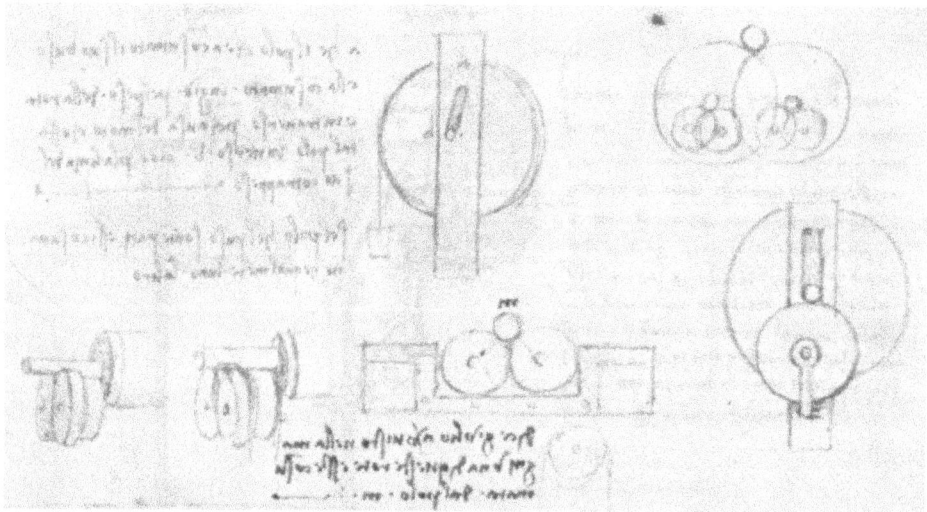

Figure 5.22. Section of the *Codex Madrid I* (Leondardo da Vinci, 1494), showing sets of anti-friction rollers. (Digital scan provided by Cornell University Library, Kinematic Models for Design–Digital Library.)

Indeed, once again we encounter Hooke's easily ruffled feathers, which once again delays a breakthrough from being disseminated... The invention of anti-friction rollers dates back to before the 1500s; for instance, Leonardo da Vinci drew sets of such rollers in his *Codex Madrid I* in 1494 [190]: see figure 5.22.

Nevertheless, on 6 March 1693, Huygens seems to have abandoned this approach and moved on to the new design of figure 5.20e [191] (see also figure 5.23a). In his notes, he wrote, '*libratio isochrona melior praecedente*', that is, '*isochronous oscillation, better than previous [model].*' In essence, his original perfect marine balance consisted of a pendulum and a balance composed of a large three-armed balance (shown in figure 5.23a) with a small pendulum attached to it. This design proved impractical, however, so the second (1695) version of his perfect marine balance contained a two-armed balance or 'foliot' instead, which Huygens had first designed in 1694. He proposed to suspend a small metal shape **ABDE** to the pivot axis **C**, with edges **DB** and **DE** curved such that a small weight p suspended by a thread or ribbon from point **D** would trace these curves throughout any oscillation. The force exerted upon p relative to the axis **C** at any given time would then be proportional to the angular deviation from the equilibrium position of the metal tongue. Huygens demonstrated that the curve satisfying the proportional force condition is the involute of the circle eDf (with centre **C**) and that the weight p, when oriented vertically, describes a parabola (see figure 5.23b) and, hence, that this contraption would perform isochronous oscillations.

Since there is no further mention of Van Ceulen as Huygens' clockmaker in any of his letters or other surviving manuscripts after September 1685 [192], it is most likely that the first model of his perfect marine balance was actually constructed by Bernard van der Cloesen (1640s–1736) [193], clockmaker from The Hague, although

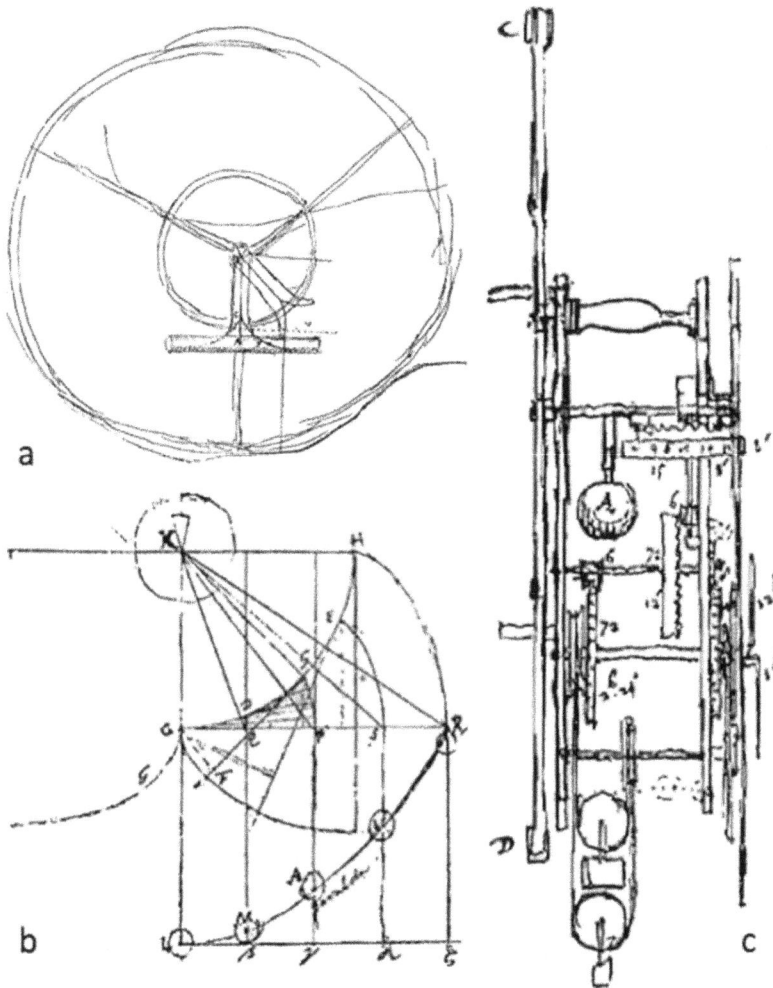

Figure 5.23. Huygens' final sketches related to his balancier marin parfait: a. development of the concept; b. mathematical (geometric) basis; c. working model, including the long balance **CD** typical of this type of clock. (Huygens C 1693 *Oeuvres Complètes de Christiaan Huygens* **XVIII** pp 562–69.)

Huygens himself enacted most of the final adjustments. This clock was weight-driven, with the large balance beating seconds.

Modern calculus allows us to see fairly easily that Huygens' perfect marine balance was indeed meant to oscillate isochronously. Consider figure 5.24 for technical details. As before, I will follow Crommelin's derivation [194] to derive the clock's isochronicity. Let us consider its weight L displaced to a point **A**. It is straightforward to show that the point **P**, where the pendulum's cord leaves the arc of the metal shape **GP**, is always located at the same height as point **C**. **CP** describes the tangent of the circular swing arc in point **C** and is the normal to the arc at point **P**.

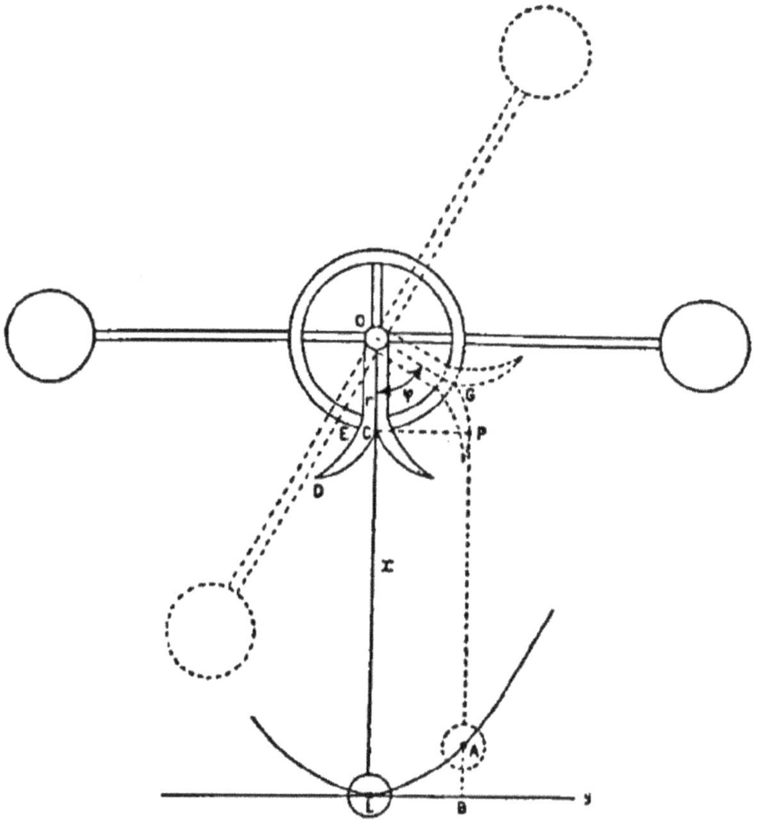

Figure 5.24. Technical details of Huygens' perfect marine balance corresponding to figure 5.23a.

It follows that **CP** is as long as the arc travelled by the bob from **C** to **G**, so that [195]

$$\text{arc } \mathbf{GP} = \frac{\mathbf{CP}^2}{2r},$$

where $r = \mathbf{OC}$. With reference to the outer circle, $\mathbf{GP} = \mathbf{AB} \equiv x$, so that we can transfer the pendulum's motion to the system defined by the axes **LB** and **LO**. Therefore, $\mathbf{CP} = \mathbf{LB} \equiv y$, where

$$x = \frac{y^2}{2r}; \quad y^2 = 2rx.$$

The curve needed for isochronous operation is therefore a parabola. The force moment M exerted in point **P** is $M = mg\mathbf{CP} = mgy$ (for mass m) and $y = \mathbf{CP} = $ arc $\mathbf{CG} = \varphi r$, so that $M = mg\varphi r$. The angle of rotation is φ, so that

$$\frac{M}{\varphi} = mgr = \text{constant}.$$

The resulting oscillations are hence isochronous. We can also calculate the expected period, using for the potential energy

$$E_{\mathrm{p}} = xmg = \frac{y^2 mg}{2r}; \quad y = \varphi r,$$

and therefore

$$E_{\mathrm{p}} = \frac{1}{2} mgr\varphi^2; \quad A = mgr.$$

With the expression for the kinetic energy,

$$E_{\mathrm{k}} = \frac{1}{2} I \left(\frac{d\varphi}{dt} \right)^2; \quad B = I = \text{constant},$$

we can now derive the system's period,

$$P = 2\pi \sqrt{\frac{I}{mgr}},$$

which is identical to the period of the compound pendulum.

Huygens' perfect marine balance included a new equation-of-time cam [196], a kidney-shaped rotating mechanical piece that ensured adherence to the annual equation-of-time corrections. His brother Constantijn Jr, then Secretary to the Dutch 'stadholder' (head of state) William III, disseminated Huygens' new design to both the renowned clockmaker Tompion and Daniel Quare (1648/9–1724) in 1694 or 1695. Quare, the English clockmaker, had become well-known for his invention of a repeating watch movement in 1680. Both clockmakers implemented Huygens' new equation-of-time design in their 'Royal equation clocks.' Interestingly, in his final years, Huygens seems to have encountered his former nemesis, Robert Hooke, in Tompion's workshop, but without either man recognizing the other, as we learn from an entry in Hooke's *Diary* of 30 March 1693: Hooke recorded that he had '*Met Mr Zulich* [Huygens] *at Tompions, but knew him not.*' If only...

Upon hearing of Huygens' success, Landgraf Karl von Hessen-Kassel (1654–1730), most likely from his son, Crown Prince Friedrich von Hessen-Kassel (1676–1751), requested that Huygens construct a similar clock for his own use, but Huygens declined [197]. Nevertheless, Huygens instructed his clockmaker to construct a second perfect marine balance, which may well have ended up in the Landgraf's possession (or perhaps Huygens had a change of mind) [198]. For the second clock, in March 1695 he had Van der Cloesen convert an existing equation-of-time clock to meet his new design specifications [199], but additional details remain beyond our historical reach. He told his brother Constantijn Jr that he meant to provide further details and a demonstration to Tompion:

'You will have spoken to him [Tompion], I believe, of my new horological invention, which I will [soon] describe and demonstrate. I have converted an old

3-foot pendulum, which also shows solar time, without the need for [a correction for] the equation of time' [200].

Huygens was keen to test his new device (figure 5.23c) in practice [201], but death intervened. He passed away on 8 July 1695, before being able to achieve his lifelong goal of manufacturing a sufficiently accurate timepiece for use at sea. He did not manage to publish his new clock design before his death, although he had clearly intended to make it public [202]. He had also continued to seek improvements of his first few perfect marine balance examples, but none was eventually constructed [203].

Sully constructed his own marine longitude clock almost 30 years later based on Huygens' perfect marine balance design. After he had presented his device to the *Académie Royale des Sciences* in April 1723, the *Académie*'s report noted that ...

'during the first trial made at the Observatory, the clock did not vary more than nineteen seconds in eight full days.'

Meanwhile, Newton continued to doubt that a sufficiently accurate marine clock could ever be constructed. Although he had never been to sea, Newton had a lot of experience with methods of longitude determination at sea. I will discuss these developments in their context in chapter 6. Yet, in response to a petition offered to the British Parliament on 25 May 1714 requesting support for the development of a sufficiently accurate means to determine one's longitude at sea, Newton is said to have commented that ...

'[o]ne is by a watch to keep time exactly; but, by reason of the motion of a ship, and the variation of heat and cold, wet and dry, and the difference of gravity in different latitudes, such a watch hath not yet been made.'

In 1721, following seven years of service on the British Board of Longitude, he lamented in a letter to Josiah Burchett, the secretary of the Admiralty, that ...

'a good watch may serve to keep a reckoning at Sea for some days and to know the time of a celestial Observ[at]ion; and for this end a good Jewel[6] may suffice till a better sort of Watch can be found out. But when Longitude at sea is lost, it cannot be found again by any watch.'

References and notes

[1] 1665-08-01: Moray, Robert—Huygens, Christiaan; *Oeuvres Complètes de Christiaan Huygens* **V** No 1436 (pp 426–8)
[2] 1665-09-18: Huygens, Christiaan—Moray, Robert: *Ibid.* **V** No 1466 (pp 485–7)
[3] 1665-12-24: Huygens, Christiaan—Moray, Robert: *Ibid.* **V** No 1508 (pp 549–51)

---

[6] Jewels—specifically diamonds, rubies and sapphires—were used at the ends of the axles of the faster-moving gears.

[4] Mahoney M S 1980 Christian Huygens: The Measurement of Time and of Longitude at Sea *Studies on Christiaan Huygens* ed H J M Bos, M J S Rudwick, H A M Snelders and R P W Visser (Lisse: Swets & Zeidlinger) pp 234–70; footnote 62

[5] Huygens C 1673 *Oeuvres Complètes de Christiaan Huygens* **XVIII** 489

[6] *Ibid.* **XVIII** 496; figure 16

[7] Chapter 3, footnote 2

[8] Hooke R 1678 *De Potentia Restitutiva* (London: '*Printed for John Martyn, Printer to the Royal Society, at the Bell in St. Paul's Church-Yard.*')

[9] 1691-03-02: Leibniz, Gottfried Wilhelm—Huygens, Christiaan; *Oeuvres Complètes de Christiaan Huygens* **X** No 2664 (pp 49–52)

[10] 1691-03-26: Huygens, Christiaan—Leibniz, Gottfried Wilhelm; *Ibid.* **X** No 2667 (pp 55–8)

[11] Huygens C 1675 *Ibid.* **VII** 408

[12] Leopold J H 1980 *Christiaan Huygens and his instrument makers Studies on Christiaan Huygens* ed H J M Bos, M J S Rudwick, H A M Snelders and R P W Visser (Swets & Zeidlinger: Lisse) 228

[13] *Ibid.* **VII** 409–16

[14] See also Leopold J H 1980 *op. cit.* 227–8

[15] Huygens C 1675–1676 *Ibid.* **VII** No 2008 (p 410); transl Mahoney M S 2004 Drawing mechanics *Picturing Machines 1400–1700* ed W Lefèvre (Cambridge, MA: MIT Press) 281–308

[16] See Mahoney M S 2004 *Ibid.* p 301 (incorrectly referenced in his Note 41)

[17] *Ibid.*

[18] 1675-09-10: Thuret, Isaac—Huygens, Christiaan; *Ibid.* **VII** No 2055 (p 499)

[19] 1675-02-15: Colbert, Jean-Baptiste—Huygens, Christiaan; *Ibid.* **VII** No 2011 (pp 419–20)

[20] 1675-09: Huygens, Christiaan—Perrault, Claude; *Ibid.* **VII** No 2054 (pp 497–8)

[21] 1675-01-30: Huygens, Christiaan—Oldenburg, Henry; *Ibid.* **VII** No 2003 (pp 399–400)

[22] 1675-02-12: Oldenburg, Henry—Huygens, Christiaan; *Ibid.* **VII** No 2009 (pp 416–7)

[23] 1675-02-20: Huygens, Christiaan—Oldenburg, Henry; *Ibid.* **VII** No 2013 (pp 422–4)

[24] Duplicate available in *Phil. Trans. R. Soc.* **112** (25 March 1675)

[25] Reprinted as 1675-02: Huygens, Christiaan—Gallois, Jean; *Ibid.* **VII** No 2014 (pp 424–5); App. to No 2013: 1675-02-20: Huygens, Christiaan—Oldenburg, Henry; *Ibid.* **VII** 422–4

[26] 1675-02-09: Huygens, Christiaan—Colbert, Jean-Baptiste; *Oeuvres Complètes de Christiaan Huygens* **VII** No 2006 (pp 405–6)

[27] Plomp R 1999 A Longitude Timekeeper by Isaac Thuret with the Balance Spring Invented by Christiaan Huygens *Ann. Sci.* **56** 379–94

[28] 1675-08-09: Huygens, Christiaan—Huygens, Constantijn Jr.; *Oeuvres Complètes de Christiaan Huygens* **VII** No 2045 (pp 483–5)

[29] 1675-11-21: Huygens, Christiaan—Oldenburg, Henry; *Oeuvres Complètes de Christiaan Huygens* **VII** No 2078 (pp 542–3); transl Plomp R 1999 *op. cit.*

[30] Gould R T 1923 *The Marine Chronometer: Its History and Development* (London: J. D. Potter) 30

[31] See Matthews M R 2000 *Time for Science Education: How Teaching the History and Philosophy of Pendulum Motion Can Contribute To Science Literacy* (New York: Springer) 153

[32] de Hautefeuille J 1675 *Factum, touchant Les Pendules de Poche* etc., petition to the *Parlement de Paris* to block Huygens' patent registration (*Oeuvres Complètes de Christiaan*

Huygens **VII** 439–453); signed 20 December 1675, published February or March 1676 (*Ibid.* **VII** 452 Note 22)

[33] de Hautefeuille J 7 July 1674, letter to the *Académie Royale des Sciences*

[34] See 'Espinasse M 1956 *Robert Hooke* (Berkeley and Los Angeles, CA: University of California Press) 66–7

[35] *Ibid.*

[36] Birch Th 1757 *The History of the Royal Society of London* **III** 179. The date of Huygens' letter and that of the Royal Society's meeting appear discrepant. However, note that the Julian calendar was still in use in England at that time, while France—where Huygens was based—had already adopted the modern Gregorian calendar.

[37] Inwood S 2003 *The Man Who Knew Too Much: The Strange and Inventive Life of Robert Hooke (1635–1703)* (London: Macmillan) 200

[38] 1675-02-20: Huygens, Christiaan—Oldenburg, Henry; *Oeuvres Complètes de Christiaan Huygens* **VII** No 2013 (pp 422–4)

[39] Oldenburg's earlier support of Hooke's work is evidenced, for instance, in a letter he wrote to Huygens on 17 October 1665, 1665-10-17: Oldenburg, Henry—Huygens, Christiaan; *Oeuvres Complètes de Christiaan Huygens* **V** No 1479 (pp 500–2), where we read, '*For the impediments which Mr. Hooke has met with in the working of his [lens grinding] machine, I must first speak with the inventor before describing them to others, several circumstances having escaped me.*'

[40] Purrington R D 2009 The First Professional Scientist: Robert Hooke and the Royal Society of London *Science Networks: Historical Studies* **39** (Basel/Boston/Berlin: Birkhäuser) p 92; note, however, that already on 28 January 1674, Hooke noted in his *Diary* '*Oldenburg treacherous and a villain.*' On 25 March 1675, he intimated that he '*suspected Hills whispers with Oldenburg,*' and on 8 April of that year, he worried about '*Lord Brouncker & Oldenburg, discovered their design,*' followed by a complaint about a '*Grubendolian Caball at Arundel House*' on 15 October 1675. ('Grubendol' was a commonly used anagram of 'Oldenburg.')

[41] Purrington R D 2009 *op. cit.* 93

[42] Birch Th 1757 *op. cit.* **II** 112

[43] Birch Th 1757 *op. cit.* **III** 190

[44] Purrington R D 2009 *op. cit.* 92

[45] Espinasse M 1956 *op. cit.* 67

[46] Spratt Th 1667 *The History of the Royal Society of London, for the Improving of Natural Knowledge* (London) 247

[47] Giles J 2006 Dealer unearths Hooke's Royal Society notes *Nature* **439** 638–9

[48] Adams R and Jardine L 2006 Report: The Return of the Hooke Folio *Notes and Records of the Royal Society* **60** 235–9

[49] Birch Th 1757 *op. cit.* **III** 148

[50] Inwood S 2003 *op. cit.* 200–1

[51] Hooke R 1675 *Diary* **I** p 151

[52] *Ibid.* **I** pp 163–4

[53] Gunther R T 1930 *Early Science in Oxford* **VIII** 149–50

[54] 1675-07-11: Huygens, Christiaan—Oldenburg, Henry; *Oeuvres Complètes de Christiaan Huygens* **VII** No 2040 (pp 477–8)

[55] Gunther R T 1930 *op. cit.* **VIII** p 208

[56] Oldenburg H 1676 *Phil. Trans. R. Soc.* **128** 710

[57] Waller R 1705 *The Posthumous Works of Robert Hooke* (London: Smith and Walford)

[58] See 'Espinasse M 1956 *op. cit.* 63–5

[59] Patterson L D 1952 Pendulums of Wren and Hooke *Osiris* **10** 277–321

[60] Andrewes W J H 1996 Even Newton Could Be Wrong: The Story of Harrison's First Three Sea Clocks, *The Quest for Longitude*, ed W J H Andrewes (Cambridge MA) pp 190–1 (Note 59)

[61] 1675-07-11: Huygens, Christiaan—Oldenburg, Henry; *Oeuvres Complètes de Christiaan Huygens* **VII** No 2040 (pp 477–8)

[62] 1675-08-10: Huygens, Christiaan—Oldenburg, Henry; *Ibid.* **VII** No 2048 (pp 488–90)

[63] 1675-02-15: Colbert, Jean-Baptiste—Huygens, Christiaan; *Ibid.* **VII** No 2011 (pp 419–20)

[64] *Réponse des Etats de Hollande et de Westfrise à une requête (inconnue) de Chr. Huygens au sujet de la détermination des longitudes*, 25 September 1675, *Ibid.* **XVIII** 523

[65] 1675-09-27: Staten van Hollant en Westfriesland—Huygens, Christiaan; *Ibid.* **XVIII** 524

[66] Huygens C 1679 *Memoire concernant l'Académie Royale des Sciences, Ibid.* **VIII** No 2185 (pp 196–9)

[67] *Histoire de Louis XIV, depuis la mort de Mazarin jusqu'à la paix de Nimègue en 1678*, par M. Pellisson de l'Académie françoise. A Paris, chez Rollin fils, m.dc.xlix. Avec approbation et privilège du Roy. 3 Vol. in-12° (see *Oeuvres Complètes de Christiaan Huygens* **VIII** No 2185 Note 1)

[68] Transl Plomp R 1999 *op. cit.*

[69] 1682-10-01: Huygens, Christiaan—Gallois, Jean; *Oeuvres Complètes de Christiaan Huygens* **VIII** No 2279 (pp 393–5); transl Plomp R 1999 *op. cit.*

[70] Gould R T 1923 *op. cit.* 24

[71] Resolutiën van de Bewinthebberen van de Oostindische Comp.[ie] ter Camer tot Amsterdam, 31 December 1682, *Oeuvres Complètes de Christiaan Huygens* **XVIII** 533–5

[72] This letter is no longer available, but we know of its existence because of Gallois' response: 1683-01-07: Gallois, Jean—Huygens, Christiaan; *Ibid.* **VIII** No 2287 (pp 405–6)

[73] 1675-03-21: Oldenburg, Henry—Huygens, Christiaan; *Ibid.* **VII** No 2016 (pp 426–7)

[74] 1675-05-01: Huygens, Christiaan—Justel, Henri; *Ibid.* **VII** No 2026 (pp 456–7)

[75] Application pratique aux horloges de différents mouvements vibratoires plus ou moins isochrones. Avertissement, *Oeuvres Complètes de Christiaan Huygens* **XVIII** 508

[76] Huygens C 1658 *Horologium (The Hague)* 1

[77] Palmer W 1716 *A great improvement in watch-work; which may be of great use at sea, for discovering the longitude. Humbly offer'd to the Consideration of the Learned. With some remarks on another way of discovering the longitude. To which is added, an advertisement Relating to divers other Useful Inventions. By William Palmer, a Lover of Mechanical Motion* (York: Grace White and Charles Bourne)

[78] Plank S 1714 *An introduction to the only method for discovering longitude. Humbly presented to both Houses of Parliament, and to those Worthy Persons that are to judge of it* (London)

[79] Huygens C 1683 L'application de décembre 1683 des vibrations de torsion aux horloges marines (pendulum cylindricum trichordon), *Oeuvres Complètes de Christiaan Huygens* **XVIII** 527–33

[80] Crommelin C A 1938 Pendulum Cylindricum Trichordon van Christiaan Huygens *Nederlands Tijdschrift voor Natuurkunde* **5** 314–8

[81] *Ibid.*

[82] 1683-12-12: Huygens, Christiaan—Fullenius, Bernhardus; *Ibid.* **VIII** No 2327 (pp 474–8)

[83] Resolutiën van de Bewinthebberen van de Oostindische Comp.[ie] ter Camer tot Amsterdam, 27 July 1684, *Ibid.* **XVIII** 533–5

[84] *Ibid.* Note 1

[85] 1685-08-23: Huygens, Christiaan—Huygens, Constantijn; *Ibid.* **IX** No 2394 (pp 20–2)

[86] 1685-09-03: Hudde, Johannes—Huygens, Christiaan; *Ibid.* **IX** No 2396 (p 24)

[87] 1685-09-06: Huygens, Christiaan—Huygens, Constantijn; *Ibid.* **IX** No 2397 (p 25); see also 1685-10-01: Huygens, Constantijn—Huygens, Christiaan; *Ibid.* **IX** No 2400 (p 30)

[88] Resolutiën van de Bewinthebberen van de Oostindische Comp.[ie] ter Camer tot Amsterdam, 30 August 1685 *Ibid.* **XVIII** 534–5

[89] 1685-09-09: Huygens, Christiaan—Huygens, Constantijn; *Ibid.* **IX** No 2398 (pp 26–7)

[90] *Ibid.*

[91] 1685-09-11: Huygens, Constantijn—Huygens, Christiaan; *Ibid.* **IX** No 2399 (pp 28–9)

[92] 1685-10-03: Huygens, Christiaan—Hudde, Johannes; *Ibid.* **IX** No 2401 (pp 30–1)

[93] 1685-10-01: Huygens, Constantijn—Huygens, Christiaan; *Ibid.* **IX** No 2400 (p 30)

[94] 1685-10-03: Huygens, Christiaan—Huygens, Constantijn; *Ibid.* **IX** No 2401 (pp 30–2)

[95] 1685-10: Huygens, Christiaan—Hudde, Johannes; *Ibid.* **IX** No 2406 (p 37)

[96] 1685-10-26: Huygens, Christiaan—Hudde, Johannes; *Ibid.* **IX** No 2407 (p 37)

[97] 1686-04-23: Huygens, Christiaan—Helder, Thomas; Instructie en onderwijs aangaande het gebruijk der Horologiën tot het vinden der Lengde van Oost en West, *Ibid.* **IX** No 2423 (pp 55–76)

[98] 1685-10-26: Huygens, Christiaan—Hudde, Johannes; *Ibid.* **IX** No 2407 (p 37)

[99] Huygens Instituut voor Nederlandse Geschiedenis, *The Dutch East India Company's shipping between the Netherlands and Asia 1595–1795: Details of voyage 1526.4 from Texel to Batavia* http://resources.huygens.knaw.nl/das/detailVoyage/92585 [accessed 10 October 2016]

[100] 1688-04-24: Huygens, Christiaan—Hudde, Johannes; *Ibid.* **IX** No 2517 (pp 267–268); App. II: No. 2519, Rapport aengaende de Lengdevindingh door mijne Horologien op de Reys van de Caep de B. Esperance tot Texel A°. 1687, *Ibid.* **IX** pp 272–291

[101] Huygens Instituut voor Nederlandse Geschiedenis, *op. cit.: Details of voyage 5820.5 from Masulipatnam* http://resources.huygens.knaw.nl/das/detailVoyage/96729 [accessed 10 October 2016]

[102] 1687-10-03: Huygens, Christiaan—de Graef, Abraham; *Ibid.* **IX** No 2488 (pp 222–3)

[103] Letter dated 18 April 1687; VOC Archief (The Hague: Algemeen Rijksarchief) 4023 folio 42ff

[104] 1688-04-24: Huygens, Christiaan—Hudde, Johannes; *Oeuvres Complètes de Christiaan Huygens* **IX** No 2517 (pp 267–8); App. II: No. 2519, Rapport aengaende de Lengdevindingh door mijne Horologien op de Reys van de Caep de B. Esperance tot Texel A°. 1687, *Oeuvres Complètes de Christiaan Huygens* **IX** pp 272–91

[105] Note that the discrepancy originally reported by Huygens was 25 minutes of arc; De Volder's corrections led to the adjusted value adopted here.

[106] Huygens C 1690 Addition *Ibid.* **XXI** pp 466–7

[107] *Gregory manuscripts*, 1627–1720, held at the Royal Society: GB 0117 MS/247

[108] 1689-09-03: Huygens, Christiaan—Huygens, Constantijn; *Oeuvres Complètes de Christiaan Huygens* **IX** No 2544 (pp 333–6) Note 1

[109] Schliesser E and Smith G E 2000 Huygens' 1688 Report to the Directors of the Dutch East India Company on the Measurement of Longitude at Sea and the Evidence it offered

against Universal Gravity *Archive for the History of the Exact Sciences* http://philsci-archive.pitt.edu/5510/, p 30 Note 69 [accessed 15 October 2016]

[110] Tachard G 1688 *A Relation of the Voyage to Siam Performed by Six Jesuits, sent by the French King, to the Indies and China, in the year, 1685. With their Astrological Observations, and their Remarks of Natural Philosophy, Geography, Hydrography, and History* (London: A. Churchil) 49–59

[111] Riccioli G B 1661 *Geographiae et Hydrographiae reformatae libri duodecim quorum argumentum sequens pagina explicabit* (Bologna) 378

[112] 1688-02-23: Vossius, Isaäc—van Beuningen, Coenraad; *Oeuvres Complètes de Christiaan Huygens* **IX** No 2518 (pp 269–72)

[113] 1688-04-24: Huygens, Christiaan—Hudde, Johannes; *Ibid.* **IX** No 2517 (pp 267–8)

[114] *Ibid.*; Appendix II: No. 2519, Rapport aengaende de Lengdevindingh door mijne Horologien op de Reys van de Caep de B. Esperance tot Texel A°. 1687 *Ibid.* **IX** 272–91

[115] 1688-04-30: Hudde, Johannes—Huygens, Christiaan; *Ibid.* **IX** No 2521 (p 294)

[116] 1689-07-22: de Volder, Burchard—Vereenigde Oostindische Compagnie; *Ibid.* **IX** No 2547 (pp 339–43)

[117] *Ibid.* 341 (last paragraph)

[118] Huygens C 1669 Instructions Concerning the Use of Pendulum-Watches, for Finding the Longitude at Sea *Phil. Trans. R. Soc.* **4** 937–53

[119] Schliesser E and Smith G E 2000 *op. cit.*

[120] Huygens C 1688 Rapport aengaende de Lengdevindingh door mijne Horologien op de Reys van de Caep de B. Esperance tot Texel A° 1687 *Oeuvres Complètes de Christiaan Huygens* **IX** 285

[121] *Ibid.*; elsewhere, he refers to the cartographer as 'Rembrandtz'.

[122] Schliesser E and Smith G E 2000 *op. cit.:* App. 1

[123] 1688-04-24: Huygens, Christiaan—Hudde, Johannes; *Oeuvres Complètes de Christiaan Huygens* **IX** No 2517 (p 268)

[124] Huygens C 1688 *op. cit.* 287

[125] Huygens C 1688 Résultats de quelques expéditions maritimes *Oeuvres Complètes de Christiaan Huygens* **XVIII** 640; figure **129**

[126] Schliesser E and Smith G E 2000 *op. cit.*: App. 1

[127] *Ibid.* 7–9

[128] 1687-05-01: Huygens, Christiaan—de la Hire, Philippe; *Oeuvres Complètes de Christiaan Huygens* **IX** No 2455 (pp 130–3)

[129] Schliesser E and Smith G E 2000 *op. cit.*

[130] Huygens C 1690 Discours de la cause de la pesanteur *Oeuvres Complètes de Christiaan Huygens* **XXI** 464; transl Schliesser E and Smith G E 2000 *op. cit.*

[131] 1686: Huygens, Christiaan—Helder, Thomas; *Oeuvres Complètes de Christiaan Huygens* **IX** No 2520 (pp 292–3)

[132] Schliesser E and Smith G E 2000 *op. cit.*

[133] Huygens C 1688 *op. cit* 275–76

[134] 1686: Huygens, Christiaan—Helder, Thomas; *Oeuvres Complètes de Christiaan Huygens* **IX** No 2520 (pp 292–3) Note 1

[135] Nationaal Archief, *VOC—opvarenden: Gegevens van Johannis de Graaff uit Amsterdam*, http://vocopvarenden.nationaalarchief.nl/detail.aspx?ID=1622417 [accessed 18 October 2016]

[136] *Ibid., Gegevens van Pieter van Laer uit Amsterdam,* http://vocopvarenden.nationaalarchief. nl/detail.aspx?ID=1569816 [accessed 18 October 2016]

[137] *Ibid., Gegevens van Jillis Meijbos uit Groeningen,* http://vocopvarenden.nationaalarchief.nl/ detail.aspx?ID=1617580 [accessed 18 October 2016]

[138] *Ibid., Gegevens van Evert Verbrugge uit Amsterdam,* http://vocopvarenden.nationaalarchief. nl/detail.aspx?ID=1633683 [accessed 18 October 2016]

[139] Huygens Instituut voor Nederlandse Geschiedenis, *op. cit.: Details of voyage 1607.1 from Texel to Ceylon,* http://resources.huygens.knaw.nl/das/detailVoyage/92667 [accessed 18 October 2016]

[140] Huygens Instituut voor Nederlandse Geschiedenis, *op. cit.: Details of voyage 502.3 from Ceylon to Veere,* http://resources.huygens.knaw.nl/das/detailVoyage/96811 [accessed 18 October 2016]

[141] 1690-12-21: de Graaff, Johannes—Huygens, Christiaan; *Oeuvres Complètes de Christiaan Huygens* **IX** No 2646 (p 578)

[142] 1690-12-23: de Graaff, Johannes—Huygens, Christiaan; *Ibid.* **IX** No 2647 (p 579)

[143] 1690-12-26: de Graaff, Johannes—Huygens, Christiaan; *Ibid.* **IX** No 2650 (p 582)

[144] 1690-12-29: de Graaff, Johannes—Huygens, Christiaan; *Ibid.* **IX** No 2653 (p 584)

[145] Huygens C 1693 Verklaeringh en aenmerckingen op het Journael van Jo. de Graef en 't geen ontrent de Horologien is voorgevallen in de laetste proeve der Lengdevindingh A° 1690, 1691 en 1692, *Ibid.* **XVIII** 643–51

[146] 1693-02-14: de Graaff, Johannes—Huygens, Christiaan; *Ibid.* **X** No 2789 (pp 396–8)

[147] Huygens C 1693 *op. cit*

[148] 1691-12-17: de Graaff, Abraham—Huygens, Christiaan; *Ibid.* **X** No 2718 (pp 205–6)

[149] 1691: de Graaff, Johannes; Meybos, Gilles; van Laer, Pieter—Vereenigde Oostindische Compagnie; *Ibid.* **X** No 2720 (pp 207–8)

[150] Huygens C 1693 *op. cit.*

[151] 1694-04-19: Huygens, Christiaan—de Volder, Burchard; *Ibid.* **X** No 2802, 2803 (pp 442–4)

[152] 1693-02-10: Huygens, Christiaan—de Graaff, Johannes; *Ibid.* **X** No 2786 (pp 389–90); see also Huygens C 1693 *op. cit.*

[153] Schliesser E and Smith G E 2000 *op. cit.:* Appendix 2; also Huygens C 1693 *op. cit.*

[154] 1692-11-19: de Graaff, Johannes—Huygens, Christiaan; *Ibid.* **X** No 2774 (p 341)

[155] 1693-04-06: de Volder, Burchard—Huygens, Christiaan; *Ibid.* **X** No 2800 (pp 435–6)

[156] 1693-04-19: Huygens, Christiaan—de Volder, Burchard; *Ibid.* **X** No 2803 (pp 443–4); App. to No 2802

[157] *Ibid.;* see also Schliesser E and Smith G E 2000 *op. cit.:* App. 2, footnote 128

[158] Schliesser E and Smith G E 2000 *op. cit.:* App. 2, footnote 128

[159] Huygens C 1693 *op. cit.*

[160] *Ibid.* 647

[161] *Ibid.* 649

[162] 1693-03-06: Huygens, Christiaan—Vereenigde Oostindische Compagnie; *Oeuvres Complètes de Christiaan Huygens* **X** No 2796 (pp 423–4)

[163] 1693-03-06: Huygens, Christiaan—van de Blocquery, Josias; *Ibid.* **X** No 2795 (pp 422–3)

[164] 1693-03-23: Huygens, Christiaan—de Volder, Burchard; *Ibid.* **X** No 2798 (pp 433–4)

[165] 1658-09-24: Huygens, Christiaan—Hodierna, Giovanni Batista; *Ibid.* **II** No 518 (pp 223–5)

[166] 1693-03-24: Huygens, Christiaan—de Volder, Burchard; *Ibid.* **X** No 2798 (pp 433–4)

[167] 1693-04-06: de Volder, Burchard—Huygens, Christiaan; *Ibid.* **X** No 2800 (pp 435–6)

[168] Schliesser E and Smith G E 2000 *op. cit.*: App. 2

[169] 1693-04-06: de Volder, Burchard—Huygens, Christiaan; *Oeuvres Complètes de Christiaan Huygens* **X** No 2800 (pp 435–6)

[170] 1693-04-19: Huygens, Christiaan—de Volder, Burchard; *Ibid.* **X** No 2803 (pp 443–4)

[171] 1693-03-24: Huygens, Christiaan—de Volder, Burchard; *Ibid.* **X** No 2798 (pp 433–4)

[172] Schliesser E and Smith G E 2000 *op. cit.*: App. 2, footnote 139

[173] 1693-04-19: Huygens, Christiaan—de Volder, Burchard; *Oeuvres Complètes de Christiaan Huygens* **X** No 2803 (pp 443–4)

[174] *Ibid.*

[175] Schliesser E and Smith G E 2000 *op. cit.*: App. 2

[176] 1693-03-06: Huygens, Christiaan—Vereenigde Oostindische Compagnie; *Oeuvres Complètes de Christiaan Huygens* **X** No 2796 (pp 423–4)

[177] 1693-04-19: Huygens, Christiaan—de Volder, Burchard; *Ibid.* **X** No 2803 (pp 443–4)

[178] Huygens C 1693 Application pratique aux horloges de différents mouvements vibratoires plus ou moins isochrones: La 'libratio isochrona melior praecedente' de mars 1693, *Ibid.* **XVIII** 569

[179] Huygens C 1683–1684 **III**. Premier projet de 1683 ou 1684 du 'Balancier marin parfait' de 1693 *Ibid.* **XVIII** 536–8

[180] 1693-09 Huygens, Christiaan—Editors of the *Acta Eruditorum*; *Ibid.* **X** No 2823 (pp 512–5)

[181] *Ibid.* Note 16

[182] Huygens C 1693 **V**. Le 'Balancier marin parfait' de janvier–fevrier 1693 *Ibid.* **XVIII** 546–61

[183] Huygens C 1659 Projet de 1659 d'une horloge à pendule conique *Ibid.* **XVII** 88–9

[184] Crommelin C A 1947 Les horloges de Christiaan Huygens *Journal Suisse d'Horlogerie et de Bijouterie* **LXXII** 189–204

[185] Huygens C 1693 *Oeuvres Complètes de Christiaan Huygens* **XVIII** 546–61

[186] *Machines et Inventions approuvées par l'Académie Royale des Sciences depuis son établissement jusqu'à présent; avec leur Description, par M. Gallon*, 1735 (Paris: G. Martin)

[187] Berthoud F 1802 *Histoire de la Mesure du Temps par les Horloges* pp 365ff

[188] Anonymous (Poniz Ph) 2013 *Recently discovered: Huygens' perfect marine balance longitude clock 'BMP2'*, article released as part of 'marketing' a clock formerly known as the *Sully Equation of Time Clock* at the *Rockford Time Museum*, www.uhren-muser.com/cn/documents/News_7-2013_US.pdf [accessed 3 November 2016]

[189] Bryson A 1859 Notes on clock and watch making; with descriptions of several antique timekeepers deposited in the Museum *Proc. Soc. Antiquaries Scotl.* **3** 430–6

[190] da Vinci L 1494 *Tratado de Estatica y Mechanica en Italiano (Codex Madrid)* **I**; National Library Madrid No. 8937

[191] Huygens C 1693 **VI**. La 'Libratio isochrona melior praecedente' de mars 1693 *Oeuvres Complètes de Christiaan Huygens* **XVIII** 562–70

[192] Huygens C 1685 Application pratique aux horloges de différents mouvements vibratoires plus ou moins isochrones *Oeuvres Complètes de Christiaan Huygens* **XVIII** 509 Note 5

[193] *Ibid.* 516 Note 4

[194] Crommelin C A 1947 *op. cit.*

[195] *Ibid.*

[196] Lloyd H A 1943 Some Notes on Very Early English Equation Clocks, and Joseph Williamson's claim to have made them all!, *Horol. J.*; cited by Piggott K 2011 *A Royal 'Haagseklok,' Appendix Five, Alexander Bruce's English and Dutch Longitude Sea-Clocks*

Rediscovered, www.antique-horology.org/piggott/rh/appendix5.pdf [accessed 3 November 2016]

[197] 1694-10-01: Huygens, Christiaan—de Roisey, Alexander Rolas; *Oeuvres Complètes de Christiaan Huygens* **X** No 2878 (pp 684–5)

[198] Anonymous (Poniz Ph) 2013 *op. cit.*

[199] 1695-03-04: Huygens, Christiaan—Huygens, Constantijn; *Oeuvres Complètes de Christiaan Huygens* **X** No 2891 (pp 708–10)

[200] *Ibid.*

[201] *Ibid.*

[202] 1695-03-14: Marquis de l'Hospital, Guillaume François Antoine—Huygens, Christiaan; *Ibid.* **X** No 2892 (pp 711–3)

[203] Crommelin C A 1947 *op. cit.*

Chapter 6

The merits of horology versus astronomy

Huygens' death in 1695 signified the end of an era in which the focus of solving the longitude problem was largely driven by mechanical improvements in clock design. It would take the better part of the next two decades before the focus returned to this specific pursuit. With Huygens no longer in the picture and Hooke disappointed by the perceived opposition he had encountered within the Royal Society, Isaac Newton entered centre stage, a position he would maintain for most of his life. Newton had rapidly become the most respected British scholar at the start of the 18th Century; he took over as President of the Royal Society in 1703. At the end of the previous chapter we already saw that Newton was rather vocal in his doubts that any manmade timepiece could deliver the definitive solution to the longitude problem. Instead, he advocated a return to astronomy as a viable means to reach closure.

6.1 The nature of gravity

The 1686–1687 VOC sea trial of Huygens' clocks was not merely a victory of the clockmaker's skills. Its ramifications were much more far-reaching; the voyage's scientific insights caused a seismic shift in our understanding of the nature of the Earth's gravity. As an unintended consequence, it pitted the brightest minds of the time, those of Huygens and Newton, against each other—and all this on the basis of one of the simplest measurements, the length of the seconds pendulum and its variation with latitude.

Huygens and Newton initially considered the nature of gravity independently. In 1669, Huygens presented his initial thoughts *On the Cause of Gravity* to the *Académie des Sciences* in Paris [1], in discussions with the mathematicians Gilles Personne de Roberval and Bernard Frénicle de Bessy (c. 1604–1674), a group which also included the astronomer Jacques Buot (Buhot; before 1623–1678). Observations obtained by De Roberval, the physicist Edme Mariotte (c. 1620–1684), and Du Hamel implied, according to the French scholars engaged in these discussions, that gravity involved attraction. In 1690, Huygens combined this old presentation from 1669 with a section he wrote in 1686–1687 and an *Addition* in response [2] to Newton's *Principia* into his *Discourse on the Cause of Gravity* [3], which he appended to his *Treatise on Light*.

6.1.1 Huygens and Newton at loggerheads

In the section of his *Discourse* which he claimed to have written before having read Newton's *Principia*, Huygens made the arguments that the Earth's shape is non-spherical and that the Earth's rotation is the *only* cause of variations in the planet's gravity with position. These suggestions caused an immediate conflict with Newton's gravitational theory, since the latter had argued that the Earth's gravity could be decomposed into a so-called 'inverse-square' celestial gravity component and a component referred to as 'universal gravity.' The inverse-square gravitational strength was thought to depend on the distance to the centre of a gravitational potential well (such as the centre of a celestial body), whereas universal gravity represents the gravitational force exerted by every particle of matter on every other particle. Whereas Huygens was prepared to accept the inverse-square gravitational component, he was wholly unconvinced that Newton had demonstrated the reality of universal gravity. Huygens' reluctance to accept Newton's theory was driven by his adherence to gravity as a 'mechanical philosophy,' with which universal gravity was incompatible. Huygens preferred to explain the gravitational force by means of mechanical processes such as pressure, rather than through 'action at a distance' in the way that Newton saw it.

Huygens' theory of gravity, which he developed between 1669 and 1690, was based on basic philosophical principles which had first been proposed in the 1630s by Descartes, who proceeded to publish about them in 1644 [4]. Descartes had suggested that empty space cannot exist and, thus, that space must be filled with a substance he referred to as 'æther.' Because of the high density of these æther particles, they cannot move freely; Descartes took this to imply that their motion must be circular, which in turn made him suggest that the æther would be structured as a collection of vortices, as illustrated in figure 6.1:

'*Now that we have, by this reasoning, removed any possible doubts about the motion of the Earth, let us assume that the matter of the heaven* [Descartes referred to each vortex as a 'heaven'], *in which the Planets are situated, unceasingly revolves, like a vortex having the Sun as its centre, and that those of its parts which are close to the Sun move more quickly than those further away; and that all the Planets ... always remain suspended among the same parts of this heavenly matter. For by that alone, and without any other devices, all their phenomena are very easily understood*' [5].

The centrifugal motion of the æther particles would result in a concentration of matter (mostly composed of 'fine' matter) towards the outer edges of the vortices, except for what Descartes called 'rough' matter. Owing to its greater moment of

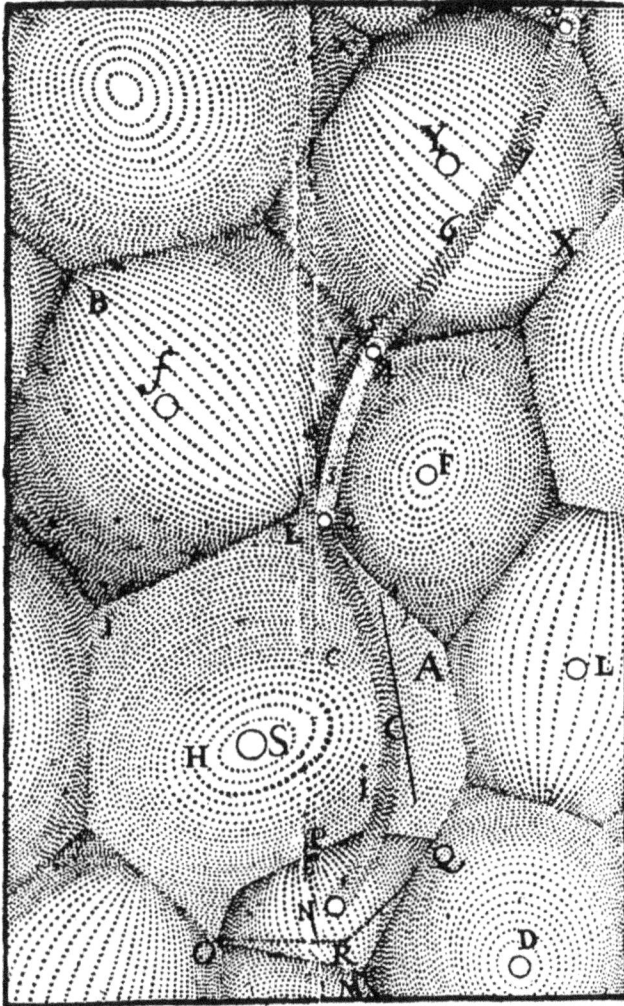

Figure 6.1. Æther vortices around stars (suns) and planets. (Renatus des-Cartes 1644 *Principia Philosphiæ* (Amsterdam: Ludovicus Elzevirius, 1644.)

inertia, rough matter would be forced towards the centres of the vortices, a type of inward pressure resembling the effects of gravity [6].

Huygens' model was a more exact, mathematically supported theory of gravity. Instead of assuming circular motion, he assumed that æther particles were moving randomly and isotropically, that is, statistically similarly in any direction, but that they were slowed down at the outer edges of the vortices. Once again, in this model the fine particles in the outer vortex regions would cause a net pressure force on the rough particles, driving the latter to the centres of the vortices. Huygens' calculations also showed him that the centrifugal motion was exactly counterbalanced by the centripetal force.

Newton's theory, on the other hand, did not rely on a mechanical explanation, but it is based on the idea of gravitational attraction, the 'inverse-square law,' combined with the idea that any mass in the Universe exerts an attractive force on any other mass. Specifically, in its modern form, Newton's theory of universal gravity states that the gravitational force between two bodies is proportional to the product of their masses, say m_1 and m_2, and inversely proportional to the square of the distance, r, between them (the inverse-square law):

$$F_1 = F_2 = G\frac{m_1 m_2}{r^2},$$

where $G = 6.67408 \times 10^{-11}$ m^3 kg^{-1} s^{-2} is Newton's gravitational constant, which is the same anywhere in the Universe. Although G appears in Newton's theory of universal gravity, it took until 1797–1798 for the British experimental and theoretical physicist Henry Cavendish (1731–1810) to measure its value based on the 'Cavendish experiment,' which used a torsion apparatus to yield the 'specific gravity' of the Earth (or, equivalently, the Earth's mass) [7] instead of G. In modern units, Cavendish determined $G = 6.754 \times 10^{-11}$ m^3 kg^{-1} s^{-2}.

Newton postulated that the gravitational force is an attractive force, which works instantaneously at any distance and has an infinite range. It could thus account for both the downward force described by Galileo, which keeps us on Earth, and the orbital motion of the planets in our solar system discussed by Kepler.

Newton's breakthrough, popularized by the well-known story about a falling apple, was initially committed to paper by his friend and biographer William Stukeley (1687–1756): see figure 6.2. Stukeley recalled a meeting in the shade of some apple trees in Newton's garden, when the scientist started to contemplate the effects of gravity on Earth:

> 'After dinner, the weather being warm, we went into the garden and drank thea, under the shade of some apple trees… he told me, he was just in the same situation, as when formerly, the notion of gravitation came into his mind. It was occasion'd by the fall of an apple, as he sat in contemplative mood. Why should that apple always descend perpendicularly to the ground, thought he to himself…' [8]

6.1.2 Geographic dependence of the length of the seconds pendulum

Contrary to Huygens, Newton argued that a rotating, oblate Earth could not generate enough gravity to explain the observed shortening of the seconds pendulum at the Equator. The accuracy of empirical measurements of the length of the seconds pendulum thus became the crux of the disagreement between the most important natural philosophers of the late 17th Century.

In theory, Huygens and Newton agreed, the Earth's rotation, combined with its slightly flattened shape, reduced the net gravitational component at the Equator compared with those at the Poles. As a consequence, the length of the seconds pendulum should become shorter as one travels towards lower latitudes. However,

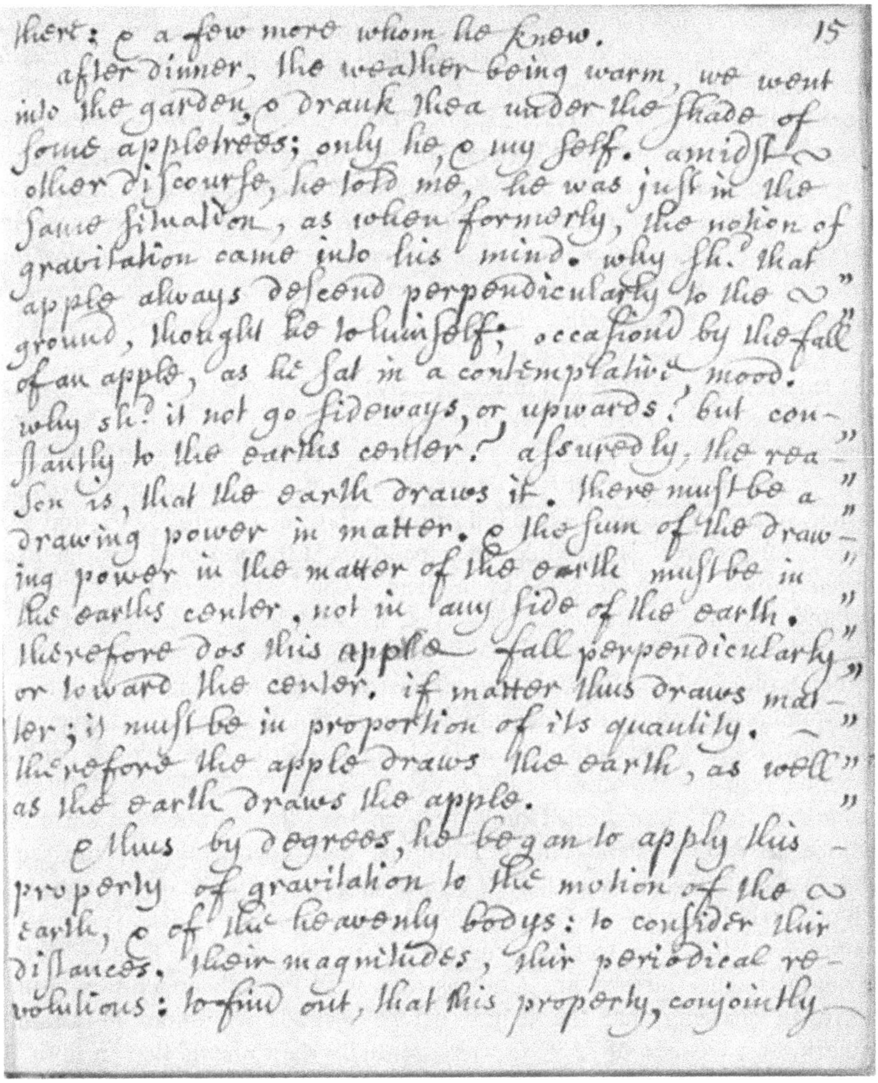

Figure 6.2. Excerpt of William Stukeley's 1752 biography of Newton, where the apocryphal story of the famous falling apple first came to life. Note that the word 'apple' replaces an earlier instance of the word 'earth.' (© The Royal Society.)

as we saw in chapter 4.1.4, existing measurements of the geographic dependence of the lengths of seconds pendulums prior to the 1686–1687 VOC sea trial of Huygens' clocks were inconsistent. In fact, their apparently random nature baffled contemporary scientists: in May 1687, prior to the completion of the VOC expedition, Huygens expressed his confusion in a letter to Philippe de la Hire:

'*Among my papers, you will find a short essay on the cause of gravity, to which I would like to add some reflections on what Mr. Richer and others have observed, in relation to the different pendulum lengths in different climates, but having seen that Mr. Picart, on the Isle of Tycho Brahe* [Uraniborg, Denmark], *claims to have found the same length as at Cape Cet[t]e* [Sète, on the French Mediterranean coast] *and Paris, and that the observation of Mr. Varin, which is referred to in Mr. Mariotte's treatise on the motion of water and other fluids, is not proportional to that of Mr. Richer, I do not know what to believe about this phenomenon. That is why I ask you to advise me as soon as possible whether you have any other information that convinces you that there is indeed this variation in nature, which seems to me very likely, although I can also give reasons if it is not there'* [9].

The length of the seconds pendulum, l_{sec}, in Paris was well-established. Huygens derived a length of 3 Paris feet 8½ *lignes* (l_{sec} = 3.059 Paris feet; 1 Paris foot = 32.48 cm) for his seconds pendulum [10], corresponding to a one-second fall length of ½ $\pi^2 l_{sec}$ = 15.096 Paris feet [11]. Huygens, Richer, and Picard measured highly consistent values for the one-second fall length in Paris of 15.096, 15.099, and 15.096 Paris feet, respectively [12], while an expedition [13] sponsored by the French *Académie Royale des Sciences*, led by 'le Sieur Varin,' a mathematics teacher from Paris, and which also included Deshayes, as well as the mathematician and hydrography teacher Guillaume de Glos (who joined the other two scientists at a later stage), obtained 15.098 Paris feet, all measures that were internally consistent. Away from Paris, the available measurements were not as clear-cut, however. As we already saw, Richer's measurements in 1672 of the length of the seconds pendulum in Cayenne, near the Equator, implied a shortening by 1¼ *lignes* and a one-second fall length of 15.055 Paris feet. Confusingly, in 1682 the team led by Varin on an expedition to Gorée Island and Guadeloupe concluded [14] that the seconds pendulum had to be shortened by 2 *lignes*, corresponding to a one-second fall length of 15.029 Paris feet, on Gorée Island, 10° further north than Cayenne, while a correction to 15.027 Paris feet was required in Guadeloupe, more than 16° north of the Equator in the Caribbean… Contrast this with Picard's conclusion from the early 1670s, showing that the length of the seconds pendulum in Uraniborg (Denmark, at a latitude of 55.9°N) was essentially the same as that in both Paris and Cape Sète (43.4°N), corresponding to a one-second fall length of 15.096 Paris feet. We already discussed the discrepant results obtained by Mouton in Lyon, which suggested an inexplicably short fall length of 15.02 Paris feet.

Meanwhile, Huygens had estimated theoretically that the centrifugal effect at the Equator owing to the Earth's rotation should be $^1/_{289}$th of the gravitational force, leading to a quantitative prediction [15] of the shortening of the seconds pendulum which could now be tested based on the expedition's measurements. I note that Huygens consistently performed his calculations of the effects of the Earth's rotation while implicitly assuming a spherical shape. In fact, he concluded that—assuming that in the absence of rotation the surface gravity is uniform anywhere on the planet —the rotation effect implied by the measurements from the VOC expedition indicates that the Earth must have an oblateness of $^1/_{578}$, compared with a value of $^3/_{689}$ proposed by Newton in the first edition of his *Principia* (1684–1687).

The oblateness implied by Huygens' calculations of November/December 1687 hardly mattered for the rotational effect of gravity: instead of $^1/_{289}$th of the gravitational force, the effect would need to be adjusted to $^1/_{288.5}$th [16].

Schliesser and Smith [17] assessed the implications of Huygens' calculations for the credibility of the pendulum-length measurements available to him; they concluded that the effects of rotation alone would imply a shortening of the pendulum length at Cayenne and at Gorée that was $^4/_{10}$ and 1¼ of a *ligne* less than the lengths found by Richer's and Varin's expeditions, respectively. Moreover, they pointed out that Picard should have been able to determine that the seconds pendulum had to be longer by ⅓ of a *ligne* in Uraniborg than at Cape Sète, but he clearly did not. Given Huygens' significant displeasure with the apparent carelessness with which some of the available measurements had been obtained, particularly those reported by Richer (see chapter 4.1.4), he was not convinced that the Earth's rotation would have a measurable effect:

'... *because other observations made in various regions regarding this unequal length of pendulums did not turn out that well according to expectations, I previously had the thought that maybe this effect of the rotation of the Earth was nullified by some other natural cause, or it was rendered irregular*' [18].

6.1.3 The 1686–1687 VOC expedition

Given these inconsistencies, a systematic exploration of the length of the seconds pendulum and its variation with latitude was urgently needed, specifically using the same experimental set-up. Huygens was clearly in an excellent position to pursue such an analysis based on the measurements obtained on his behalf during the VOC's return expedition from the Cape of Good Hope to its Texel anchorage. Indeed, this appears to have been one of the main motivations to undertake the 1686–1687 VOC-sponsored expedition.

De Graaff's frequent measurements using Huygens' clocks on the homebound voyage in 1687 allowed the latter to carefully calculate the difference in longitude between both reference locations. As we already saw in chapter 5.2.3, Huygens additionally combined Tachard's 1685 longitude determination of the Cape with respect to that of Paris with the corresponding difference between Paris and Texel from Riccioli's *Geographiae* to conclude that the longitude determinations enabled by his clocks' measurements showed that the *only* variation in the gravitational force with geographic position on Earth was owing to the Earth's rotation [19]: the cumulative correction (loss) to his clocks' measurements over the course of the 117-day voyage from the Cape to Texel amounted to a mere 68 seconds (but see below). Note, however, that Huygens' available evidence that the Earth's rotation was the only effect that he needed to correct his clocks for was, strictly speaking, only valid for the comparison between the longitudes of the Cape and Texel [20], and not for the comparison of the *Alcmaer*'s navigators' course based on dead reckoning and that implied by his clocks.

Turning now to Newton's proposed inverse-square law of gravity, Huygens also considered the variation in the gravitational force implied if the inverse-square law applied both to astronomical objects as well as at and below the surface of the Earth. He concluded that the Earth's radius at the Poles would be $^1/_{578}$ shorter than that at the Equator, resulting in a shortening of the seconds pendulum by a factor of $^1/_{289}$:

> 'This is nearly the same difference that is produced for daily motion or centrifugal force. Thus a clock with the same length pendulum would run slower at the Equator than at the Pole by twice what it would be slowed by the motion of the Earth; and so this daily difference at the Equator would be about 5 minutes. And at the other parallels it would everywhere be more than twice what it was previously. But I strongly doubt that experience confirms this large of a variation, since I have observed, in the voyage that I mentioned that the first equation alone suffices, and it [this great variation] would give more than twice too great a difference around the middle of the course between the route of the vessel calculated with the pendulum and the route estimated by the Mariners. ... And, to provide the reason why the second variation would not occur, I say that it would be strange if the gravity near the surface of the Earth did not follow in precisely the same way as in the higher regions' [21].

Newton's theory of universal gravity—including the effects of variations in the gravity owing to the Earth's slightly flattened shape—predicted a somewhat greater shortening of the length of the second pendulum than that implied by Huygens' calculations based on the Earth's rotation alone. Again, Huygens' best evidence against Newton's theory was based on the precisely determined longitude difference between the Cape of Good Hope and the VOC's anchorage at Texel, which strongly implied to him that the Earth's rotation alone was responsible for the variation in the length of the seconds pendulum with latitude. He thus rejected Newton's claim of universal gravity based on the empirical evidence available to him [22].

Newton had implied, in his *Principia*, that the only clear distinction between inverse-square and universal gravity could be provided by measurements of the variations in the Earth's surface gravity with latitude—and this was indeed apparently borne out by Huygens' experimental results. This implication was also realized by Leibniz and expressed in a letter to Antonio Schinella Conti (1677–1749)—the Italian philosopher and physicist who acted as intermediary in the Leibniz–Newton calculus controversy which fully erupted in 1711—in November or December 1715:

> 'I am strongly in favour of the experimental philosophy but Mr. Newton is departing very far from it when he claims that all matter is heavy (or that every part of matter attracts every other part) which is certainly not proven by experiments, as Mr. Huygens has already asserted; gravitating matter could not have itself that weight of which it is the cause and Mr. Newton adduces no experiment or sufficient reason for the existence of a vacuum or of atoms or for the general mutual attraction' [23].

Huygens himself did not publish the strong conclusions I just summarized (and which have been highlighted in much more quantitative detail by Schliesser and Smith). Nevertheless, in his letter to Hudde which accompanies his report on the 1686–1687 VOC expedition, the implications are clear:

'*As regards the purported effect of the Earth's rotation, you may have seen what has recently been suggested by Professor Newton in his treatise* Philosophiæ Naturalis Principia Mathematica, *positing several hypotheses that I cannot approve of and that lead to different conclusions than my reckonings give, ...*' [24].

because, as he stated in the introductory text of the *Addition* to his *Discourse*,

'*... as for the different lengths of the pendulums in different regions, which he* [Newton] *has also addressed, I believe to have, by the average of these clocks, a clear confirmation not only of the effect of the motion of the Earth, but also of the measures of these lengths, which agrees very well with the calculation I have just given*' [25].

Of course, despite all of Huygens' elaborate objections, with the benefit of hindsight and our modern understanding of physics, we now know that Newton's theory was correct after all, and Huygens' alternative theory of gravity has faded into the depths of history. The correction required to make a pendulum clock run consistently while sailing from the Poles to the Equator is approximately 1½ times greater than the effects caused by the Earth's rotation alone. Schliesser and Smith's analysis of the records from the 1687 return expedition provide a plausible explanation as to why Huygens was unable to reconcile his calculations with Newton's [26].

As it turns out, Huygens' clock ran approximately 5¼ minutes behind upon the *Alcmaer*'s return to the Texel anchorage, which would have translated into a geographic position to the East of its actual location. At the same time, the cumulative (negative) correction to the clock's measurement that day owing to the Earth's rotation, 11 minutes and 34 seconds, was 5 minutes and 34 minutes smaller than it should have been, thus resulting in a smaller East–West correction than warranted and almost offsetting the effect caused by the clock's slow running.

Whereas the VOC expedition had provided enormously valuable measurements which would form the basis of future efforts of longitude determination at sea, as well as novel (although incorrect) scientific insights into the nature of the Earth's gravity, this single expedition was just the starting point. The prevailing uncertainties, both systematic and statistical, were simply too significant to reach firm conclusions. Better calibration of both sea-going clocks and geographic locations was urgently needed; intermediate calibration points should be included on any future voyages—yet none of these recommendations were taken into account by the relevant authorities...

6.2 Newton's early contributions to resolving the longitude problem

Although he had never been to sea, Newton was well versed with methods of longitude determination at sea. He had worked on using the Moon's path across the sky for this purpose while writing the first edition of his *Principia*, in an attempt to combine the best aspects of astronomy and accurate timekeeping. Newton understood that the success of the lunar distance method for longitude determination stood or fell with one's accurate understanding of the complex lunar motions. The Moon's orbit spans an arc of approximately its own width every hour, so that large numbers of careful observations are required to accurately trace its movement across the sky. However, despite the significant efforts he undertook in preparation for the three editions of his *Principia*, he did not manage to obtain a satisfactory solution to the problem.

6.2.1 John Flamsteed, first Astronomer Royal

Let us first consider Newton's efforts in a contemporary context. On 4 March 1675, King Charles II had signed a royal warrant to appoint John Flamsteed as *'our astronomical observer,'* with an annual salary of one hundred pounds (taxed at ten percent), in addition to a house to live in. He was also required to provide an education to two boys a month at the Royal Mathematical School at Christ's Hospital in London as part of his duties. In 1673, Sir Jonas Moore, the English mathematician and surveyor, and the famous diary writer Samuel Pepys had jointly founded the Royal Mathematical School. It was established specifically to train 40 boys annually in navigation techniques, so that they could serve the King at sea, in either the Royal Navy or the English merchant navy. Moore, the driving force behind the subsequent establishment of the Royal Observatory at Greenwich, was tasked with communicating Charles' II instructions to the *Astronomer Royal*, namely, ...

> *'forthwith to apply himself with the most exact Care and Diligence to the rectifying* [correction of] *the Tables of the Motions of the Heavens, and the places of the fixed Stars so as to find out the so-much desired longitude of places, for the perfecting the art of navigation'* [27].

It is unclear whether the title *Astronomer Royal* was formally bestowed upon Flamsteed, or if he simply assumed it. Following his appointment by the King, he usually added the letters *M. R., Mathematicus Regius*, after his name [28]. Flamsteed was instrumental in modernizing the state of positional astronomy. Until his appointment, positional astronomy relied on the tried-and-tested yet sparsely covered catalogue compiled by Tycho Brahe (1546–1601) or incarnations of Johannes Kepler's *Rudolphine Tables*, which had first been published in 1627 and are named after the Holy Roman Emperor Rudolph II (1552–1612). Brahe's catalogue only contained some 1000 stellar positions. These had been observed without the aid of a telescope, since the instrument had not yet been invented during the Danish astronomer's life. Flamsteed had used his own observations to show that Brahe's catalogue was prone to errors.

Flamsteed, a self-taught astronomer from the north of England, had attracted the attention of a number of Fellows of the Royal Society in 1669–1670, most importantly of its President and Secretary, Brouncker and Oldenburg. His achievements in astronomy and mathematics also caught the attention of John Collins (1625–1683), the mathematician, with whom Flamsteed soon afterwards started a cordial exchange of letters. He visited London in 1670, where he struck up a friendship with Moore.

Flamsteed's contact with Moore served him well: in 1674 the mathematician proposed to set him up at a private observatory, which he intended to finance and establish at Chelsea College under the auspices of the Royal Society. However, events soon overtook their intentions. That same year, a French explorer calling himself Le Sieur de Saint-Pierre, most likely Jean-Paul Le Gardeur (1661–1738), claimed to have solved the longitude problem using astronomical data [29], specifically based on application of the lunar distance method.

Meanwhile, the King received increasing numbers of proposals from a wide variety of scholars and opportunists claiming to have solved the longitude problem. Upon having received proposals from Le Sieur de Saint-Pierre and from a certain 'old' Henry Bond (c. 1600–1678), author of *The Longitude Found* (1676), mathematician, and teacher of navigation from London, on 15 December 1674 Charles II issued a warrant appointing Brouncker, Pell (chair), Hooke, as well as Wren, Seth Ward, the Bishop of Salisbury and Savilian Professor of Astronomy at Oxford, Sir Charles Scarborough, the physician and mathematician, Sir Samuel Morland (1625–1695), mathematician, inventor and Groom of the Bedchamber to Charles II, and the politician, Colonel Silius Titus (1623–1704), Groom of the Bedchamber, to a Royal 'Longitude Commission' to scrutinize these proposals:

'... *that he* [St. Pierre] *hath found out the true knowledge of the Longitude, and desires to be put on Tryall thereof; Wee having taken the Same into Our consideration, and being willing to give all fitting encouragement to an Undertaking soe beneficiall to the Publick ... hereby doe constitute and appoint you* [the Commissioners], *or any four of you, to meet together ...*

And You are to call to your assistance such Persons, as You shall think fit: And Our pleasure is that when you have had sufficient Tryalls of his Skill in this matter of finding out the true Longitude from such observations, as You shall have made and given him, that you make Report thereof together with your opinions there-upon, how farre it may be Practicable and usefull to the Publick' [30].

In retrospect, Bond's 1674 proposal based on magnetic declination [31] resulted in the only Royal bounty ever awarded, [32] consisting of a life annuity of £50. Hooke recorded his disbelief in Bond's magnetic theory (see also chapter 6.3.1) in his *Diary*, although it apparently provided him with a popular discussion item for years to come:

'Met at Colonel Titus's in Hatton Garden. With Lord Sarum. Dr Pell examined Bond's theory. Found it ignorant and groundless and fals but resolved to speak favorably of it. He supposed the magnetic poles of the earth to be somewhere in the air and to be left behind by the motion of the earth...' [33].

Moore appears to have played merely a supporting role in these efforts. However, he arranged that Flamsteed was appointed as assistant to the Royal Commission in February 1675. Flamsteed subsequently undertook an observational programme to test the viability of Saint-Pierre's proposal. Although he acknowledged the method's potential, he highlighted its practical infeasibility given the lack of detailed tables of the positions of the so-called 'fixed stars' relative to the Sun's annual path and the Moon's orbital parameters, despite Newton's attempts to describe the Moon's orbit based on his new theory of gravity. Flamsteed stated that he would need years of observational data, obtained with large instruments fitted with telescopic sights, to make the lunar distance method a practical reality.

6.2.2 Establishment of a Royal Observatory at Greenwich

Through Moore's intervention, Flamsteed found himself visiting the King, whom he impressed by emphasizing the competing efforts by Picard at the Paris Observatory, which focussed in particular on addressing the longitude problem. He pointed out that his French counterparts had established a Royal Observatory for their work, in 1667, prompting the King to decree that England should have a Royal Observatory as well, specifically to pursue positional astronomy. The competition in France instead focussed on cataloguing the ephemerides of Jupiter's satellites as a potentially viable method of longitude determination. In fact, Flamsteed later pointed out that in 1674, ...

'[a]n accident happened that hastened, if it did not occasion, the building of the [Royal] Observatory. A Frenchman, that called himself Le Sieur de St. Pierre, having some small skill in astronomy, and made [sic] *an interest with a French lady,* [34] *then in Favour at the Court, proposed no less than the discovery of Longitude'* [35].

At the suggestion of Wren, the Royal Observatory's architect, the King issued an additional warrant on 22 June 1675, decreeing that the Observatory was to be established at Greenwich.

Until the formal inauguration of the Royal Observatory on 10 July 1676, Flamsteed performed his astronomical observations from the Tower of Moore's residence in London, which formed the basis—jointly with his earlier observations from Derby in northern England—for the first edition of his *Historia Coelestis Britannica*. He continued his efforts in positional astronomy from the Queen's House in Greenwich Park until he could move into the newly established Observatory.

One of Flamsteed's first tasks as Astronomer Royal was to prove that the Earth rotates on its axis. Although this had been assumed by Copernicus when he first

Figure 6.3. The Octagon Room at the Royal Observatory in the 1670s. The Tompion clocks can be seen in the centre; they are labelled A and B. Note the oval-shaped bobs seen above the clock faces. Engraving: Francis Place, 1676. (© National Maritime Museum, Greenwich, London.)

proposed his new model of the solar system, it had never been proven. Nevertheless, this same assumption was inherent to all methods proposed for the determination of the longitude.

Beyond the title, his salary, and a place to live, Flamsteed was only furnished with an iron sextant. His friend Moore provided him with two pendulum clocks made by Tompion—but which had not been taken care of very well [36]—with very long pendulums of 13 feet for improved accuracy. They were suspended *above* the clocks' movements (see figure 6.3) and beat once every two seconds. In addition, Tompion had designed the clocks such that they could go a full year on one winding.

Flamsteed supplemented this rudimentary equipment with his own quadrant and two telescopes he had brought along to Greenwich himself. He borrowed a second quadrant from the Royal Society on 25 January 1676, which he used until 26 September 1679, '*when the ill nature of Mr. Hooke forced it out of my hands*' [37]. He added,

> '*Yet I lost nothing by it; for it was so ill contrived by him that I could not make it perform better than my first. And now he obliged me to think of fitting up one of my own, of 50 inches radius, wherein by peculiar contrivances I had avoided all the inconveniences I had met with in his. This gives an observed height to half a minute: and now, by it, I am sure of the observed times by three seconds; which I could not have expected from either of my other instruments.*'

Judging from Flamsteed's description of Hooke's alleged efforts to construct a quadrant for the former's use, it appears that neither man held the other in high regard [38]: in 1676 Hooke had, in fact, provided Flamsteed with a 10-foot mural quadrant, but the instrument proved unworkable.

During the year it took him to complete the Earth's rotation experiment, Flamsteed used Sirius, the bright Dog Star, as his time reference. He needed an independent time keeper to transform the sidereal time obtained from successive transits to the mean solar time maintained by his clocks, with the day-to-day difference of the transit time of 3 minutes and 56 seconds owing to the Earth's orbital motion around the Sun. In a letter of 12 July 1677 to his regular correspondent, the mathematician and astronomer Richard Towneley, he wrote, …

'our clocks kept so good a correspondence with the Heavens that I doubt it not but they would prove the revolutions of the Earth to be isochronical, which if guaranteed it will follow that the Equation of Days, which I have demonstrated in the Diatribe is altogether agreeable to the Heavens. I can now make it out by three months' continuous observations, though to prove it fully it is requisite the clocks be permitted to go a whole year without any alterations' [39].

Indeed, by March 1678 he had proven the isochronicity of the Earth's rotation, his 'Equation of Natural Days:'

'My theory of the Equation of Days I looked upon but as a dream at first because one part on which it was founded, viz. the isochronicity of the Earth's revolutions, was only supposed, not demonstrated by me; but the clocks have proved that rational conjecture a very truth' [40].

6.2.3 Equipment woes and flaring tensions

A lack of access to suitable equipment continued to plague Flamsteed's performance; repeated requests to the Government for financial assistance were met with promises, which however never materialized. Hence, Flamsteed constructed his own measuring devices, initially including only a sextant, which he later supplemented with a mural arc at his own expense. Wall-mounted in the plane of the local meridian, the instrument consisted of a graduated arc fitted with a telescopic sight. Although the arc was completed by 1681, he did not erect it until 1683, fearing that it was too weak ('slight'). He turned out to be right, so that he was forced to continue his work on the basis of only his sextant.

Whereas Flamsteed obtained numerous relative celestial positions for stars visible from the Observatory, his absolute calibration was still limited by the accuracy of Brahe's catalogue. Nevertheless, Flamsteed compiled an approximate catalogue using his first mural quadrant to obtain the altitudes at the meridian of a few of the principal stars, while employing the sextant he had been furnished with to determine their mutual separations on the sky. Despite these difficulties, his scientific contemporaries kept pressuring him to publish his tables, asking him '*why he did not print his observations*'. Flamsteed lamented that he could not '*make bricks without straw,*'

'*Some people, to make me uneasy, others out of sincere desire to see the happy progress of my studies, not understanding amid what hard circumstances I lived, called hard upon me to print my observations. I had often answered that I had not any instrument for taking the meridional distances of the stars from the vertex, which such exactness as I could their intermutual distances from each other; that I wanted these to connect them; and that without them my labours would appear lame and imperfect*' [41].

In 1688 he acquired a small amount of additional funds from his father's inheritance, some of which he invested in the construction of a new, much stronger mural arc at the Observatory (see figure 6.4). Although he had been assured by George Legge, First Baron Dartmouth (c. 1647–1691), Master General of the Ordnance since 1682, that his expenses for the new mural arc would be reimbursed, to his significant disappointment and chagrin reimbursements never materialized. As such, he was forced to continue to teach private pupils to supplement his limited income. With the instrument's inauguration on 11 September 1689, Flamsteed's observations became much more precise, eventually leading to the extensive and celebrated *British Catalogue*.

His first call following verification of the arc's position was to determine the equinox, that is, the date on which day and night are approximately equally long, the Observatory's latitude, the obliquity of the ecliptic plane—that is, the inclination of Earth's equatorial plane with respect to its orbital plane—other fundamental measurements required to correctly determine the positions of the fixed stars, and

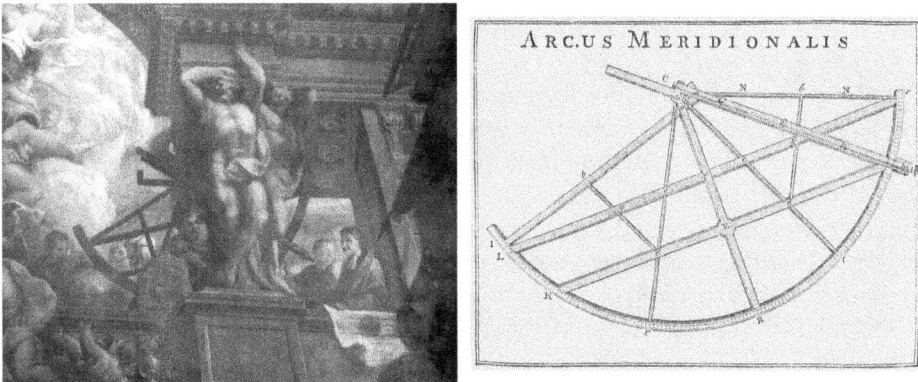

Figure 6.4. (*Left*) Flamsteed (on the right) with Thomas Weston, merchant adventurer, at the eyepiece of the Mural Arc. Detail from the ceiling of the Painted Hall in the Old Royal Naval College, Greenwich, by Sir James Thornhill. Commissioned in 1707 the painting of the ceiling was completed in 1712 (© Graham Dolan; reproduced with permission); (*right*) Flamsteed's Mural Arc. Reproduced from William Cudworth's *Life and Correspondence of Abraham Sharp* (1889).

true solar, lunar, and planetary motions. Newton used the latter as the basis of his lunar distance method of longitude determination, but it took a number of years before Flamsteed was prepared to make his results public.

Flamsteed's mural arc, used alongside an accurate pendulum clock (shown in figure 6.3), allowed him to measure zenith distances of objects crossing the local meridian, as well as the sidereal time when this happened. In turn, these measurements enabled him to calculate the objects' right ascensions and declinations, that is, their accurate positions in the sky. Flamsteed's observations were eventually published post-humously in the second volume of his *Historia Coelestis Britannica* (1725), edited by his widow Margaret. During his lifetime, he had put off publishing his observations until completed, despite being pressured to do so by Newton and the English astronomer and polymath Edmond Halley (1656–1742), Flamsteed's successor as Astronomer Royal, among others. Around 1683, Halley actually contributed observations of occasional lunar occultations of some bright stars, as well as their 'appulses' (close encounters on the sky). He had hoped to use these measurements as accurate timing devices, although the method turned out to be too cumbersome for practical application. Newton was particularly keen to see Flamsteed's observations in print. A mere two years after the mural arc had been inaugurated, Newton impressed upon Flamsteed the urgent requirement for accurate celestial positions. In a letter dated 10 August 1691, Newton wrote,

'*I hope it will not be long before you publish your catalogue of the fixed stars. In my opinion, it will be better to publish those of the first six magnitude observed by others, and afterwards, by way of an appendix, to publish the new ones observed by yourself alone, than to let the former stay too long for the latter*' [42].

However, Flamsteed was not to be persuaded. In his response of 24 February 1692, he explained the rather significant progress he had made since having installed his new meridian arc and asked, almost indignantly,

'*Tell me now sincerely (for I know you will do it) if you think it would be prudently done of me to leave off where I am, whilst I have strength and vigour (God be praised) to prosecute them? Would it, I say, be wisely done of me, to cease my designated observations of the constellations that yet remain to be taken or completed, to transcribe what I have done for the press, and to attend to it for 12 months, to gain a little present reputation? Would not even those men, who ask so peevishly why I do not print them? Would not they tell me I might have staid another year or two, for all their idle talk, and given them the whole complete?*' [43].

In fact, as Francis Bailey already pointed out, a draft version of this letter showed that this latter passage had originally been phrased much more assertively, showing—by reading between the lines—that he was embroiled in an ongoing disagreement, in particular with Halley:

'Tell me now sincerely whether you think it would be discreetly done of me, to leave the prosecution of my work, which has thriven thus successfully under my hands in 27 months, to satisfy the clamours of unreasonable and malicious people, or not? I make you judge, with Sir Christopher Wren and Mr. Caswell, whether I ought to publish some part of my observations before the rest; or only those stars in Tycho's catalogues, and I leave others, more remarkable than they, for an additional work. Tell me, if you please, whether it would be prudently done of me to leave my remaining constellations for a year, to copy papers and put them in order for the press, to gratify one only calumniating libertine, and stop his mouth, and those he has filled with his trifling and envious suggestions?' [44]

The language in this letter becomes more forcefully indignant of the way in which a certain 'friend' of Newton and Flamsteed (that is, Halley) has been treating the latter. Between October 1694 and September 1695, both men engaged in an extensive exchange of letters about Flamsteed's observations in which they frequently discussed lunar and planetary theories—although I note that Newton's style was not always very courteous [45,46]—on the mutual understanding that their correspondence was for their eyes only. Newton was indeed very keen to acquire Flamsteed's observations, with the specific aim to publish a second edition of his *Principia*, to which he planned to add more details of the theory of the motion of the Moon and the planets. In his preface to the collected correspondence between Newton and the young mathematician Roger Cotes (1682–1716), who had recently been appointed as the first Plumian Professor of Astronomy and Experimental Philosophy, Edleston phrased Newton's frustrations as follows:

'If Flamsteed, the Astronomer-Royal, had cordially co-operated with him in the humble capacity of an observer in the way that Newton pointed out and requested of him... the lunar theory would, if its creator did not overrate his own powers, have been completely investigated, so far as he could do it, in the first few months of 1695, and a second edition of the Principia *would probably have followed the execution of the task at no long interval'* [47].

In the spring of 1696, Newton was appointed to the post of Warden of the Mint and moved from Cambridge to London, where he continued his scientific discussions with Flamsteed. Rumours of a forthcoming new edition of the *Principia* surfaced occasionally. For instance, in February 1700 Leibniz wrote to the Scottish philosopher Thomas Burnett of Kemney (1656–1729) that ...

'I have learnt (I forget where) that he [Newton] *will give further details on the movements of the Moon: and I've also been told that there will be a new edition of his* Principia' [48].

Apparently, however, the discussions between Flamsteed and Newton turned sour fairly soon after Newton's move to London, and the situation somehow became explosive by late 1698 [49,50]. Interestingly, the tenseness between both men developed over a very short period, given that Newton had visited Flamsteed at the Royal Observatory late on the evening of 4 December 1698 [51] to acquire an additional dozen of computed lunar positions (which, in retrospect, contained a number of errors which were later corrected by Flamsteed [52]).

By this time, Flamsteed was swayed to publish the results of his many years of careful observations. In November 1704, the *Referees*, a committee of the Royal Society—composed of Newton in his role as President of the Royal Society, Wren, the polymath John Arbuthnott (1667–1735), the mathematician David Gregory, and the composer and scholar Francis Roberts (Robartes; c. 1649–1718), who subsequently became a Member of Parliament for Bodmin (Cornwall)—recommended that the results of Flamsteed's labours of the last three decades be published. The printing costs of Flamsteed's *Observations* and the *Catalogue* were borne by Prince George of Denmark, a patron of science himself.

The first volume of Flamsteed's work contained only his observations made with the sextant. The *Referees* threw many obstacles onto Flamsteed's path to publication of the second volume, which was meant to include his observations with the mural arc. Meanwhile, Prince George died on 28 October 1708, which put a stop to any plans to get the observations published, since Flamsteed's main funding source was no longer available. He was relieved, it seems, that Newton would hence no longer pressure him to publish his results, so that he could continue to perfect his catalogue:

'… and, since Sir Isaac Newton has put a full stop to the press, [I] shall not urge it forward again till I see a good fund settled, and secured, for carrying it on without any danger of impediment, or obstruction from him or any of his tools' [53].

However, developments unexpectedly got ahead of him: while Flamsteed was working on the completion of his observational data, he was privately informed that his catalogue was actually in press. Indeed, Flamsteed had provided Newton with a sealed copy of his catalogue, which was meant as a deposit rather than as the basis of a printed publication! Moreover, his annoyance was increased manifold when he received a letter dated 14 March 1711 from Arbuthnott, apparently on behalf of the Royal Society (that is, pushed forward by Newton and Halley), demanding that Flamsteed hand over those parts of his catalogue which were as yet lacking and that he (Arbuthnott) had been instructed by Queen Anne (1665–1714; reigned from 1702) to supervise the publication of the *Historia Coelestis* and ensure its completion [54].

Meanwhile, Halley had been put in charge of printing Flamsteed's manuscript, but he took it upon himself to make changes and 'corrections' to the *Catalogue*—apparently out of spite, if we may believe Francis Bailey's account [55]. Flamsteed's rebuttal clearly shows his indignation:

'I have now spent 35 years in the composing and work of my catalogue; which may, in time be published for the use of her Majesty's subjects, and ingenious men all the world over. I have endured long and painful distempers by my night watches and day labours. I have spent a large sum of money above my *appointment, out of my own estate, to complete my catalogue, and finish my astronomical works under my hands. Do not tease me with banter, by telling me that these alterations are made* to please me, *when you are sensible nothing can be more displeasing nor injurious, than to be told so. Make my case your own, and tell me ingenuously [sic] and sincerely, were you in my circumstances, and had been at all my labour, charge, and trouble, would you like your labours to have* surreptitiously *forced out of your hands, conveyed into the hands of your declared, profligate enemies, printed without your consent, and spoilt, as mine are, in the impression? Would you suffer your enemies to make themselves judges of what they really understand not? Would you not withdraw your copy out of their hands, trust no more in theirs, and publish your own works rather at your own expense, than see them spoilt, and yourself laughed at, for suffering it?'* [56].

Exasperated and clearly wounded in his pride as a careful observer, he continued,

'I see no way to prevent the evil consequences of Dr Halley's conduct, but this. I have caused my servant to take a new copy of my catalogue; of which I shall make as much to be printed of as Dr Halley has spoilt, and take care of the correction of the press myself, … and that all the last proof sheets may be sent to Greenwich, at my own charge, …, and not printed off till I have seen a proof without faults. After which, I will proceed to print the remaining part of the catalogue as fast as my health, and the small help I have, will suffer me. But if you like not this, I shall print it alone, at my own charge, …; for I cannot bear to see my own labours thus spoilt, to the dishonour of the Nation, Queen, and People. If Dr Halley proceed, it will be a reflection on the President of the Royal Society; and yourself will suffer in your reputation, …'

However, neither this personal appeal, nor a petition to the Queen on 16 April 1712 [57], were to any avail: 400 copies of the fruits of his hard labour with the mural arc, although incomplete, were printed in a single volume in 1712, in a significantly '*garbled and incorrect manner.*' The *Referees* responded that the Queen herself had urged them to proceed, but it appears that the Queen's order was only obtained *after* the seal on the preliminary version of the catalogue had been broken [58]. Flamsteed resolved to publish his own catalogue, at his own expense, which prompted him to request the return of Newton's copy of both his catalogue and the 175 sheets of observations that formed the basis of Halley's volume, which he had provided to Newton as a 'sacred deposit' a few years earlier. Newton did not oblige. Legal wrangling ensued, while tensions ran high.

Eventually, once the officers at Court had been changed following the death in 1715 of the Earl of Halifax—Newton's stalwart supporter at the Court—Flamsteed's luck turned. He petitioned the Lords of the Treasury and managed to have the 300 remaining copies of his 'garbled and incorrect' catalogue delivered to him [59]. He promptly destroyed the 'spoilt' catalogues (but not the published observations made with his sextant) by burning them, so …

'that none might remain to show the ingratitude of two of his countrymen, who had used him worse than ever the noble Tycho was used in Denmark' [60].

He diligently set out to republish the correct catalogue, but he never saw it in print since death intervened. The second and third volumes of his *Historia Coelestis* contained more than 28 000 observations made with the mural arc between 11 September 1689 and 27 December 1719, which provided 2935 stellar positions more accurately than any previous compilation had done. The remainder of the second volume—on which Flamsteed had been working at the time of his passing—and all of the third volume were prepared by his assistants, Joseph Crosthwait and Abraham Sharp. All three volumes eventually saw the light of day in 1725, six years after Flamsteed's death. Although the catalogue was not entirely free of errors, the tabulated stellar, lunar, and planetary positions were sufficiently accurate to make the lunar distance method a viable navigation approach. However, one still needed to correct the motions of the Sun and the Moon and somehow measure the Moon's position while at sea.

Although Flamsteed's primary task was to improve the tables of stellar positions and characterize the orbits of the Moon and the planets, he also worked diligently on rectifying the tables of the motions of Jupiter's satellites, which he started to employ in practice from the early 1680s. He published the ephemerides of Jupiter's moons for 1684–1687 in the *Philosophical Transactions of the Royal Society*. The journal did not appear in 1688, 1689, and 1690. However, Flamsteed provided his unpublished ephemerides observations directly to Huygens in December 1689 [61], so that the latter could correct the longitude measurements taken on the VOC's expedition to the Cape of Good Hope in 1686–1687.

In addition, in Flamsteed's 1697 *Account of the Beginning, Progress, and present State of our Improvements and Deficiencys in the Doctrine and Practice of Navigation* to Pepys, the astronomer outlined how he had used observations of Io from 1689 to derive that the difference in longitude between Greenwich and Hoai-gnan-fu, present-day Huai'an (Jiangsu province) on the east coast of China, was of order 10 degrees, or some 600 miles, less than had been thought previously:

'Tis in Vain to talke of the Use of findeing the Longitude at Sea, except you know the true Longitude and Latitude of the Port for which you are designd ...

... Observations of the Moon or Satellite Eclipses in any port wherever they came, which being compared with those made at the Greenwich Observatory or elsewhere by Ingenious and accurate men, would give us the true Longitude of those Ports, whereby the faults of our present Mapps and Sea Charts (of which our Navigators complain so much) will be corrected and halfe the Buisness of Navigation perfected.'

Flamsteed's personal history is rather depressing. Despite his great efforts at compiling an accurate catalogue of both stellar positions and lunar, solar, and planetary motions, the eventual product was riddled with errors introduced by his editors. As a consequence, Flamsteed's reputation has been tarnished beyond repair; not even Baily's carefully reconstructed timeline managed to cast Flamsteed in a more favourable light. It is also a tale of a collision of personalities, of Flamsteed the perfectionist and Newton the visionary, who was keen to reach his grander goals by, perhaps, sacrificing some accuracy or by employing preliminary results to arrive at the new insights he was so keen to achieve.

6.2.4 The lunar distance method

Newton was a strong advocate of the lunar distance method of longitude determination, a method involving angular measurements on the sky of the distances ('lunars') between the Moon and a range of celestial objects. The method's first publication can be traced back to the German mathematician Johann(es) Werner's (1486–1522) *In hoc opere haec continentur Nova translatio primi libri geographiae Cl. Ptolomaei* (Nuremberg, 1514). Werner appears to have been inspired by a letter dated some time in 1499–1500 from Amerigo Vespucci, the Italian explorer, to his patron Lorenzo di Pierfrancesco de'Medici, in which he claimed to have determined his longitude based on observations of celestial objects, on 23 August 1499. In a second, 16th Century copy of a fragmentary letter known as the *Ridolfi Fragment*, discovered in 1937 by Professor Marques Ridolfi in the Conti archives [62] and possibly dating from 1502, Vespucci wrote to de'Medici,

'*Briefly to support what I assert, and to defend myself from the talk of the malicious, I maintain that I learnt this [my longitude] by the eclipses and conjunctions of the Moon with the planets; and I have lost many nights of sleep in reconciling my calculations with the precepts of those sages who have devised the manuals and written of the movements, conjunctions, aspects, and eclipses of the two luminaries and of the wandering stars, such as the wise King Don Alfonso in his* Tables, *Johannes Regiomontanus in his* Almanac, *and Blanchinus, and the Jewish rabbi Zacuto in his* Almanac, *which is perpetual; and these were composed in different meridians: King Don Alfonso's book in the meridian of Toledo, and Johannes Regiomontanus's in that of Ferrara, and the other two in that of Salamanca. It is certain that I found myself, in a region that is not uninhabited but highly populated, 150 degrees west of the meridian of Alexandria, which is eight equinoctial hours. If some envious or malicious person does not believe this, let him come to me, that I may affirm this with calculations, authorities, and witnesses. And let that suffice with respect to longitude; for, if I were not so busy, I would send you full detail of all the many conjunctions which I observed, but I do not wish to become tangled in this matter, which strikes me to be the doubt of a literary man, and not one which you have raised. Let that suffice*' [63].

Werner's method was subsequently discussed at length by Petrus Apianus (1495–1552), the German humanist, mathematician, and printer, in his first major work, *Cosmographicus seu descriptio totius orbis* (Landshut, 1524): see figure 6.5 (left). He

Figure 6.5. (*Left*) Folio from the *Cosmographia Petri Apiani* (1524) showing the lunar distance method in practice. (Credit: Smithsonian Institution Libraries; https://timeandnavigation.si.edu/multimedia-asset/using-lunar-distances-to-find-longitude) (*right*) Title page of Apianus' *Introductio Geographica* (Ingolstadt, 1533).

even illustrated the method on the title page of his *Introductio Geographica* (Ingolstadt, 1533), as shown in the right-hand panel of figure 6.5. In his second major publication, *Astronomicon Caesareum* (1540), he explored the use of solar eclipses instead. Figure 6.5 shows Apianus' explanation of the lunar distance method, including the cross staff that was to be used for this purpose. Although the cross staff was useful for measuring latitudes, its accuracy was insufficient for use with the lunar distance method at the time, for which one needed better lunar almanac tables and the development of the sextant. This latter development took until the early 1730s to mature, with the instrument's first practical implementation by the English mathematician John Hadley (1682–1744) and Thomas Godfrey (1704–1749), an inventor based in the American colonies, although Newton's unpublished papers actually already contain the salient details to operate a sextant.

In 1634, the French mathematician and astronomer Jean-Baptiste Morin (1583–1656) claimed that he had discovered a new way of finding longitude, commenting that …

'I do not know if the Devil will succeed in making a longitude timekeeper but it is folly for man to try' [64].

Cardinal Richelieu, King Louis XIII's chief minister, consequently established a commission to examine this claim. The commission concluded that Morin's approach, which was based on measuring absolute time by determining the position of the Moon relative to the stars and which took into account atmospheric refraction as well as the lunar parallax (small positional shifts depending on the observer's location), was in essence a more sophisticated version of the lunar distance method. Morin and the commission were in dispute for five years. Richelieu died in 1642. In 1645, his successor, Cardinal Jules Raymond Mazarin (1602–1661), gave Morin 2000 *livres* for his efforts, but the method was still deemed impractical. In fact, it took the invention of the telescope, the pendulum clock, the micrometer screw, and logarithms in the 17th Century for the lunar distance method to become practically viable.

Despite Newton's strong support of the lunar distance method, he never achieved the accuracy required for the method to be used as a viable means of longitude determination. In 1697, Halley published a summary of Newton's *Principia*, in which he specifically referred to lunar observations for longitude determination at sea. In 1699, Newton discussed the problem of determining one's longitude at sea with the British Admiralty secretary, Josiah Burchett (1666?–1746). The next year, in February 1700, he decided to try and improve his understanding of the Moon's motions through observations, which he eventually published in 1702 in both English and Latin, but he did not manage to derive a comprehensive orbital model. It is clear that Newton was struggling to make sense of the observations he had obtained, but his subsequent modelling efforts did not yield the desired results. Although he announced that he had constructed an innovative reflecting octant to measure distances between the Moon and stars while at sea, and hence to help him in his modelling efforts, a description of the instrument was only found among his papers by Halley long after Newton's death; it was eventually published in 1744.

6.3 Human ingenuity

6.3.1 Terrestrial magnetism to the rescue?

Meanwhile, Halley had been pursuing an independent method he hoped to employ for longitude determination at sea, commonly referred to as the 'magnetic declination.' He considered whether careful observations of magnetic deviations on Earth could be used to achieve that aim. Although the Earth's magnetic field was only poorly understood at the time, it was commonly known that the directions of magnetic and geographic North deviated from one another systematically across the globe. Halley, as well as many others who had proposed similar approaches before him since the 16th Century [65–67], hoped that if this East–West pattern of deviation could be linked to terrestrial longitude and mapped accurately, it might be employed to determine one's longitude.

Contemporary navigators were convinced that if they could read the compass and see the stars on a clear night, they could determine their longitude by measuring the distance between the magnetic and the geographic North Poles, indicated by the compass needle and the direction of the North Star. When sailing along any

East–West parallel in the northern hemisphere, the apparent distance between the magnetic and the geographic North Poles changes systematically. Halley expended significant efforts mapping and modelling the Earth's magnetic field, resulting in maps showing 'halleyan' or isogonic lines.

In 1683, Halley published his first of many scientific articles on the Earth's magnetic field. He devised a model of the Earth which, although later proven incorrect, implied that the behaviour of the terrestrial magnetism had its origin deep in the planet's core—which we now believe to be true. In 1696, he suggested that the Earth was composed of an outer shell surrounding an independently moving inner core; both components would produce their own magnetic dipoles (that is, north and south poles), but the motion of the inner core was responsible for the observed behaviour of the terrestrial magnetic field.

In 1693, supported by the Royal Society [68], Benjamin Middleton, Fellow of the Royal Society, and Halley had petitioned the British government to provide a vessel for an expedition ...

> 'to discover what may be learnt ... [of] the variations of the Magneticall Needle ... in that vast Tract of Sea betweene America and China, being neare halfe the Globe...' [69]

Middleton would finance the voyage, as long as the government would provide the ship. Although their proposal was quickly—and in full—approved by Queen Mary II, who was 'graciously pleased to incourage the said undertakeing' [70] the Nine Years' War with France (1688–1697) intervened—or perhaps the relationship between Halley and Middleton changed [71]—and forced them to delay their departure on the *H.M.S. Paramour* (*Paramore*) from the English port of Deptford until 20 October 1698. In addition, the British Admiralty had limited the scope of Halley's expedition to the North and South Atlantic Oceans only.

The voyage of the *Paramore*, a 52-foot, three-masted 'pink' specifically built for the expedition, was the first ever voyage commissioned for strictly scientific purposes only. Despite having no experience whatsoever in navigation, the Royal Navy selected Halley as the vessel's captain; he was given his commission as 'Master and Commander without Instructions' on 4 June 1696, taking command of the ship on 19 August 1698:

> 'Mr Hally has gott a ship from the government, in which he has sett sail to goe round the globe on new discoverys, and the rectifying of geography...,'

according to James Gregory, the Scottish mathematician, astronomer, and Fellow of the Royal Society, to the Reverend Colin Campbell in May 1699. His inexperience was not lost on the crew, resulting in problems of insubordination in relation to his perceived incompetence as commander. Having failed to replace the officers giving him trouble in the West Indies, Halley commandeered the ship back to its home port by July 1699 to initiate legal proceedings against his officers; the Court only mildly rebuked the crew, to Halley's great dissatisfaction [72].

Soon afterwards, on 24 August 1699, Halley recommissioned the *Paramore*, setting sail the following September with a temporary commission as Captain in the Royal Navy. This second Atlantic voyage lasted for a year, until 6 September 1700, during which Halley obtained extensive observations of the Earth's magnetic field, covering the latitude range from 52°N to 52°S—within 200 miles of the Antarctic continent. Halley published his findings in the form of the *General Chart of the Variation of the Compass* (1701), the first isogonic map of the North and South Atlantic Oceans, still used as an important reference by geophysicists: see figure 6.6.

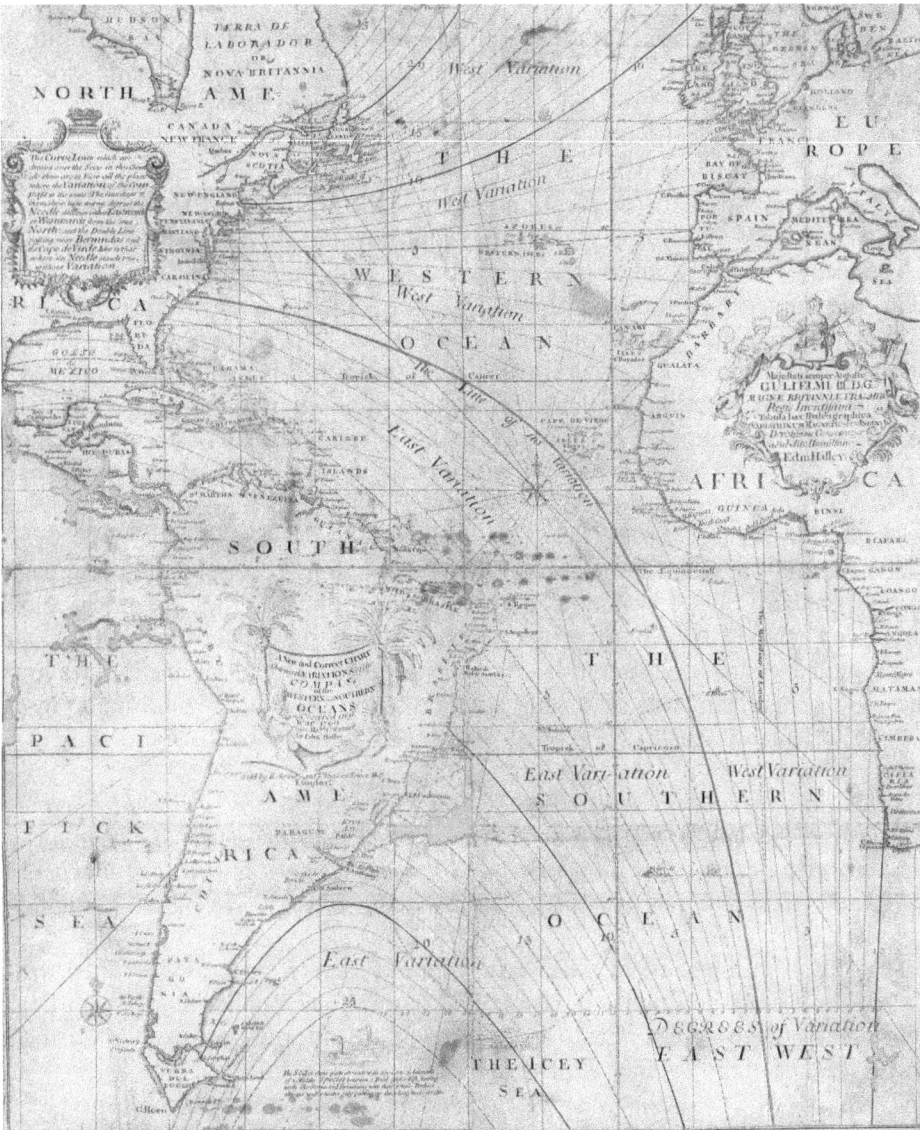

Figure 6.6. Halley's first isogonic map, 1701. (Courtesy of the John Carter Brown Library at Brown University.)

Nevertheless, he was eventually forced to abandon his efforts to use magnetic-field measurements for longitude determination, since local variations from large-scale trends, combined with the inherent inaccuracies of contemporary compasses, rendered the approach unreliable. He continued to pursue this line of research, however, although efforts to use geographic variations in the terrestrial magnetic field eventually ceased, since they were overtaken well into the 18th Century by the development of reliable and accurate marine timepieces.

6.3.2 Alchemy

In addition to the scientifically well-justified methods we have discussed so far, the 17th Century also witnessed opportunistic proposals for longitude determination that were in line with the *Zeitgeist* but which we would discard as outright wacky today. One such proposal, commonly referred to as the 'wounded dog theory,' first appeared on an anonymous pamphlet in 1688, *Curious Enquiries Being Six Brief Discourses* (London). It involved the application of a so-called 'powder of sympathy' (*pulvis sympatheticus*), an alchemical remedy that somehow cured wounds at a distance.

The idea of curing wounds at a distance was losing traction in the 17th Century under the influence of the Scientific Revolution which was in progress across the European intellectual landscape. Until then, it had formed part of the notion of 'sympathetic magic:' a concoction known as 'weapon salve' should be applied to the bladed weapon that had inflicted the wound rather than to the wound itself. The weapon and the wound were thought to be joined magically, since one caused the other. Weapon salve was first introduced in the pseudo-Paracelsian book *Archidoxis Magica*, published in 1570. The philosophy behind its use goes back, at least in spirit, to the holistic approach of the Swiss–German philosopher, alchemist, and physician Paracelsus (Philippus Aureolus Theophrastus Bombastus von Hohenheim; 1493–1541), often considered the founder of the field of toxicology. Paracelsus believed that man and heaven were intimately linked through a Divine spirit, and hence that one should take a holistic approach in all of one's endeavours. The salve's ingredients would usually include blood from the wound made by the weapon to which it was to be applied. Paracelsus' own recipe instructed one to ...

'*[t]ake of moss growing on the head of a thief who has been hanged and left in the air; of real mummy; of human blood, still warm—of each one ounce; of human suet, two ounces; of linseed oil, turpentine, and Armenian bole—of each two drachms. Mix all well in a mortar, and keep the salve in an oblong, narrow urn*' [73].

The idea of using a dry powder rather than a salve is generally credited to the dynamic and versatile English diplomat and natural philosopher Sir Kenelm Digby (1603–1665). Digby purported to have received his information from a Carmelite monk while visiting Florence (Italy) in 1622:

'*It was a religious Carmelite, that came from the Indies and Persia to Florence, he had also been at China, who having done many marvailous cures with his Powder, after his arrival to Toscany, the Duke said he would be very glad to learn it of him. It was the father of the Great Duke, who governs now. The Carmelite answered him, That it was a secret which he had learnt in the oriental parts, and he thought there was not any who knew it in Europe but himself, and that it deserved not to be divulged... But a few moneths after I had opportunity to do an important courtesie to the said Fryer, which induced him to discover unto me his secret, and the same year he returned to Persia; insomuch that now there is no other knows this secret in Europe but my self...' [74]*

With his cure in powder form, Digby hoped to secure the priority of invention, at least in Europe, and explain the underlying principles in terms of the corpuscular theory which was prevalent at the time. This way, he managed to avoid accusations of invoking the occult forces attributed to weapon salves: since weapon salves worked without direct contact, the cure was deemed to be occult, thus running the danger of suggesting devils at work.

In 1658, Digby gave a speech [75] to a 'famous assembly,' presumably a precursor to the *Académie des Sciences*, in Montpellier (southern France), emphasizing the efficacy of his powder of sympathy and establishing himself as the leading authority on the substance. He recounted in detail how he allegedly cured an injury his historian friend James Howell (c. 1594–1666) had sustained while intervening in a duel as far back as 1624, and which did not heal well:

'I asked him then for any thing that had the bloud upon it, so he presently sent for his garter, wherewith his hand was first bound: and as I called for a Bason of water, as if I would wash my hands; I took a handfull of Powder of Vitriol, which I had in my study, and presently dissolved it. As soon as the bloody garter was brought me, I put it within the Bason, observing in the interim what Mr. Howel did, who stood talking with a Gentleman in a corner of my Chamber, not regarding at all what I was doing: but he started suddenly, as if he had found some strange alteration in himself; I asked him what he ailed? I know not what ailes me, but I find that I feel no more pain, me thinks that a pleasing kind of freshnesse, as it were a wet cold napkin did spread over my hand, which hath taken away the inflammation that tormented me before; I replyed since then that you feel already so good an effect of my medicament, I advise you to cast away all your playsters, onely keep the wound clean, and in a moderate temper, twixt heat and cold.'

Figure 6.7 illustrates the ideas behind the application of a powder of sympathy from a contemporary source, the *Theatrum Sympatheticum Auctum* (1662) [76]. The illustration accompanies the Latin translation of Digby's 1658 speech [77]. In true baroque fashion the text includes a lengthy explanatory poem in Latin, of which the full translation in English [78] is provided below (for the numbers in square brackets, refer to the round panels in figure 6.7).

Figure 6.7. Application of a powder of sympathy (*Theatrum Sympatheticum Auctum*, 1662 p 125; Reproduced with permission, Herzog August Bibliothek, Wolfenbüttel, 30.4 Med.: VD17 23:290712A.)

Behold, a wondrous cure of wounds!
Which Celsus himself, the artisan, did not know,
And whatever in the world that more ancient healer favoured,
Perhaps (by Hercules!) even he had failed to note.
With Pamphilus lend your ears, you who do not know these things,
And look with lynx's eyes at the front of this book.
A cloth is washed in water, which is red with blood [2]
From the gore of a recently wounded man.
But the powder was previously dissolved in the water,
The cloth is then dried. What more?
The pain of the wound ceases and is removed.
Each of these things benefits from a temperate warmth,
For if they are done with heat and blazing fire,
The wound is scorched with a fiery burning.
To many the author seems here mystical,
And the spirit astral. Very many are they, besides,
Who ascribe the force and power to the occult quality of a lurking sorcerer,
And go on and on about the Anima Mundi.

But Digby is another Oedipus before this sphinx,
And presumes something different than the rest,
And everywhere unties the knots of enigmas,
With which intricate Nature abounds.
For he would have that all the air is full of light,
And that light presides over everything,
And carries atoms with itself across long intervals of space,
And after a while, carries them back,
Just like a ball bounced forcefully from a wall,
Draws out quite a few corpuscles.
Nor is it dissimilar to a cloth on which liquid was poured,
Which, when it is warmed by a nearby fire,
Produces a mist, or something like it.
And thus we see that a lamp is reflected,
In the morning when the radiance of Phoebus,
Illuminates the earth, drunk with a dewy moisture.
A thick air rises, seeking the heavens,
And forms a little cloud which thereupon is carried,
From the orbit of the sun until it disappears from sight.
And nothing, the author continues, is therefore empty of air,
As he proves by the corpuscles already mentioned [3],
And by baby vipers which are closed-up,
In a glass prison but thrive,
And feed upon the atoms with great profit,
For day by day they grow larger.
But, Argus, you perceive a knot is lurking here:
Why does a wart, when the rays of the Moon are seen,
No longer oppress the hand with its presence?
But proceed, Muse, and turn the pages!
It is still, asserts our author, that these corpuscles,
However much they be divided, will never find an end [4].
Wherefore he narrates very many things,
About these corpuscles and atoms,
By which a canny dog finds his prey, and about hammered gold,
In order that these examples may sparkle with light.
However (he later adds), these are not attracted,
In just one way, but oftentimes they emulate filtration,
Whereby a fluid moves upwards toward the pole [5],
Or often they obey the laws of fire which attracts what is wet.
But more strongly are they glued together,
Whose nature most agrees,
As a burden and scale add their own weight [6],
So that they may correspond in all things.
But of things whose nature is very discrepant,
It is the greatest effort to draw them together.

He points to that wet Chaos which, he briefly relates,
Some people keep in a glass vessel,
To indicate what is the order and place,
Of the four natural things, the Grand Sophos,
The prime elements which thereupon he names:
Figure, Quantity, Imagination,
These are believed to be of the greatest weight.
Indeed! Great is the force of Imagination;
Figure and quantity are exceptionally powerful!
Hence if somebody yawns, another will yawn as well,
Hence the harmony of lutes;
Hence will you dance if bitten on the foot by a tarantula.
Lastly the author adds to these what he maintains thus:
Namely, that a body which attracts these spirits,
Attracts also whatever is adjacent to them.
Thus when milk boils over and spills,
And experiences the force of fire on the coals,
The cow suffers from tremendous pain,
And is racked with a fiery heat in her udders.
And thus the ulcerated foot of an ox,
Feels relief or is afflicted with pain,
According to how much the breeze of a beneficial spirit soothes,
Or a malignant one aggravates it.
Thus also at the lightening bolts of Mars,
Does fiery Vulcan sink down on the anvil [7],
And along with the falling, black soot,
The anger of the raging fire vanishes.
And many things, which narrow meter cannot hold,
Digby, worthy of all praise, inserts.
He concludes that in the treatment of the wound,
The sun and light attract spirits of the gore from the bandage,
Along with corpuscles of the powder.
But meanwhile, he says, the wound is exhaling spirits,
That are full of heat and fiery,
And which are always drawing new air to themselves,
And when thus the spirits of the gore and powder,
Finally arrive, those spirits, as they are returned to themselves,
And their origin, they dig in their heels.
And each enjoys his own proper seat.
Therefore, so long as the horde of powder corpuscles,
Is mixed tightly with the rest,
And every current from the wound,
Passes through each other, Lo! A cure arises!
Even the surgeon was summoned as a joke,
With Momus shouting: Go away, come back after the feast!

Digby invoked seven principles to explain the nature of his cure, all rooted in a mechanistic philosophy and contemporary alchemy. First, sunlight attracts the spirits of the blood on the weapon. Meanwhile, the warmth of the injured man's heart pushes out the atoms, which *'march of themselves a good way in the air, to help there by the attraction of the Sun and of the Light'* [79]. The spirit of the vitriol (sulphuric acid or ferrous sulphate), the essential ingredient in the powder, joins the atoms of the blood from the weapon in the air, whereas the wound exhales hot, fiery spirit of the blood into the ambient air. This would diffuse into an air current, which eventually comes in contact and diffuses with the air carrying the spirit of vitriol from the weapon. This way, the atoms of the blood on the weapon find their origin, and since the atoms of the blood and the spirits of the vitriol are joined, …

'both the one and the other do jointly imbibe together within all the corners, fibres, and offices of the veins which lie open about the wound of the party hurt, which hereby are comforted, and in fine imperceptibly cured' [80].

With the benefit of both historical hindsight and our current understanding of medicine, we can understand why these sympathetic cures sometimes worked. Indeed, their application prevented any of the frequently used drugs to touch the wounded part, thus preventing infections, while the bandages where keeping the dirt out. And most importantly, the injury was allowed to cure itself.

In a carefully researched historical reconstruction [81], Hedrick considered Digby's claim of priority, as well as the truthfulness of his 1658 account. She provides clear, abundant yet circumstantial evidence that the original sympathetic cure in powder form must be attributed to the courtier and Member of Parliament Sir Gilbert Talbot (c. 1607–1695), who was a Fellow of the Royal Society from its inception. It was Talbot who was called as authority on the powder of sympathy by his contemporaries at the Royal Society, rather than Digby, in a series of meetings between 5 and 26 June 1661, when the Fellows explored the substance's efficacy—Talbot's superior credibility is clearly in evidence, given that Digby was also present at these meetings. Therefore, while Digby should be credited with pushing powders of sympathy to the forefront in the public mind, he was clearly not regarded by the scientific elite as the substance's originator. Nevertheless, history shows that his public relations campaign has been extraordinarily successful in linking his name to the matter.

In this late-17th Century context, where alchemy had not yet been fully discarded, the 1688 *Curious Enquiries* pamphlet was not immediately written off as a quack idea. The author rather cleverly linked his suggested means of longitude determination to Digby's well-known efforts to give the idea a semblance of authority. Specifically, he suggested, rather cruelly, that a ship's captain could take along on his voyage a wounded dog, with the wound's discarded bandage left in the ship's home port. A trusted timekeeper would be tasked with dipping the bandage into the powder at a set time every day, causing the animal to howl in pain, since the cure was certainly not painless. In turn, this would provide a direct and accurate measure of the time on shore, which hence could be used to calculate the vessel's longitude at sea.

> 'Sir Kenelm Digby in his Discourse of the Sympathetick Powder, tells us how he made Mr. Howel start, upon his putting a Bloody Garter, with which Mr. Howel's wounded hand had been bound up withal, into a Bason of Water mixed with that Powder: If such a starting cold be made to any Inferiour creature at a great distance, and by often doing it, it would not in two or three months lose its power, we might at Sea with great Ease and Pleasure know when the Sun was upon the Meridian at London, or any other appointed time; and consequently by the difference of Time, the difference of Longitude' [82].

There are no records as to whether the theory has ever been put to the test, or of its efficacy—imagine the cruelty involved in keeping the wound open on long ocean voyages, irrespective as to whether or not the cure would even work over the enormous distances implied... In fact, the pamphlet may well have been meant tongue in cheek rather than to be taken literally: the *Curious Enquiries* is thought to have been part of the pseudonymous *Poor Robin*'s series of satirical writings, almanacs, and chapbooks; the early contributions to this series were attributed to William Winstanley (c. 1628–1698), English poet and biographer. Satirical or not, the author did not go quite as far as to formally endorse the method, although he pointed out that this would not be any more gruesome than the practice routinely adopted prior to 1595 of having a crew member take direct readings of the Sun to aid in their ship's navigation, since ...

> 'before the Back-Quadrants were Invented, when the Forestaff was most in use, there was not one Old Master of a Ship amongst Twenty, but what a Blind in one Eye by daily staring in the Sun to find his Way' [83].

The original configuration of the quadrant, which involved a cross staff, required the user to look directly into the Sun to determine its height above the horizon. Following the invention of the back-quadrant in 1595 by John Davis (c. 1543/50–1605), the famous English explorer and navigator, readings were taken with one's back to the Sun and based on measuring the position of the shadow of the quadrant's upper vane as it was cast on an horizon vane.

On the same pamphlet extolling the virtues of the wounded dog theory, the author also reviewed and rejected a number of alternative methods of longitude determination,

> 'Many Men have had many Whimseys about this Longitude, and such indeed they have only proved: I have read formerly in some Old (I have forgot now what) Book, That if you fill a Glass of Water full to the Brim, and watch the time of the New, or Full Moon, the Water of it self at that very instant of time, would run over. Well, thinks I, by this at least we can discover our Longitude at Sea twice a Month; and very probably may give a good guess at it daily, by

observing the Increase and Decrease of the Water, and allowing for the Condensation and Rarefaction of the Air, observed by a Thermometer; but when I tried at several Full Moons, I never observed the Water to stir; so then I concluded that to be false.'

Given the great importance and increasing urgency of solving the longitude problem, learned societies, individual scholars, opportunists, and even clergymen latched onto the bandwagon of the times, all vehemently promoting their own interests. A particularly inventive pamphlet [84], which was clearly meant for moral guidance, was self-published in 1699, and once again a year later, by the Reverend Samuel Fyler (1638–1703), rector of the parish of Stockton in the southern English county of Wiltshire. Fyler phrased his proposed method in the form of a dialogue between himself, the author, and 'an intelligent seaman.' Cleverly, the pamphlet's title drew in those interested in finding a solution to the scientific problem at hand, but the author's aim was clearly to emphasize why *'the longitude of eternity in [one's] thoughts'* was what ultimately mattered.

Nevertheless, Fyler's method was indeed novel, but it was wholly impractical: he proposed that, with an intimate knowledge of the night sky, one could identify sequences of stars from the horizon to the zenith, one such meridian for each hour. He expected that the leading astronomers of the day could then easily draw maps and calculate timetables showing when each of these meridians would be visible straight overhead on the reference meridian, the most commonly used of which at that time intersected the Canary Islands, exactly 20° West of Paris. Armed with these materials, any navigator at sea with access to a reasonably accurate timepiece could then observe which of those meridians was visible straight overhead at midnight. Clearly, Fyler thought, the sailor's longitude would then follow directly from the difference in time between the local meridian and the Canary Island's prime meridian. Unfortunately, however, this approach relied on circular reasoning and required access to more astronomical observations than were available at any of the contemporary observatories on Earth, even on a clear night...

6.4 Developments leading up to the 1714 Longitude Act

6.4.1 Disaster off the Isles of Scilly

The sense of urgency to resolve the longitude problem is often said to have reached fever pitch following the naval disaster of 23 October 1707, when four English navy ships under the command of the Admiral of the Fleet, Sir Cloudesley Shovell (c. 1650–1707) on the *H.M.S. Association*, shipwrecked on the rocks of the Isles of Scilly, just off the southwestern tip of England. The English fleet's navigators were collectively of the opinion that they were located well West of the Île d'Ouessant, an extension of France's Brittany peninsula, but upon continuing their voyage north that fateful, foggy evening, they rapidly realized to their dismay that they were instead much closer—in fact, too close—to the Isles of Scilly.

Having struck the rocks around 8 p.m., the *Association* sank in a mere three to four minutes, and none of the 800 men on board were saved. In fact, Admiral

Shovell washed ashore at Porth Hellick Cove on St. Mary's Island, the largest of the Isles of Scilly, and local legend has it that he was still breathing but only semi-conscious. By some accounts, he was smothered in the sand by a local woman for the emerald ring on his finger which he had been given by his close friend, Captain James Lord Dursley (after 1679–1736), Earl of Berkeley [85]. This tale was made public after the woman had confessed to her act and produced the ring on her deathbed, several decades later. Reportedly, the local clergyman sent the ring back to Dursley. However, historians disagree about the reliability of the murder story, since it is unknown whether the ring was ever recovered and the tall tale originates from an unverifiable deathbed confession [86].

Three other large ships, *H.M.S. Eagle*, *H.M.S. Romney*, and the fire ship *H.M.S. Firebrand* also found their watery grave at the same time. The Scilly disaster, responsible for the demise of some 2000 men, is often directly linked to the passing of the British Longitude Act in 1714, seven years after the event, but careful scrutiny of the available historical records shows that there is little evidence of such a causal connection. The disaster's news coverage was factual, with the *Daily Courant* of 1 November 1707 reporting …

> '*an Account, that Sir Cloudsly Shovel with about 20 Sail of Men of War coming from the Streights, having made an Observation the 21st, lay the 22d from 12 to about 6 in the Afternoon; but the Weather being very hazy and rainy and Night coming on dark, the Wind being S.S.W, they Stearing E by N, supposing they had the Channel open, were some of them upon the Rocks to the Westward of Scilly before they were aware, about 8 a Clock at Night. Of the* Association *not a Man was sav'd … The Captain and 24 Men of the* Firebrand *Fire-Shop were saved, as were also all the Crew of the* Phoenix. *'Tis said the* Rumney *and* Eagle, *with their Crews, were lost with the* Association.'

Similar reports appeared elsewhere [87], all blaming the disaster on the foggy weather conditions and not on difficulties related to the navigators' longitude determination.

Nevertheless, another local legend on the Isles of Scilly has persistently reared its head since it first surfaced in 1780. It suggests that a local sailor on board the *Association*, a native of the Isles of Scilly, knew the waters well and recognized that they were on course to the Scilly rocks. Although navigation was the purview of the senior commanders, he risked his life to warn Admiral Shovell of the impending danger. Because of the sailor's insistence and his low station, Shovell became incensed by this insubordination and had him hanged from the yardarm for inciting mutiny:

> '*About one or two aft. noon on the 23rd [22nd] Octr Sir C. call'd a council & examd ye Masters wt lat. they were in; all agreed to be in that of Ushant on ye coast of France, except Sr W. Jumpers Mr of ye Lenox, who believ'd 'em to be nearer Scilly, & yt in 3 hours should be up in sight of, [wch unfortunately happen'd] but Sr Cloud. listened not to a single person whose opinion was*

contrary to ye whole fleet. [They then alter'd their opinion and thought 'emselves to be on ye coast of France, but a lad on board ye _ _ said the light they made was Scilly light, tho' all the ships crew swore at & gave him ill language for it; howbeit he continu'd in his assertion, and wt they made to be a saile and a ship's lanthorn prov'd to be a rock and ye Light aforementioned, wch rock ye lad call'd ye Great Smith, of ye truth of which at day-break they was all convinced.]' [88].

Again, historians doubt the veracity of this account, given the absence of contemporary supporting records and the almost occult nature of the story [89,90]. The legend may have found its origin in the discussions regarding their ship's exact location commonly expressed with great concern by sailors and commanders alike when entering the treacherous English Channel.

Contrary to many popular accounts, no 'public outcry' over navigational inadequacies followed the disaster, beyond a national outpouring of grief for the demise of so many able young men.[1] Indeed, careful consideration of historical records implies that the 1714 Longitude Act does not appear to have resulted from the Isles of Scilly disaster. Any reference to navigational difficulties did not appear until the eve of the British Parliament's vote on the Longitude Act, in a flyer published on 10 June 1714 by the mathematicians William Whiston (1667–1752) and Humphry Ditton (1675–1715). Whiston had succeeded Newton as the Lucasian Professor of Mathematics at the University of Cambridge in 1702. The men presented 11 reasons why their bill should be supported, with the tenth suggesting that ...

'it will prevent the Loss of abundance of Ships and Lives of Men; as it would certainly have sav'd all Sir Cloudsly Shovel's Fleet, had it been then put in Practice' [91].

It seems, therefore, that the Isles of Scilly disaster was merely brought up as a lobbying instrument. Nevertheless, the unavailability of any method for accurate geographic position determination at sea was never far from the minds of those concerned with trading across the world's oceans. As a case in point, in relation to his 1683 voyage to Tangier (North Africa), Pepys—in his role as Royal Navy official —lamented in his diary that ...

'[i]t is most plain, from the confusion all these people are in, how to make good their reckonings, even each man's with itself, and the nonsensical arguments they would make use of to do it, and disorder they are in about it, that it is by God's Almighty Providence and great chance, and the wideness of the sea, that there are not a great many more misfortunes and ill chances in navigation than there are' [92].

---

[1] Shovell's body was taken to Plymouth and embalmed before being transported to London by road. He lay in state in his London home before his state burial in Westminster Abbey.

The impression that the Isles of Scilly disaster was probably only recalled as an effective lobbying tool is further borne out by reference to a petition signed by 'Captains of Her Majesty's Ships, Merchants of London, and Commanders of Merchant-Men' that had been submitted to the House of Commons. It is thought that Whiston and Ditton were the driving force behind this petition, given that they had already petitioned Parliament in April 1714, lobbying for it to offer large financial rewards for the same purpose. In addition, in a letter to *The Guardian* newspaper published on 14 July 1713, and a second letter in *The Englishman* of 10 December of the same year, they had first suggested that Parliament should offer sustained support to finally solve the longitude problem. The men promoted their own scheme, but they did not mention the Isles of Scilly disaster. We are therefore left to conclude that details of the disaster were apparently only revived when Members of Parliament were already considering whether or not to support a possible Longitude Act.

6.4.2 Whiston and Ditton's proposal

Although Whiston and Ditton's June 1714 petition stated that ...

> '*Sir Isaac Newton's own Paper, delivered into the Committe [sic], gives hopes, that the Known Method by the Theory of the Moon, which is hitherto not exact enough, may, upon due Encouragement, in time be brought to Perfection,*'

they specifically pursued a method which, in apparent disagreement with Newton's convictions, was ...

> '*easy to be understood and practis'd by Ordinary Seamen, without the Necessity of any puzzling Calculations in Astronomy.*'

The men's main longitude scheme initially put forward was inspired by a combination of unrelated events and developments. During a leisurely afternoon conversation, Ditton mentioned to Whiston that ...

> '*[t]he nature of Sounds would afford a method, true at least in Theory, for the discovery of the Longitude*' [93],

since the time difference between the moment when a sound is made and when it is heard would be equivalent to the difference in longitude between the locations where the sound is made and where it is heard. Whiston concurred, saying that, while in Cambridge, he had heard guns being fired 90 to 100 miles (145–160 km) away, off Beachy Head on the English southern coast, presumably during the Battle of Beachy Head on 10 July 1690, one of the naval battles associated with the Nine Years' War between the French and a coalition of the English and the Dutch. He had also come to know that explosions from the artillery of the Dutch Wars carried to '*the very middle of England, at a much greater distance.*'

Their idea was to have signal vessels stationed at precisely known locations across the Atlantic Ocean, 600 miles apart, so that marine navigators could estimate their distance from these stationary beacons by comparing the known time of the expected signal to the actual shipboard time at the moment the signal was heard. Unfortunately, sounds do not carry at sea reliably enough for accurate location finding, so that this idea was doomed to fail from the outset. However, Whiston was additionally inspired by a fireworks display he had witnessed on the River Thames in London on 7 July 1713, which celebrated the Peace of Utrecht, a series of treaties which eventually helped end the War of the Spanish Succession (1702–1715).

Their reinvented proposal, published in *The Guardian* the following week, therefore suggested to tackle the longitude problem by measuring the time delay between seeing a star shell fired at a given time each day, say, at local midnight, and hearing its sound [94]. The shells would be timed to explode at an altitude of 6440 feet, the limit of the technology available at that time, which Whiston reckoned could be seen from 100 miles. The men made a crucial miscalculation, however, in asserting that the Atlantic was nowhere more than 300 fathoms (approximately 550 metres) deep, so that the light ships could be anchored at fixed positions fairly easily. Nevertheless, any irregularities could be worked out easily once the scheme was operational, or so they thought. In reality, the average depth of the Atlantic Ocean is about 2000 fathoms (3700 m), with occasional recesses down to 3450 fathoms (6300 m). The exact positions of the vessels could be obtained on the basis of absolute position calibrations using the ephemerides of Jupiter's satellites or by observing the times of solar or lunar eclipses.

As a case in point, figure 6.8 shows Whiston's map of locations where one would be able to see a star shell in the sky shot up from a mortar on Shooter's Hill, then a location in the countryside east of London. Whiston selected Shooter's Hill as his suggested firing location because of its optimal conditions, including the altitude, the relatively open countryside location, easy accessibility, and proximity to the resources needed for his experimental work to determine the speed of sound and the height to which the star shells could be propelled. In fact, from 1716 artillerymen used to fire their guns on the hill and occasionally helped out with the scientific experiments designed by Whiston and his contemporaries. However, in practice the location was not as pastoral as Whiston had imagined, but it was frequented by highwaymen, robbers that cornered unsuspecting travellers: '*I never was so rob'd in all my Life*,' claimed one victim [95].

A Parliamentary committee chaired by the Liverpool Member of Parliament William Clayton was tasked with looking into reaching closure on the longitude problem. It called witnesses in response to Whiston and Ditton's petition, including Newton. Newton gave evidence to the House of Commons committee on 2 June 1714, stating that for the determination of the longitude at sea several approaches had been proposed, which might work in theory but which were difficult to execute in practice. He discussed four of the prevailing proposals in particular, including methods based on accurate timekeeping, or on the eclipses of Jupiter's moons, the lunar distance method, and '*a new method proposed by Mr. Ditton.*' Newton criticized all four methods, particularly the timekeeping approaches:

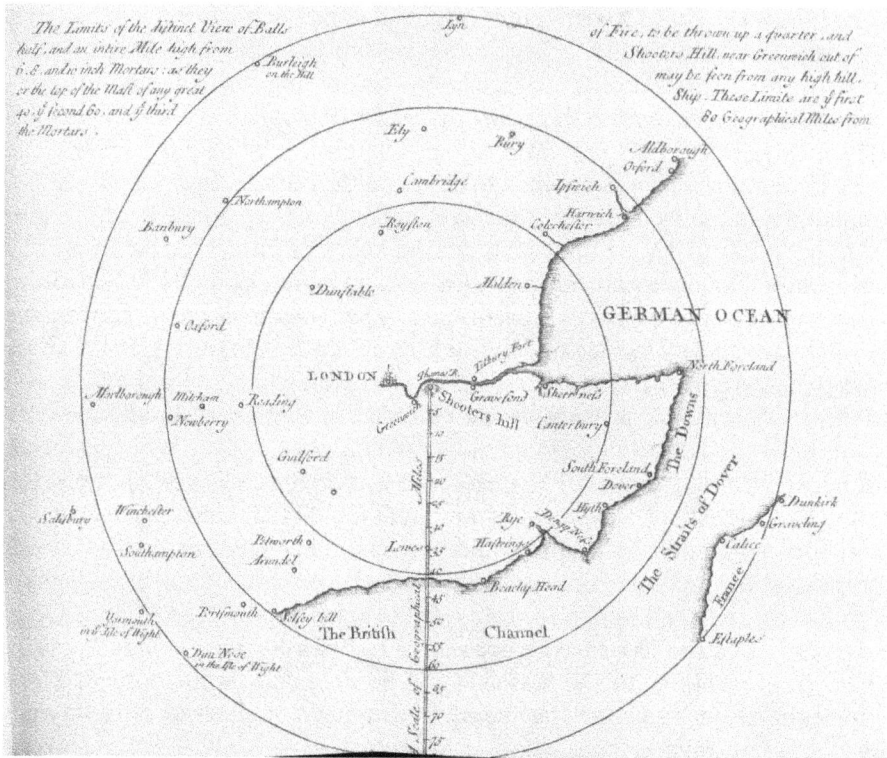

Figure 6.8. Plate from William Whiston, *The longitude discovered*, 1738, engraving, 19.4 × 25.5 cm (© National Maritime Museum, Greenwich, London.)

'One [method] is by a Watch to keep time exactly. But, by reason of the motion of the Ship, the Variation of Heat and Cold, Wet and Dry, and the Difference of Gravity in different Latitudes, such a watch hath not yet been made' [96].

Yet, his evidence made the committee produce a report that stated ...

'that it is the opinion of this committee that a reward be settled by Parliament upon such person or persons as shall discover a more certain and practicable method of ascertaining the longitude than any yet in practice; and the said reward be proportioned to the degree of exactness to which the said method shall reach.'

Whiston and Ditton's suggestion that Parliament encourage the development of novel, more accurate methods of longitude determination was thus endorsed by Newton, Samuel Clarke (1675–1729), the philosopher, the English mathematician Roger Cotes, and Halley. Indeed, Whiston and Ditton had specifically suggested Halley to undertake the measurements to support the practicality of their idea:

'And we take leave to Recommend the Learned Savilian Professor of Geometry at Oxford, Dr Halley, as the fittest Person in the World for the Tryal, and

Practice, and Improvement of this Method; and do hereby Declare, that we are willing that he go equal Shares with us in the Reward, if he pleases to undertake so Useful a Work, and the Publick please to make that Reward equivalent to the great Dignity and Importance of the Discovery' [97].

The deliberations of the House of Commons committee eventually led to adoption and approval of the Longitude Act by both Houses of Parliament. On 9 July 1714, Queen Anne came to the House of Lords to give her royal assent to 29 Acts of Parliament, including *An Act for providing a Public Reward for such Person or Persons as shall discover the Longitude at Sea*, with as one of its main goals to safeguard *'the Safety and Quickness of Voyages, the Preservation of ships, and the Lives of Men.'* As a result, 22 'Commissioners for the Discovery of the Longitude at Sea,' popularly known as the 'Board of Longitude,' were initially appointed to adjudicate claims (see table 6.1)[2].

Newton strongly believed that astronomy would eventually provide the solution to the longitude problem. His arguments affected the terms of the reward offered, in the sense that the size of the reward would match the accuracy achieved. The Act offered to pay up to £2000 for experimental work deemed promising and a reward of £10 000 to anyone who suggested a demonstrably practical method that was successful to determine one's longitude at sea with an accuracy down to one degree of a great circle, equivalent to 60 nautical miles (69 miles) at the Equator. The prize money was to be increased to £15 000 for accuracies down to 40 minutes of arc, and £20 000 for methods with the highest accuracy, that is, of half a degree.

The Longitude Act did not foresee in establishing an actual Board of Longitude[3] or any formal procedure for submitting proposals and demonstrations; all interested parties were expected to get in touch with the Commissioners or the Admiralty through published pamphlets, broadsheets, or newspapers, by letter, or in person. The quest for a solution to the longitude problem gripped the nation, particularly in the years immediately following the passing of the Longitude Act [98–100].

Newton was a strong proponent of pursuing approaches based on astronomical considerations. He had argued that the Act's proposed rewards should be offered once a method had been tried successfully on a single voyage to a specified port across the ocean. The condition that a reward might be awarded after a single successful voyage suggests that Newton's arguments advocating the use of astronomy—implying a universal method rather than a machine—had gained traction with the committee.

However, trying to pre-empt developments, Whiston and Ditton suggested in their petition of June 1714 that the efficacy of their method had already been

---

[2] Although the 1714 Longitude Act only specifically refers to the Savilian professor of 'mathematics,' both the Savilian professors of Geometry and Astronomy were treated as acceptable Commissioners, thus resulting in a final tally of 23 Commissioners.

[3] A Board was established eventually, chaired by none other than Newton, but its first meeting did not occur until 30 June 1737, when the Commissioners discussed John Harrison's first timepiece, H1.

Table 6.1 Commissioners indicated in the 1714 Longitude Act by office and by name.

1. *Ex officio* Commissioners[a]	Identity as of July 1714[c]
The Lord High Admiral of Great Britain, in his role as the First Commissioner of the Admiralty	Lieutenant-General Viscount Thomas Wentworth, 1st Earl of Strafford KG, 3rd Baron Raby (1672–1739)
The Speaker of the Honourable House of Commons	Sir Thomas Hanmer, 4th Baronet (1677–1746)
The First Commissioner of the Navy	Sir John Leake Knt (1656–1720)
The First Commissioner of Trade	Francis North, 2nd Baron Guilford PC (1673–1729)
The Admirals of the Red, White, and Blue Squadrons	
• Admiral of the Fleet (Red)	Sir John Leake Knt (1656–1720)
• Admiral of the White	Sir James Wishart (1659–1723)
• Admiral of the Blue	Sir John Norris PC (1670/1 or 1674–1749)
The Master of Trinity House	Lieutenant-General Viscount Thomas Wentworth, 1st Earl of Strafford KG, 3rd Baron Raby (1672–1739)
The President of the Royal Society	Sir Isaac Newton PRS (1642–1726)
The Royal Astronomer of Greenwich	John Flamsteed FRS (1646–1719), 1st Astronomer Royal
The Savilian, Lucasian, and Plumian Professors of Math in Oxford and Cambridge	
• The Savilian professor of astronomy at Oxford[b]	John Keill FRS (1671–1721)
• The Lucasian professor of mathematics at Cambridge	Nicholas Saunderson FRS (1682–1739)
• The Plumian professor of astronomy and experimental philosophy at Cambridge	Roger Cotes FRS (1682–1716)

2. Commissioners indicated by name	Notes
The Right Honourable Thomas Earl of Pembroke and Montgomery	Thomas Herbert, 8th Earl of Pembroke and 5th Earl of Montgomery KG PC PRS (c. 1656–1733); English and later British statesman
Philip Lord Bishop of Hereford	Philip Bisse FRS (1667–1721)
George Lord Bishop of Bristol	George Smalridge (Smallridge; 1662–1719)
Thomas Lord Trevor	1st Baron Trevor PC (1658–1730); British judge and politician, Attorney General and later Lord Privy Seal
Sir Thomas Hanmer Baronet, Speaker of the Honourable House of Commons	4th Baronet (1677–1746)

Francis Robarts, *Esq.*	Member of Parliament/statesman, scholar, and composer (FRS; c. 1649–1718)
James Stanhope, *Esq.*	1st Earl Stanhope PC (c. 1673–1721); British statesman and soldier
William Clayton, *Esq.*	English merchant and Member of Parliament (after 1650–1715)
William Lowndes, *Esq.*	Secretary to the Treasury of Great Britain/ Member of Parliament (1652–1724)

[a] Note that multiple *ex officio* appointments were commonly held simultaneously by the same individuals; some of these offices, particularly those of the white and blue admirals, are poorly recorded. [b] It would be prudent to also include Edmond Halley FRS here, the Savilian professor of geometry at Oxford, since the Savilian professors of astronomy and geometry were both considered suitable Commissioners. [c] FRS/PRS: Fellow/President of the Royal Society; KB: Knight of the Bath; KG: Knight of the Garter; Knt: Knight; PC: Privy Council of Great Britain.

confirmed in practice by the likes of Newton and Halley, and it did not need to rely on astronomical insights:

'*V. Because its great Use at Land and in Geography is indisputable, and was distinctly observ'd by Sir Isaac Newton and Dr Halley, upon the first Proposal of this Method to them. And we beg leave to say, that this Use alone is so great and extensive, that if there were no other, it would highly deserve the Encouragement of the Publick.*

VI. Because another great Use is also undoubted, viz, for all Places in the Narrow Seas, and within about 100 Miles of all Shores and Islands; that is, for all Places where Ships are in the greatest Danger; as Sir Isaac Newton own'd to the Committee; so that if this Method extended no farther, yet it would highly deserve the publick Encouragement.

VII. Because there is little or no Reason to doubt of its Use at any Place at Sea, even where Ships are allowed to be in the least Danger; since in the most doubtful Case of All Sir Isaac Newton has, in his Paper delivered to the Committee, proposed a most effectual Remedy for the same; as will be clearly understood when the Method itself is known to the World.'

Although it is said that the principle of Whiston and Ditton's method [101] was indeed applied successfully to finding the difference in longitude between Paris and Vienna, and between the Observatories at Greenwich and Paris [102], Arbuthnott ridiculed the idea in a letter of 17 July 1714 to Jonathan Swift (1667–1745), the well-known satirist, essayist, and Dean of St Patrick's Cathedral in Dublin (Ireland). He bluntly declared that their approach anticipated a burlesque proposal of his own:

'*Whetstone [a deliberate misspelling of Whiston] has at last published his project of the longitude; the most ridiculous thing that ever was thought on. But a*

*pox on him! he has spoiled one of my papers of Scriblerus, [103] which was a
proposition for the longitude not very unlike his, to this purpose; that since there
was no pole for east and west, that all the princes of Europe should join and build
two prodigious poles, upon high mountains, with a vast lighthouse to serve for a
polestar. I was thinking of a calculation of the time, charges, and dimensions.
Now you must understand his project is by lighthouses, and explosion of bombs
at a certain hour'* [104].

However, Joseph Addison (1672–1719), the English writer and politician, clearly
held a different opinion, stating that the Whiston and Ditton scheme *'deserves a
much higher Name than that of a Project, if our Language afforded any such Term'*
[105,106]. These contradictory developments clearly provoked both Swift and
William Hogarth (1697–1764), the painter, to mock the proposal. Swift, as an
active member of the Scriblerus Club of satirists, took further punitive action
by publishing a song riddled with unsavoury rhymes ridiculing Whiston and
Ditton [107]:

ODE FOR MUSIC.
ON THE LONGITUDE.

RECITATIVO
The longitude miss'd on
By wicked Will. Whiston;
And not better hit on
By good Master Ditton.

RITORNELLO
So Ditton and Whiston
May both be bep-st on;
And Whiston and Ditton
May both be besh-t on.
Sing Ditton
Besh-t on;
And Whiston
Bep-st on.
Sing Ditton and Whiston,
And Whiston and Ditton,
Besh-t and Bep-st on,
Bep-st and Besh-t on.

DA CAPO

Despite their success in determining the longitude differences on land, the Longitude
Board did not see fit to pursue the idea any further [108] for numerous practical
reasons, not least the requirement for international cooperation to maintain a fleet of

stationary vessels across the World's oceans. The men had considered these aspects, however, and they suggested that the signal ships should receive legal protection from trading nations:

> '*And it ought to be a great Crime with every one of them, if any other Ships either injure them, or endeavor to imitate their Explosions, for the Amusement and Deception of any.*'

However, numerous other practical obstacles stood in the way of success. In fact, Flamsteed had already noted that the rockets ...

> '*will be of little or no use at sea because of the short duration of the appearances and that the seamen will want to know where and at what Moment to look*' [109],

while Newton commented that, similarly to using a set of chronometers, the method was ...

> '*rather for keeping an Account of the Longitude at Sea, than for finding it, if at any time it should be lost*' [110].

In addition, cost concerns and the logistics involved in keeping such a huge network of manned, stationary vessels functioning made the idea a non-starter.

Contrary to popular accounts suggesting that his disappointment in this outcome, aggravated by Swift's mockery, accelerated Ditton's death on 15 October 1715, the Longitude Board had not yet reached a conclusion on the men's proposal at the time of his death [111], so that it is more likely that he died of '*a putrid fever*' [112]. Interestingly, at some point around this time Ditton actually considered using a chronometer for the purposes of longitude determination, and he event sent one of his clock designs to Leibniz, but he discarded this approach as impractical in his pamphlet [113].

Meanwhile, and encouraged by the passing of the Longitude Act, Whiston continued to pursue the rich reward on offer by suggesting practical methods of longitude determination, including approaches based on the direction of magnetic declination, for instance. However, such pursuits, which extend well into the 18th Century, are beyond the scope of this text; they have been covered comprehensively and authoritatively elsewhere [114,115]. Given the increasing importance of accurate navigation across the World's oceans, Whiston and Ditton were not alone in their pursuit of a practically viable method for longitude determination. Well before the passing of the Longitude Act, both serious and less serious proposals from many 'projectors' had seen the light of day, from serious scholars like Huygens, Hooke, and Newton to the 'longitude lunatics' so vividly depicted in Hogarth's masterwork, *A Rake's Progress* (see figure 6.9). John Flamsteed, as Astronomer Royal an *ex officio* Commissioner of Longitude, complained of the '*Swarme of hopefull Authors*' who tried to contact him [116]. As early as 1714, he

Figure 6.9. Interior of Bedlam (Bethlem Royal Hospital), from *A Rake's Progress* by William Hogarth, 1735. This image shows Tom Rakewell (centre, front, shirtless, and bald) fallen so low that he has ended up in Bedlam, London's notorious mental asylum. The poor soul in the background is trying to solve the longitude problem, an oblique reference to Whiston and Ditton's method proposed in 1713–1715.

wrote to Abraham Sharp, his former assistant, about the letters, pamphlets, and 'pretenders' reaching out to him:

> '*A couple of young Non conformist preachers from Worksop in the North of Derbyshire* [more than 150 miles away] *came thither to have my approbation of some Method they had to propose for finding the Longitude at sea, one I shall tell you because it will make you laugh abundantly*' [117].

The practice of 'projecting' had become associated with naive, foolish, or even malicious schemes that took advantage of public gullibility, generating financial investments which never yielded a return. The quest for a solution to the longitude problem unfortunately became the ultimate example of such a project. Already in 1697, Daniel Defoe, the famous author, had little positive to say about projectors:

> '*A meer Projector then is a Contemptible thing, driven by his own desperate Fortune to such a Streight, that he must be deliver'd by a Miracle, or Starve; and when he has beat his Brains for some such Miracle in vain, he finds no remedy*

but to paint up some Bauble or other, as Players make Puppets talk big, to show like a strange thing, and then cry it up for a New Invention' [118].

The Story of Longitude is not just the pursuit of mathematical beauty and engineering prowess. It touches the deepest reaches of humanity and as such it carries a powerful message to anyone open to embracing the challenges of the time. Despite significant adversity, human ingenuity will not back down; progress, although slow at times, is unstoppable.

Time and time again.

References and notes

[1] Huygens C 1669 Debat de 1669 à l'académie sur les causes de la pesanteur *Oeuvres Complètes de Christiaan Huygens* **XIX** 628–45

[2] Huygens C 1690 Discours de la cause de la pesanteur *Addition Oeuvres Complètes de Christiaan Huygens* **XXI** 462–6

[3] *Ibid.* 443–88

[4] Descartes R 1644 Les principes de la philosophie ed V Cousin 1824–1826 *Oeuvres de Descartes* vol **3** (Paris: F.-G. Levrault)

[5] *Ibid.*

[6] Zehe H 1980 Die Gravitationstheorie des Nicolas Fatio de Duillier *Arch. Hist. Exact Sci.* (Hildesheim: Gerstenberg) **28** 1ff

[7] Cavendish H 1798 Experiments to Determine the Density of the Earth *Phil. Trans. R. Soc.* **88** 469–526

[8] Stukeley W 1752 *Memoirs of Sir Isaac Newton's Life* (London: Royal Society) 15

[9] 1687-05-01: Huygens, Christiaan—de la Hire, Philippe; *Oeuvres Complètes de Christiaan Huygens* **IX** No 2455 (pp 130–3)

[10] Huygens C 1673 *Horologium Oscillatorium* **IVB**

[11] Harper W L 2011 *Isaac Newton's Scientific Method: Turning Data into Evidence about Gravity & Cosmology* (Oxford: Oxford University Press) chapter 5

[12] *Ibid.* table 5.2 (also the source of the other length measurements in this paragraph)

[13] Dew N 2010 Scientific travel in the Atlantic world: the French expedition to Gorée and the Antilles, 1681–1683 *Brit. J. Hist. Sci.* **43** 1–17

[14] Varin, Deshayes J and de Glos G 1682 *Observations astronomiques faites au Cap Verd, en Afrique, et aux Isles de l'Amerique;* reprinted in *Mémoires de l'Académie Royale des Sciences Depuis 1666 jusqu'a 1699* (Paris, 1729) vol **VII** pp 431–59

[15] Huygens C 1688 Rapport aengaende de Lengdevindingh door mijne Horologien op de Reys van de Caep de B. Esperance tot Texel Ao *1687 Oeuvres Complètes de Christiaan Huygens* **IX** 275 No 2519

[16] Huygens C 1690 *op. cit.* 396

[17] Schliesser E and Smith G E 2000 Huygens' 1688 *Report to the Directors of the Dutch East India Company on the Measurement of Longitude at Sea and the Evidence it offered against Universal Gravity,* in: *Archive for the History of the Exact Sciences* http://philsci-archive.pitt.edu/5510/ table 2 [accessed 15 October 2016]

[18] Huygens C 1688 *op. cit.* 275

[19] Huygens C 1688 *op. cit.* 287

[20] Schliesser E and Smith G E 2000 *op. cit.* 17–9

[21] Huygens C 1688 *op. cit.* pp 475ff; transl Schliesser E and Smith G E 2000 *op. cit.* p 28

[22] 1690-09-02: Huygens, Christiaan—Papin, Denis; *Oeuvres Complètes de Christiaan Huygens* **IX** No 2617 (pp 482–7)

[23] Robinet A (ed) 1957 *Correspondance Leibniz–Clarke: présentée d'après les manuscrits originaux des bibliothèques de Hanovre et de Londres* (Paris: Presse Universitaire de France) p 43

[24] 1688-04-24: Huygens, Christiaan—Hudde, Johannes; *Oeuvres Complètes de Christiaan Huygens* **IX** No 2517 (pp 267–8)

[25] Huygens C 1690 *op. cit.* Note 4; but note that De Graaff did not provide averages of the measurements of his clocks.

[26] Schliesser E and Smith G E 2000 *op. cit.* 40

[27] Baily F 1835 An Account of the Rev. John Flamsteed, the First Astronomer Royal; compiled from his own Manuscripts and other authentic documents, never before published *The Museum of Foreign Literature, Science and Art* **28** (New York: E. Littel) 289ff

[28] *Ibid.*

[29] Forbes E G 1976 The origins of the Greenwich Observatory *Vistas Astron.* **20** 39–50

[30] British Library Add. MS. Birch 4393, f. 89; holography copy with signatures of the King and Williamson (PRO/SP44/334/27-8)

[31] Taylor E G R 1939 'Old' Henry Bond and the Longitude *Mariner's Mirror* **25** 162–9

[32] Shaw W A (ed) 1676–1679 *Calendar of Treasury Books* **5** pp 132, 287

[33] Hooke R 1676 *Diary 1672–1680* 97

[34] This was Mademoiselle Louise Renée de Penancoët de Querouaille (1649–1734), mistress of King Charles II, later Duchess of Portsmouth; cf. Flamsteed J *Notes to my state of the Observatory* in: Baily F 1835 *op. cit.* vol **33** p 51

[35] Baily F 1835 *op. cit.* 37

[36] 1677-01-17: Flamsteed, John—Moore, Jonas; *Flamsteed's Manuscripts* 36

[37] Baily F 1835 *op. cit.* p 45 Note †

[38] *Ibid.* 46

[39] 1677-07-12: Flamsteed, John—Towneley, Richard; in: Howse D 1970 *Greenwich Time and the Discovery of Longitude* (Oxford: Oxford University Press) pp 38–40; quote from Royal Society Ms 243 (Fl.)

[40] 1678-03: Flamsteed, John—Moore, Jonas; *Ibid.* p 40; citing Royal Greenwich Observatory Ms 36 Fol 54

[41] *Ibid.* 54

[42] 1691-08-10: Newton, Isaac—Flamsteed, John; in: Baily F 1835 *op. cit.*; No 14 (p 129)

[43] 1692-02-24: Flamsteed, John—Newton, Isaac; *Ibid.* No 15 (pp 129–33)

[44] Baily F 1835 *op. cit.* p 132 Note *

[45] 1695-06-29: Newton, Isaac—Flamsteed, John; in: Baily F 1835 *op. cit.*; No 30 (p 157)

[46] 1695-07-09: Newton, Isaac—Flamsteed, John; in: Baily F 1835 *op. cit.*; No 31 (pp 157–8)

[47] Edleston J 1850 *Correspondence Of Sir Isaac Newton And Professor Cotes: Including Letters Of Other Eminent Men* p x

[48] 1700-02-2/12: Leibniz, Gottfried Wilhelm—Burnett of Kemney, Thomas; Deutsche Akademie der Wissenschaften (ed) 1923 *Sämtliche Schriften und Briefe* (Berlin: Akademie-Verlag) vol **I** pp 376–7

[49] 1699–01-06: Newton, Isaac—Flamsteed, John; in: Baily F 1835 *op. cit.* No 43 (p 166)

[50] Baily F 1835 *op. cit.* pp xxxii–xxxiv

[51] *Ibid.* 65

[52] *Ibid.* xxxv Note *

[53] 1709-03-24: Flamsteed, John—Sharp, Abraham; in: Baily F 1835 *op. cit.* No 140 (p 270)

[54] 1711-03-14: Arbuthnott, John—Flamsteed, John; in: Baily F 1835 *op. cit.* No 151 (p 280)

[55] Baily F 1835 *op. cit.* pp xxxviii–xxxix

[56] 1711-04-19: Flamsteed, John—Arbuthnott, John; in: Baily F 1835 *op. cit.* No 157 (pp 283–4)

[57] 1712-04-16: Flamsteed, John—Queen Anne; in: Baily F 1835 *op. cit.* No 170 (pp 295–6); draft of a petition

[58] Baily F 1835 *op. cit.* p xlii Note †

[59] 1716-03-29: Flamsteed, John—Sharp, Abraham; in: Baily F 1835 *op. cit.* No 210 (p 321)

[60] Flamsteed J Royal Greenwich Obs. Ms. 1/32C p 92 (Cambridge: Cambridge University Library)

[61] 1689-12-23: Huygens, Christiaan—Huygens, Constantijn Jr; *Oeuvres Complètes de Christiaan Huygens* **IX** No 2555 (p 354)

[62] Conti de Ottomano Freducci, Archivio di Stato, Firenze, Italy

[63] Vespucci A 1992 *Ridolfi Fragment Letters from a New World: Amerigo Vespucci's Discovery of America* (Marsilio Classics), ed L Formisano (University of Texas Press) pp 38–9 (Engl. transl); original: Ridolfi M 1937 Fragmentaria: Una lettera inedita di Amerigo Vespucci sopra il suo terzo viaggio, *Archivo Storico Italiano* **95** 3–20

[64] Marguet F 1917 *Histoire de la longitude à la mer, au XVIIIe siècle, en France* (A. Challamel/Presse Universitaire de France) p 7

[65] Cook A 2001 Edmond Halley and the magnetic field of the earth, *Notes & Records (Roy. Soc. J. Hist. Sci.)* **55** 473–90

[66] Taylor E G R 1939 *op. cit*

[67] The story of Ralph Walker's novel compass is particularly encouraging: Salisbury H 2012 *An alternative longitude solution—Ralph Walker and magnetic variation,* www.rmg.co.uk/discover/behind-the-scenes/blog/guest-post-alternative-longitude-solution-ralph-walker-and-magnetic [accessed 18 February 2017]

[68] Royal Society 1693 *Journal Book Original* JBO/9 p 118

[69] Royal Society 1693 *Collectanea Newtonia* vol **IV** No 425; March 1693

[70] National Maritime Museum, ADM/A/1797

[71] Halley's Log, 24 March 2013, *Mr Hally has gott a ship: the origins of Halley's voyage,* https://halleyslog.wordpress.com/2013/03/24/mr-hally-has-gott-a-ship-the-origins-of-halleys-voyage/ [accessed 18 February 2017]

[72] Halley E 1982 *The Three Voyages of Edmond Halley in the Paramore, 1698–1701* ed N J W Thrower (London: Routledge/Hakluyt Soc.) pp 129–31

[73] Goclenius R 1613 *Tractatus de Magnetica Vulnerum Curatione* (Frankfurt) p 95 (orig. publ. Marburg, 1608)

[74] Digby K 1658 *A Late Discourse Made in a Solemn Assembly of Nobles and Learned Men at Montpellier in France, by Sir Kenelm Digby, Kt. &c., Touching the Cure of Wounds by the*

Powder of Sympathy, transl (1664) by R White, 4th edn (London: printed by J.G. and to be sold by Octavian Pulleyn)

[75] Digby K 1658 *Discours fait en une celebre assemblee par le chevalier Digby, chancelier de la reine de la Grande Bretagne &c., touchant le guerison des playes par la poudre de sympathie* (Paris: Augustin Courbe & Pierre Moët)

[76] Tentzel A (ed) 1662 *Theatrum Sympatheticum Auctum* (Nuremberg: Johann Andreas Endter & the Heirs of Wolfgang Endter Jr); Tentzel's name does not appear in the book, but it has been ascribed to his editorship.

[77] Digby K 1658 *op. cit.*; English transl Digby K 1664 *op. cit.*

[78] Transl http://idolsofthecave.com/cabinet/2-odd-sympathy/ 3:20–3:35 [Accessed 10 February 2017]

[79] Digby K 1658 *op. cit.* p 134

[80] *Ibid.* 133–6

[81] Hedrick E 2007 Romancing the salve: Sir Kenelm Digby and the powder of sympathy *Brit. J. Hist. Sci.* **41** 161–85

[82] Anonymous 1688 *Curious Enquiries Being Six Brief Discourses, I, Of the Longitude* (London: Randal Taylor); tentatively attributed to W Winstanley

[83] Sobel D 1995 *Longitude: The True Story of a Lone Genius Who Solved the Greatest Scientific Problem of His Time* (London: Walker Books) p 22

[84] Fyler S 1699 *Longitudinis Inventæ Explicatio Non Longa, Or, Fixing The Volatilis'd, And Taking Time On Tiptoe, Briefly Explain'd; by which rules are given to find the longitude at sea by, as truly and as exactly as the latitude is found by the star in the tayle of Ursa Minor, called the Pole Star* (London: printed for the author)

[85] Sobel D 1995 *op. cit.* pp 11–6

[86] Powell D 1957 The Wreck of Sir Cloudesley Shovell (Glasgow: Soc. for Nautical Res.) *The Mariner's Mirror* **43** 333–6

[87] Royal Museums Greenwich 2014 *The 1707 Isles of Scilly Disaster*—Part 2 www.rmg.co.uk/discover/behind-the-scenes/blog/1707-isles-scilly-disaster---part-2 [accessed 12 February 2017]

[88] Herbert E 1709; extract from a letter; passages in brackets are corrections and annotations found in the letter's margins; in: Cooke J H 1883 *The Shipwreck of Sir Cloudesley Shovell, on the Scilly Islands in 1707 from Original and Contemporary Documents Hitherto Unpublished* (Gloucester: John Bellows) www.hmssurprise.org/shipwreck-sir-cloudesley-shovell [accessed 22 February 2017]

[89] Cooke J H 1883 *op. cit.*

[90] Bottrell W 1873 *Traditions and Hearthside Stories of West Cornwall* vol **2** (Penzance: Beare and Son) p 232 www.sacred-texts.com/neu/celt/swc2/swc250.htm [accessed 22 February 2017]

[91] Whiston W and Ditton H 1714 *REASONS for a BILL, PROPOSING A Reward for the Discovery of the LONGITUDE* (London?: s.n.)

[92] Pepys S 1683, published in: Chappell E 1935 *The Tangier Papers of Samuel Pepys* (Reading: Navy Records Soc.) p 129

[93] Whiston W and Ditton H 1715 *A New Method for Discovering the Longitude Both at Sea and Land, Humbly Proposed to the Consideration of the Publick* 2nd edn (London: printed

for Mr. Whiston and Mrs. Ditton; and sold by J. Roberts near the Oxford-Arms-Inn in Warwick-Lane). Note that Ditton had passed away by the time this manuscript was published, hence the reference to 'Mrs.' Ditton.

[94] Werrett S 2016 *Day 6: Response to Figure 8,* in: Barrett K 2016 Looking for 'the Longitude' *Brit. Art Stud.* **2** https://doi.org/10.17658/issn.2058-5462/issue-02/kbarrett

[95] Shakespeare W 1714 *The History of Sir John Oldcastle,* in: *The works of Mr. William Shakespear ... Adorn'd with cutts,* ed N Rowe, 8 vols (London) vol **8** pp 181–249 (p 194)

[96] Newton I 1714 cited by: Brewster D 1831 *The Life of Sir Isaac Newton* (London: John Murray) p 256

[97] Whiston W and Ditton H 1714 *op. cit.*

[98] Thacker J 1714 *The longitudes examin'd Beginning with a short epistle to the longitudinarians, ...* (London: J. Roberts)

[99] Haldanby F 1714 *An attempt to discover the longitude at sea pursuant to what is proposed in a late act of Parliament* (London: John Morphew)

[100] Glennie P and Thrift N 2009 *Shaping the Day: A History of Timekeeping in England and Wales 1300–1800* (Oxford: Oxford University Press) p 403 Note 37

[101] Whiston W and Ditton H 1715 *op cit.*

[102] Curtis T 1837 *The London encyclopaedia, or, Universal dictionary of science, art, literature and practical mechanics; including an English Lexicon* vol **25** (London: Thomas Tegg) p 271

[103] He referred here to the *Memoirs of Martinus Scriblerus,* the collaborative publications of the Scriblerus Club of London in the early 18th Century.

[104] 1714-07-17: Arbuthnot, John—Swift, Jonathan; in: Woolley D (ed) 1999–2007 *The Correspondence of Jonathan Swift,* D. D. 4 vols (Frankfurt am Main: Lang) pp 11–12

[105] Nicolson M H and Rousseau G S 1960 *'This Long Disease, My Life:' Alexander Pope and the Sciences* (Princeton, NJ: Princeton University Press) pp 142–5, 167–8

[106] Worth C 1996 Swift's 'Flying Island': Buttons and Bomb-Vessels, *Rev. Engl. Stud.* **42** 343–60

[107] Swift J 1759 *The Works of Dr Jonathan Swift* 8 vols (Edinburgh: printed for A. Donaldson at Pope's Head) vol **6** p 125

[108] The first recorded meeting of the Board of Longitude was held on 30 July 1737; until then, the Commissioners acted mostly as independent authorities (see RGO 14/5, https://cudl.lib.cam.ac.uk/view/MS-RGO-00014-00005/7; accessed 14 February 2017)

[109] 1715-02-01: Flamsteed, John—Sharp, Abraham; in: Forbes E G, Murdin L and Willmoth F (ed) 1995–2002 *The Correspondence of John Flamsteed, the First Astronomer Royal* 3 vols (Bristol: IOP Publ. Ltd) vol **3** p 723

[110] Hall A R and Tilling L (ed) 1959–1977 *The Correspondence of Isaac Newton* 7 vols (Cambridge: Cambridge University Press for The Royal Society) vol **6** 161

[111] Whiston W and Ditton H 1715 *op. cit.*

[112] Chalmers A 1813 *The General Biographical Dictionary Containing an Historical and Critical Account of the Lives and Writings of the Most Eminent Persons in Every Nation; Particularly the British and Irish; From the Earliest Accounts to the Present Time* (London: J. Nichols and Son, etc.)

[113] Ditton H 1714 *A Discourse concerning the Resurrection of Jesus Christ. In Three Parts* (London: printed by J. Darby, and sold by Andr. Bell, and B. Lintott); see notes in the German translation

[114] Sobel D 1995 *op. cit.*

[115] Lynall G 2014 Scriblerian Projections of Longitude: Arbuthnot, Swift, and the Agency of Satire in a Culture of Invention *J. Lit. Sci.* **7** 1–18

[116] Flamsteed J 2001 *The Correspondence of John Flamsteed, The First Astronomer Royal, 1703–1719* vol **3** ed E G Forbes (Bristol: CRC Press) 712; letter 1366

[117] 1714-08-31: Flamsteed, John—Sharp, Arbraham; in: Flamsteed J 2001 *op. cit.* 3 700; letter 1360

[118] Defoe D 1697 *An Essay upon Projects* (London: Printed by R.R. for Tho. Cockerill) 33–4

Epilogue

Zero longitude

E.1 The Greenwich Meridian

The Royal Observatory at Greenwich, just outside London, has become synonymous with the modern concept of 0° longitude, commonly known as the Greenwich Meridian. You may be surprised that this choice of prime meridian as positional reference is a relatively recent development. In fact, general agreement to adopt the Greenwich Meridian as the world's most important prime meridian only dates from 1884, and it happened almost by accident.

Throughout most of history, there was no single prime meridian which everyone agreed with. Most maritime nations established their own reference meridian, often—but not always—passing through the country in question. We already encountered the reference meridians at Alexandria and Rhodes (then two principal cities in the Roman Empire) in chapter 2.1.2, those running through Toledo (Spain) versus Rome (Italy) in chapter 3.1.5, and those at Uraniborg (Denmark) and at Paris Observatory in chapter 4.1.4. We also came across differences among the meridians at Toledo, Salamanca (Spain), and Ferraro (Italy) in chapter 2.3.2, for instance.

In general, meridians used for global reference varied from the Azores to the Cape Verde and Canary Islands—specifically, the Ferro and Tenerife meridians—to reference locations, ordered here approximately from West to East, in the *United States* (Philadelphia and Washington, D.C.)[1], *Brazil* (Rio de Janeiro), *Spain* (the San Fernando meridian through Cadiz; Salamanca and Toledo), *Great Britain* (Lizard Point, Cornwall; St Paul's Cathedral, London; Royal Observatory, Greenwich), *France* (Rouen and Paris), present-day *Belgium* (Brussels and Antwerp), the *Dutch Republic* (Goes) and the *Netherlands* (the Amsterdam meridian through the Westerkerk Church, which was used to define the nation's legal time [1] between 1909 and 1937), *Norway* (Kristiania/Oslo), *Denmark* (Uraniborg and

---

[1] Washington, D.C., has played host to four different proposals for prime meridians, the Old and New Naval Observatory meridians, the White House meridian, and the Capitol meridian.

Copenhagen), *Switzerland* (Berne), *Germany* (Frankfurt, Tübingen, Ulm, and Augsburg), *Italy* (the meridian of Monte Mario passing through Rome, recomputed in 1945; Pisa, Florence[2], Bologna, Ferrara, Venice, and Naples), *Poland* (Szczecin, Krakow, and Warsaw), *Russia* (Königsberg in Prussia—present-day Kaliningrad—and the Pulkova meridian through St Petersburg), *Romania* (Oradea), *Israel* (Jerusalem), *India* (Ujjain), *Japan* ('Kairekisyo' in Nishigekkoutyou-town, Kyoto; used between 1779 and 1871), and many more, before eventually settling on the original Greenwich Meridian, that is, the Airy Transit Circle Meridian [2], followed by its later incarnations (see chapter E.4). Starting from the early 18th Century, some countries adopted the location of their national observatory as their zero meridian of choice, including Stockholm and Pulkova Observatories, for instance, and even the Great Pyramid of Giza was once mooted as reference in 1884 [3].

Given the proliferation of prime meridians in use at any given time, the French King Louis XIII issued an *ordonnance* (a decree) on 1 July 1634 in what became the first attempt to standardize the system of prime and auxiliary meridians. He announced that French scientists and navigators should use the meridian passing through the westernmost of the Canary Islands, Ferro (El Hierro). However, at the time, the longitude of the Canary Islands with respect to that of Paris was unknown to any reasonable precision. The expedition of Varin, Deshayes, and De Glos of 1681–1683 was meant to obtain a first firm measurement of Ferro island's longitude [4]. However, after significant delays associated with trade-political negotiations and setbacks, the idea of sailing to Ferro was abandoned and the expedition's focus was recast to visiting Gorée Island and the Antilles instead [5].

A more successful attempt was made in 1800 by the French astronomer and mathematician Pierre-Simon, Marquis de Laplace (1749–1827), who wrote that …

'[i]t is desirable that all the nations of Europe, in place of arranging geographic longitudes from their own observatories, should agree to compute it from the same meridian, … in order to determine it for all time to come. Such an arrangement would introduce into the science of geography the same uniformity which is already enjoyed in the calendar…'

In parallel, Otto Wilhem von Struve (Отто Васильевич Струве; 1819–1905), Director of Pulkova Observatory in St Petersburg, performed a careful study of five meridians, which he presented to the Geographical Society of Russia in 1870. Three of these were commonly used, namely those running through Greenwich, Paris, and Ferro. He also considered two additional meridians, specifically those located along North–South lines at Greenwich +30° W and the Greenwich antemeridian (Greenwich +180°). Von Struve considered the requirements of both geographers and navigators, as well as those of astronomers, in reaching his

---

[2] The Florence meridian is the antipode to a meridian line running through the Bering Strait, which was proposed as a possible neutral prime meridian at the 1884 International Meridian Conference in Washington, D.C.

recommendation. He proposed that the Greenwich Meridian be adopted, but he was also keenly aware of the political objections his proposal would face. Indeed, von Struve's study ...

'can be considered the first salvo in what turned out to be a half century of skirmishes aimed at having the world adopt Greenwich as the common meridian for longitudes' [6].

A first International Congress convened to discuss unresolved problems of geography and cosmography—the science that describes the Earth's and the Universe's general features without competing with geography or astronomy—was eventually hosted in Antwerp, Belgium, in 1871. However, since the Franco–Prussian (Franco–German) war, which was waged from 19 July 1870 to 10 May 1871, had only just concluded, the atmosphere was tense among the delegates from the 19 countries represented. Little progress was made on resolving the urgent open issues, but at least this first International Congress triggered in-depth discussions. Four years later, the second International Geographical Congress in Paris, followed by the Third International Geographical Congress held in 1881 in Venice (Italy) set the stage for an eventual successful outcome. Although most of the 29 countries represented favoured the idea of a prime meridian passing through the dedicated meridian transit telescope at the Royal Observatory in Greenwich, no significant decisions were taken at those meetings. It was hence decided to hold an international conference aimed at reaching consensus on basic issues of contention related to the concept of a Universal Day and the establishment of a globally accepted prime meridian.

Soon afterwards, the International Meridian Conference of 1884 was held in Washington, D.C., and attended by 41 delegates from 25 countries. The brainchild of Sir Sandford Fleming (1827–1915), the Scottish–Canadian engineer and inventor, the delegates eventually passed seven resolutions which cemented Greenwich's place in the history of cartography [7]:

1. *That it is the opinion of this Congress that it is desirable to adopt a single prime meridian for all nations, in place of the multiplicity of initial meridians which now exist.*
2. *That the Conference proposes to the Governments here represented the adoption of the meridian passing through the centre of the transit instrument at the Observatory of Greenwich as the initial meridian for longitude.*
3. *That from this meridian longitude shall be counted in two directions up to 180 degrees, east longitude being plus and west longitude minus.*
4. *That the Conference proposes the adoption of a universal day for all purposes for which it may be found convenient, and which shall not interfere with the use of local or other standard time where desirable.*
5. *That this universal day is to be a mean solar day; is to begin for all the world at the moment of mean midnight of the initial meridian, coinciding with the beginning of the civil day and date of that meridian; and is to be counted from zero up to twenty-four hours.*

6. *That the Conference expresses the hope that as soon as may be practicable the astronomical and nautical days will be arranged everywhere to begin at mean midnight.*
7. *That the Conference expresses the hope that the technical studies designed to regulate and extend the application of the decimal system to the division of angular space and of time shall be resumed, so as to permit the extension of this application to all cases in which it presents real advantages.*

Fleming had been heavily involved in a serious expansion of the Canadian railway network, which made him realize that adoption of some kind of universal time would allow the railways to run most smoothly. The idea of a U.S.-wide system of time zones had first been suggested to him in 1870 by Charles F Dowd (1825–1904), co-principal of Temple Grove Ladies Seminary (Saratoga Springs, New York). Fleming decided to act upon this idea, a world-wide extension of which had also been mooted by the Italian mathematician and politician Quirico Filopanti (Giuseppe Barilli; 1812–1894). Fleming consequently started to relentlessly champion the concept of a global system of universal timekeeping, specifically pointing out that civil time reckoning would be a 'cosmopolitan' system benefiting not only 'men of business' but also the 'entire family of man' [8]. Meanwhile, Cleveland Abbe (1838–1916), U.S. government meteorologist and geophysicist, also turned into a vocal proponent of timekeeping reform.

Fleming eventually proceeded to organize a conference to debate the regulation of time, the establishment of a system of longitude, and the adoption of a prime meridian. He had hoped that any prime meridian adopted by the delegates at the meeting would become the *de facto* standard of a new Universal Day, which hence would also establish a system of time zones. Events unfolded differently, however. His suggestion to link both issues caused considerable controversy and even anger among the delegates, some of whom were not actually empowered by their home countries to vote on matters of time. One key problem associated with establishing a European prime meridian, as proposed, was highlighted by the representative for the Ottoman (Turkish) Empire, who pointed out that a reference line running through Greenwich did not make much practical sense in the Middle East, for instance, since …

'the majority of our population is agricultural, working in the fields, and [they] prefer to count to sunset; besides, the hours for the Moslem prayers are counted from sundown to sundown' [9].

Nevertheless, Turkey eventually relented its opposition and adopted the universal system of time zones based on the Greenwich reference by 1920. Adopting a different approach, U.S. delegate William Frederick Allen (1846–1915), a civil engineer, Secretary of the U.S. railways' *General Time Convention* (a body which had been set up to harmonize train schedules), and Managing Editor of the *Travelers' Official Guide to the Railways*, argued that …

'[e]xactness of time reckoning is an imperative necessity in the conduct of business,'

thus moving the discussion's emphasis from agricultural requirements and religious observances to commercial arguments. With the continuing expansion of the railways and increasing exports of goods across great distances, reaching agreement on the concept of a Universal Day (and, hence, Universal Time) became a necessary commercial tool, facilitating increased connectivity of international business operations and, hence, globalization. Indeed, ...

'transnational investors used (or misused) the [conference] ... to synchronize countries to precisely coordinated capital flow' [10].

Nevertheless, the conference delegates were both unwilling and unable to make progress on any resolutions involving the establishment of a globally unified time system. Instead, they mostly concentrated on discussing the merits of establishing a universally agreed prime meridian. However, despite their misgivings, the conference's final, although non-binding, document linked Fleming's concept of the start of the Universal Day to that of the prime meridian, which was defined as the meridian running through the cross-hairs of the Airy Transit Circle (see below) at the Royal Observatory at Greenwich.

The 1884 convention gave birth to the concept of Greenwich Mean Time, GMT, that is, the mean solar time at the Airy Transit Circle. Greenwich had been favoured as the global position reference by 22 countries. One country, San Domingo (the present-day Dominican Republic), opposed this choice, while two countries—France and Brazil—abstained. France proposed to adopt a neutral line, suggesting either the Azores or the Bering Strait, but its delegates eventually abstained. The French continued to use the Paris Observatory meridian for timekeeping purposes until 11 March 1911, when they adopted GMT as the nation's civil time. Still, for decades they continued to refer to GMT as ...

'Paris Mean Time retarded nine minutes and 21 seconds,'

while they were not yet prepared to adopt the Greenwich *Meridian* as positional reference. They would, however, eventually succumb to international pressure and common sense, adopting the Greenwich Meridian reference line from 1 January 1914; we will return to the developments that led to this adoption in chapter E.3.

In response to the lack of progress at the International Meridian Conference on the establishment of the Universal Day, however, the journal *Science* commented on 31 October 1884 that ...

'[i]t seems unfortunate that Mr. Allen's resolution for local times, differing by whole hours from the universal time, was not recommended; for this would seem by all odds the simplest way of connecting local and universal times. It is already in almost universal use in this country' [11].

Indeed, the U.S. had already adopted the Greenwich Meridian as the basis for its own national system of time zones, which contributed significantly to the Greenwich Meridian's choice as global positional reference. With the increasing importance of the railways, it was no longer sufficient to rely on local times. Railway companies in both the U.S. and the U.K. introduced standard time systems to base their timetables on. In the U.K., GMT was adopted for practical reasons, since the time signals were directly transmitted to the railway system from the Royal Observatory at Greenwich via the electric telegraph.

In addition, by the late 19th Century, sea charts based on the Greenwich Meridian were used in support of almost three-quarters of the world's commerce. Specifically, in 1879 Fleming published his paper *Time-Reckoning and the selection of a prime meridian to be common to all nations* [12] demonstrating this. He presented a table '*prepared from the latest authorities within reach*,' which contained '*an estimate of the number and tonnage of steamers and sailing ships belonging to the several nations of the world*,' as well as the meridians they used for determining their prevailing longitude. Fleming had found that 95.5% of the world's ships (97.5% by tonnage) used only 11 main meridians: see figure E.1. The Greenwich Meridian was

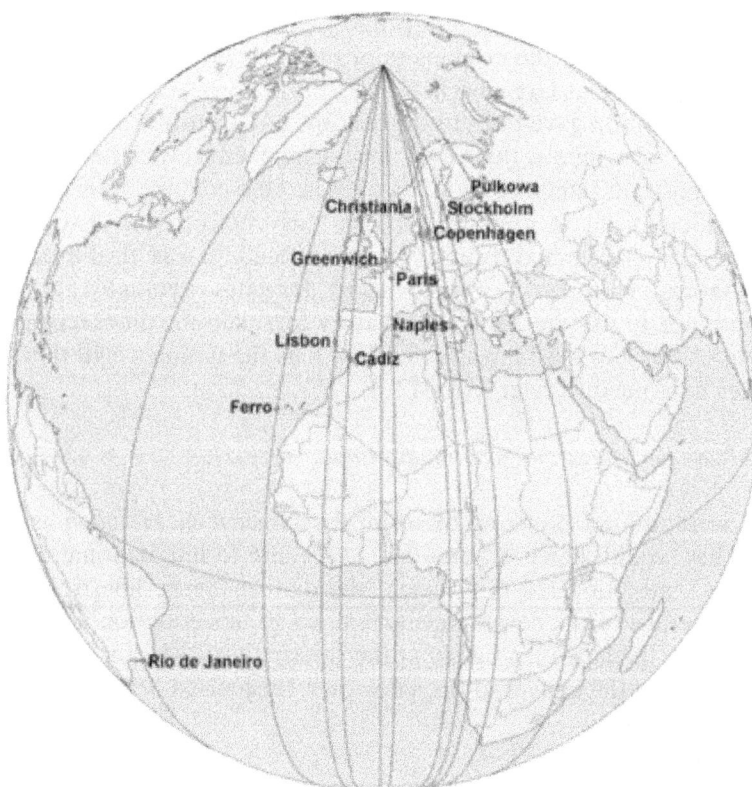

Figure E.1. In 1879, 95.5% of the world's ships used only 11 main meridians for positional reference. These meridians are labelled by their main reference cities. (© 2004 National Maritime Museum, Greenwich, London.)

popular with 65% of ships (72% by tonnage) and that of Paris by 10% (8% by tonnage). Adoption of the Greenwich meridian as 0° longitude would thus economically benefit the largest number of people globally.

Fleming's advocacy of timekeeping reform received a boost with his retirement from the railways in 1880, which allowed him to spend more time on this pursuit. In 1881, he attended and presented a related paper at the Third International Geographical Congress. His lobbying contributed to the passing of an enabling act [13] in the U.S. on 3 August 1882, *viz.*

> *Be it enacted by the Senate and House of Representatives of the United States of America in Congress assembled, That the President of the United States be authorized and requested to extend to the governments of all nations in diplomatic relations with our own an invitation to appoint delegates to meet delegates from the United States in the city of Washington, at such time as he may see fit to designate, for the purpose of fixing upon a meridian proper to be employed as a common zero of longitude and standard of time-reckoning throughout the globe, and that the President be authorized to appoint delegates, not exceeding three in number, to represent the United States in such International Conference.*

However, it was deemed that more time was needed before such a conference would likely yield positive dividends. Thus, at the October 1883 International Geodetic Association conference held in Rome, a number of resolutions were passed, in particular [14],

> *The conference proposes to the Governments to select for the initial meridian that of Greenwich ... for the reason that that meridian fulfils ... all the conditions wished for by science and because being at present the best known of all, it offers the most chances of being generally accepted.*

> *The Conference hopes that if the entire world ... [accepts] the meridian of Greenwich ... Great Britain will find in this fact ... a new step in favour of the unification of weights and measures, by acceding to the Convention du Mètre of the 20th [of] May, 1875.*

Taking the lead, on 18 November 1883, at noon on the 75th meridian (75°W), the North American railways adopted a common system of time zones for timetabling purposes. This triggered the rapid establishment of five, one-hour-wide time zones across the continent and provided the appropriate political climate for Chester Alan Arthur (1829–1886), the 21st U.S. President, to issue invitations to delegates to attend the International Meridian Conference of October 1884.

Despite the fact that the conference concluded with the publication of the seven resolutions we discussed earlier in this chapter, these resolutions did not have any teeth; they were just proposals. Implementation of these proposals by the individual governments concerned proceeded at a snail's pace, and some developments were mere symbolic gestures rather than practical implementations. During the first few years following the adoption of the resolutions, only Japan proceeded to draft legislation. In 1886, the Japanese passed in law the adoption of GMT+9 hours for their national time reckoning from the start of 1888. Elsewhere, even if it had been adopted, GMT was often used inconsistently, either by counting the number of hours from midnight (as recommended by Resolution 6) or from local noon.

Colonel Sir Thomas Hungerford Holdich (1843–1929), the English geographer and Superintendent of Frontier Surveys in British India attended the Fifth International Geographical Congress held in Berne (Switzerland) in August 1891 as delegate on behalf of the British Indian government. He voiced his concern about the state of the proposed changes:

> *'The meridian question ... has certainly advanced far enough that all English maps should possess a common origin for longitude. At present this is not so, for maps of India ... are published with a different longitude ... from the true Greenwich value ... a continuance of the present system is a grave disadvantage if we wish to persuade other nations to adopt Greenwich as the longitude of origin'* [15].

Full implementation of all of the resolutions passed by the 1884 International Meridian Conference would not occur until 1925, however. We will return to the modern timeline below, but first we take a step back and trace the history of reference meridians from their first use by Eratosthenes of Cyrene and Hipparchus of Nicea in Greek Antiquity.

E.2 Prime meridians from Greek Antiquity until the Enlightenment

Although Eratosthenes' maps from the third Century BCE have not survived the ravages of history, contemporary accounts suggest that he constructed a map that included a grid pattern, with a zero meridian that passed through his home city of Alexandria. He apparently believed that the meridian through Alexandria also passed through both the eastern Mediterranean island of Rhodes and Syene (present-day Aswan in southern Egypt), but this was incorrect. In the second Century BCE, Hipparchus subsequently adopted a reference meridian which passed through Rhodes, probably also in the mistaken notion that the same meridian passed through Alexandria and Syene. He suggested that one could determine positions East and West of this meridian by determining the local time with respect to the local time referenced at the Rhodes meridian.

Three hundred years later, during the second Century CE, Ptolemy constructed the first maps using the notion of a zero meridian in his masterwork, *Geographia*, an idea he may in fact have borrowed from his predecessor, Marinus of Tyre. Both Marinus and Ptolemy adopted the fabled Isles of the Blessed (the Fortunate Isles) as the westernmost extent of the known world, coinciding with 0° longitude.

Although the Fortunate Isles are often tentatively identified with the Canary Islands, spanning the longitude range from 13 to 18° W, Ptolemy's maps imply that their positions are, perhaps, more closely associated with the Cape Verde islands, 22 to 25° W. For the present narrative, the exact location of the Fortunate Isles is not important, as long as they are located sufficiently far west of the westernmost tip of the African continent at 17.5° W, since negative numbers had not yet been invented in the Greek sphere of influence during the lifetimes of Marinus and Ptolemy. (Negative numbers appear for the first time in *The Nine Chapters on the Mathematical Art—九章算術; Jiǔzhāng Suànshù*—which is thought to have its origins in the Chinese Han Dynasty [16], 202 BCE–220 CE.) In his *Geographia*, Ptolemy refers to the Fortunate Isles at 18° 40′ west of Venta Belgarum ('the market town of the Belgae') [17], which corresponds to present-day Winchester (England), located at a longitude of 1° 19′ W. Therefore, this places the reference meridian at the Fortunate Isles at 20° W.

Independent accounts suggest that both Marinus and Ptolemy defined their prime meridian at 2° 30′ West of the Sacred Promontory (present-day Cabo de São Vicente/Cape St Vincent in Portugal), the westernmost tip of Europe in the worldview of the ancient Greeks. That particular longitude coincided with the western edge of the Canary Islands on their maps, although modern maps place the Canary Islands approximately 8° West of Cabo de São Vicente. Based on a modern longitude determination for Cabo de São Vicente of 8° 59′ 40″ W, this would place Marinus' and Ptolemy's zero meridian at about 17° W.

With Ptolemy having set the standard, European cartographers did not manage to advance their field for well over a thousand years, when much of Europe was embroiled in war and suffering from religious persecution during the Dark Ages. During this period, leadership in cartographic developments passed over to the Islamic world, as we already saw in chapter 2.2.1. Among the choices of prime meridians made in the Arab world, an innovative choice passed through Arin, a mythical city thought to be located 10° east of Baghdad; a longitude line through Arin split the known Arab world into approximately equal halves.

Rabbi Moses ben Maimon (1135/8–1204; מֹשֶׁה בֶּן־מַימוֹן), more commonly known as Maimonides (Μαϊμωνίδης), called Mecca, Saudi Arabia, 'אמצע היישוב,' 'the middle of the habitation' (that is, the middle of the habitable hemisphere). Apparently, this was a commonly accepted convention by Arab geographers of the day. In this context, and given the importance of the direction to Mecca for Islamic religious observance, at a conference in Doha, Qatar, on 21 April 2008, Sheikh Yūsuf al-Qaraḍāwī (يوسف القرضاوي) and his Muslim cleric colleagues proposed to adopt Mecca Time as time standard, thus hoping to establish the meridian that passes through the Muslim holy city as the prime meridian.

Further East, we also already saw, in the context of Zheng He's oceanic voyages, that Chinese astronomers often relied on distance measurements with respect to their base observatory rather than using a global coordinate system. Since the fourth-Century CE publication of the *Surya Siddhanta*, a Sanskrit treatise about Indian astronomy, for timekeeping purposes [18] Indian cartographers widely employed a prime meridian passing through Avanti (present-day Ujjain in the Indian state of Madhya Pradesh) and Rohitaka, the ancient name for Rohtak in the present-day Indian state of Haryana.

Meanwhile, Ptolemy's *Geographia* continued to play a dominant role in European cartography for centuries to come. Until well into the 16th Century, many European globe and mapmakers followed Ptolemy's lead in assigning zero longitude to an imaginary line through the Fortunate Isles. Yet, simultaneously the pursuit for a 'natural' zero meridian continued unabated. One initially promising approach, pursued by luminaries including Columbus, relied on the concept of 'zero magnetic deviation' (see chapter 6.3.1), that is, the idea that the prime meridian would coincide with a line connecting the North and South Poles where the compass needle did not show any variation of the magnetic North Pole from true North. One could determine the variation of the compass from true North by 'shooting' the Pole Star and observing the direction indicated by the compass needle East or West of the Pole. In 1493, Columbus reported that his magnetic compass needle pointed due North at some point in the mid-Atlantic; his guidance was subsequently adopted in sealing the Treaty of Tordesillas (1494) to delineate the global spheres of influence of Spain and Portugal and to settle their territorial sovereignty over the newly discovered lands in the Americas (see chapter 2.3.3).

One of the first cosmographers to take forward the concept of 'zero magnetic deviation' was the Portuguese astronomer Rui (Ruy) Faleiro, the principal scientific organizer of Ferdinand Magellan's circumnavigation of the globe in 1519–1522. During this period of scientific renaissance, Gerard Mercator's early maps, including his 1538 *Orbis Imago* world map based on the recently invented double cordiform (heart-shaped) projection, had their prime meridians pass through the centre of the Canary Islands: see figure E.2. This map represented an improved version of an

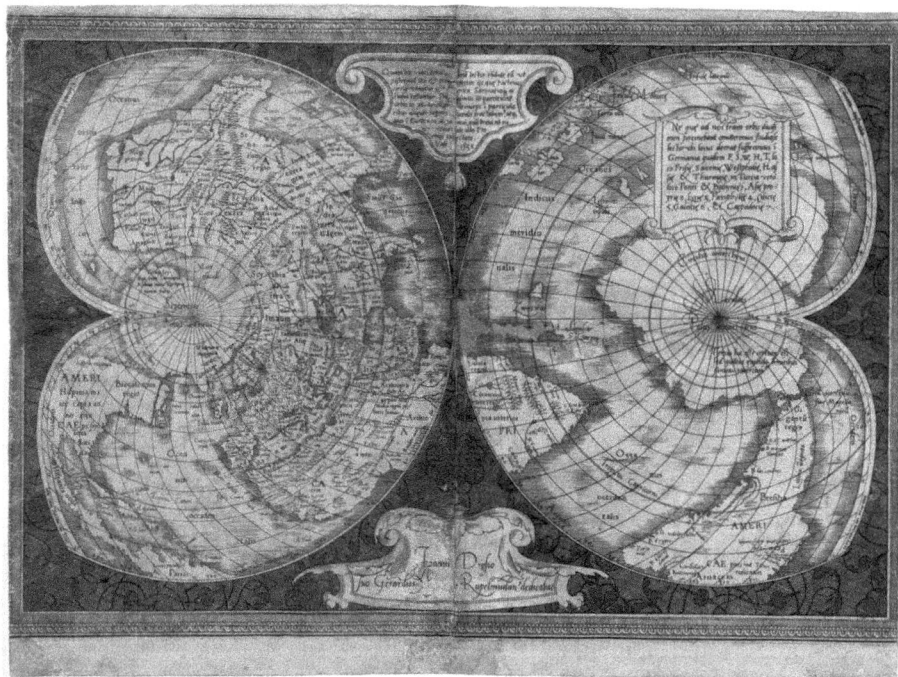

Figure E.2. Mercator's 1538 world map in double cordiform projection. (Image courtesy of the University of Wisconsin–Milwaukee Libraries.)

earlier map constructed by the French mathematician and cartographer Oronce Finé (1494–1555). The latter developed his heart-shaped projection as a balance between distortions, particularly in the Polar regions; distances from the North Pole in this map projection are correct, and so are distances along parallels.

In 1541, Mercator constructed his now-famous 41 cm globe, placing his reference meridian such that it passed through the eastern Canary Island of Fuerteventura at 14°1′ W, clearly still under the influence of Ptolemy's teachings. As early as the 1550s, he adopted the magnetic hypothesis to draw his prime meridian through the Azores [19]. Subsequently, on his famous 1569 world map (reproduced in figure 2.23), he drew the prime meridian through the Cape Verde Islands. He justified this more easterly location of the prime meridian in the map's legend, explaining that an experienced pilot, Francis of Dieppe, had assured him that the compass needle did not vary in the islands of Sal, Boa Vista, and Maio in the Cape Verde archipelago:

'Francis of Dieppe, a skillful shipmaster, asserts that movable balances, after being infected with the virtue of a magnet, point directly to the Earth's Pole in the Isles of Cape Verde: Sal, Boa Vista, and Maio. This is closely supported by those who state that this occurs at Terceira or S. Maria (which are isles of the Azores); some believe that this is the case at the most westerly of these islands which is called Corvo. Now, since it is necessary that longitudes of places should, for good reasons, have as origin the meridian which is common to the magnet and the World, in accordance with a great number of testimonies I have drawn the prime meridian through the said Isles of Cape Verde; and as the magnet deviates elsewhere more or less from the Pole, there must be a special Pole towards which magnets turn in all parts of the world, therefore I have ascertained that this is in reality at the spot where I have placed it by taking into account the magnetic declination observed at Ratisbon [Regensburg, Germany]. *But I have likewise calculated the position of this Pole with reference to the Isle of Corvo in order that note may be taken of the extreme positions between which, according to the extreme positions of the prime meridian, this Pole must lie until the observations made by seamen have provided more certain information.'*

His drawing of the magnetic meridian is confusing, however, since it does not trace a well-defined North–South longitudinal line; instead, his line passes through the Azores, specifically through the islands of São Miguel (25° 30′ W) and Santa Maria (25° 06′ W), and also through Sal (22° 56′ W), Boa Vista (22° 48′ W), and Maio (23° 10′ W). It is indeed possible that Mercator was thrown by the circuitous behaviour of magnetic isogones.

Nevertheless, the concept of 'zero magnetic deviation' continued to hold sway as late as 1594, when the English cartographer Christopher Saxton (c. 1540–c. 1610) adopted São Miguel Island (25.5°W) in the Azores as reference for his prime meridian. However, it took until 1608 before the magnetic hypothesis was finally put to rest. In the meantime, leading cartographers continued to explore the use of the compass needle as guidance for their map construction.

The Dutch–Flemish cartographer, astronomer, and navigational authority Petrus Plancius (1552–1622) was a strong proponent of adopting the western Azores, particularly the island of Santa Maria, as reference for the prime meridian, based on observations of magnetic variations. In 1594, Plancius had been awarded a patent by the Dutch Republic's Staten Generaal (the States General represent the nation's Government) for his proposed method to determine longitude ('eastfinding') at sea based on observations of magnetic-needle variations and a calibrated scale, in combination with a so-called *Astrolabium Catholicum* (a 'universal astrolabe'): see figure E.3. The instrument, only three of which are known to have survived to the present time, consisted of four components [20]:

1. the ecliptic plane, a disc which rotates within a graduated circle with hours and degrees indicated, supplemented with a shadow quadrant;

Figure E.3. Astrolabe showing evidence of the fine craftsmanship of Humfrey Cole (d. 1591). Star names are given with their magnitudes and their corresponding planetary temperaments. The mater (the hollow section of the disc) bears the markings for a *quadratum nauticum* (nautical square, used for navigation) and the back shows the universal projection as described by Gemma Frisius in his *Astrolabum Catholicum*. (Courtesy: British Museum, No. 1855, 1201.223; licensed under CC BY-NC-SA 4.0.)

2. the geographic grid in an equatorial, conformal stereographic projection (a type of projection already known to Ptolemy and Hipparchus), with a rotating ruler and a slide with an index arm and a pointer;
3. a regular astronomical astrolabe;
4. the celestial sphere in stereographic polar projection with stars shown at declinations between 90° north and 30° south, a practice that goes back to Ptolemy in Greek Antiquity.

Its operation was described in detail by its inventor, Gemma Frisius, in his treatise *Medici ac Mathematici de astrolabe catholico liber ...* (Antwerp, 1556). He used the system of meridians and parallels on the front disc alternately as a longitude–latitude grid, that is, as a grid of geographic coordinates, and as a local grid composed of zenith circles and altitude parallels [21,22]. For details, I refer the interested reader to the lucid explanation of the instrument's technical operation provided by Koeman [22].

To test his theory in practice on southern voyages of the Dutch East India Company's merchant fleet, Plancius taught junior merchant Frederi(c)k de Houtman (1571–1627) how to measure and record compass readings. De Houtman assisted Pieter Dirksz Keyser (c. 1540–1596), the chief pilot, with astronomical observations during the first Dutch expedition [23] to the East Indies, on board the flute *Hollandia*. Captained by Master Jan Dignumsz van Kwadijk, they left their Texel anchorage as part of a merchant fleet on 2 April 1595 and returned in August 1597. This voyage is now known as the *Eerste Schipvaart* (First Voyage). Non-maritime matters were the responsibility of Frederik's brother Cornelis de Houtman (1565–1599), who discovered a new sea route from Europe to Indonesia, thus facilitating Dutch dominance in the spice trade. Frederik de Houtman also joined the second expedition, known as the *Tweede Schipvaart*, of 1 May 1598 to 17 July 1599 on board the yacht *Friesland*, under the command of Master Jan Kornelisz May.

Around the same time, in 1601 the Flemish cartographer Joost de Hondt (*Lat.*: Jodocus Hondius; 1563–1612) constructed a terrestrial globe whose legend concludes as follows by referencing his adopted prime meridian:

'We have begun our longitudes not as Ptolemy did from the Fortunate Islands but from those called Azores because there the compass needle points due north.'

Hondius' prime meridian passes through São Miguel; on his earlier world map of 1598, he drew his prime meridian through the sea between the Azorean islands of Corvo and Faial.

The early 17th Century represented a radical change in the attitudes of mapmakers with respect to the adoption of prime meridians based on compass needle variations. Willem Jansz Blaeu, the famous mapmaker, went through a period of uncertainty as regards the most suitable prime meridian, as evidenced by his decision to include a number of meridians, in addition to that passing through Corvo, on his two-sheet world map of 1604, including lines passing to the west of São Miguel and through Lizard Point in Cornwall (England). However, by 1608 Blaeu had completely rejected the 'zero magnetic deviation' approach as a viable means to determine 0° longitude. He went so far as to ridicule the idea in a legend on his 1617 terrestrial globe; that globe's positional reference system used the Pico Tenerife, the highest point on the Canary Island of Tenerife, as its prime meridian, a choice he adopted from that moment onwards (see also figure 5.19).

Meanwhile, other European cartographers produced maps and globes during the 16th and 17th Centuries which also included a variety of prime meridians, even if they were not based on compass needle variations. The 24 cm wooden *Paris Green (Quirini) Globe* (c. 1513–1515) currently held by the Bibliothèque Nationale de Paris is one of the earliest post-Columbus globes to include a prime meridian; the reference line passes through the Cape Verde Islands, referred to as *Insule Portugalensium invente anno Domini 1472*. Earlier globes may have followed Ptolemy's guidance, possibly having adopted prime meridians through the Canary Islands.

A world chart from ca. 1529 made by Girolamo da Verrazzano, an Italian navigator who explored the New World on behalf of King Francis I of France, includes two sets of meridians, one in ink based on a prime meridian passing through the westernmost point of the Canary Island of La Palma. A second, in pencil without longitude markings, appears to adopt as its geographic reference either the westernmost point of Madeira or the centre of Flores in the Azores. An early world map [24] from 1542 by Alonzo de Santa Cruz (1505–1567), Principal Cosmographer to the Holy Roman Emperor Charles V (1500–1558), places the prime meridian west of the Azorean island of Faial. Interestingly, the Tordesillas Treaty Line is also included, 20° West of the map's prime meridian; see figure E.4 for a reproduction of this map.

Abraham Ortelius' (1527–1598) *Theatrum orbis terrarum* (Antwerp, 1570) is considered the first atlas in the modern sense; it included a world map in cordiform projection which was originally published in 1563 (shown in figure E.5). That map's

Figure E.4. Alonzo de Santa Cruz' world map of 1542. (Courtesy: National Library of Sweden.)

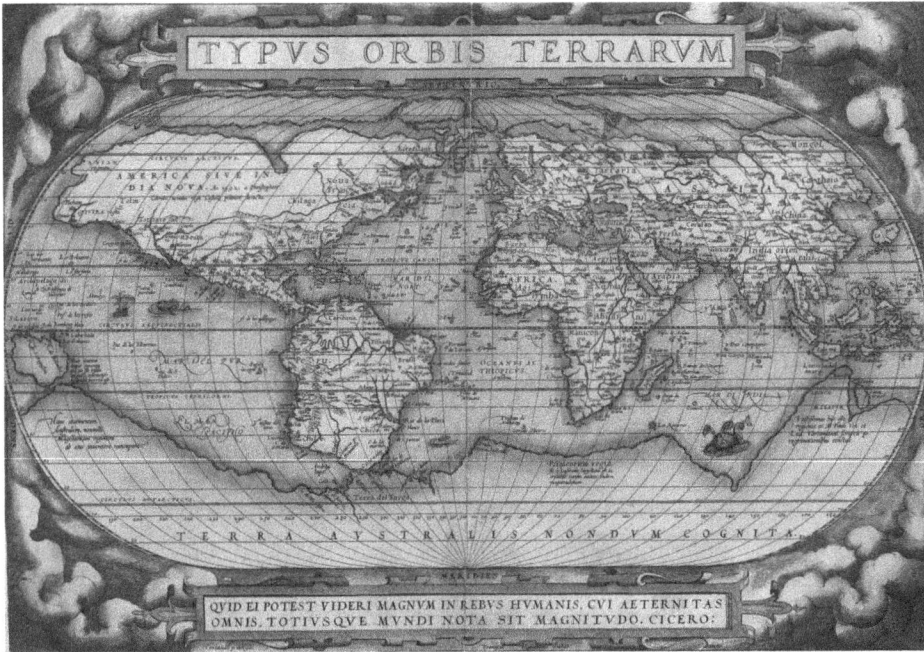

Figure E.5. Ortelius' *Typus Orbis Terrarum* world map (1570). (Courtesy: Library of Congress.)

prime meridian passes through the Azores, specifically traversing through São Miguel and Santa Maria. However, by this time, other prime meridians had already become commonplace as well, including reference lines passing through the Cape Verde and Canary Islands.

Lucas Jansz Waghenaer (1533/4–1606), a Dutch hydrographer, cartographer, and pilot, published his celebrated sea atlas, *Spieghel der Zeevaerdt*, in 1584 (published in English as *The Mariners' Mirrour* in 1588). Although only one chart in the collection includes a positional grid, the location of the prime meridian adopted is significant. It passes through the Canary Islands, between La Gomera and Tenerife. This signified an important move towards the Canary Islands as positional zero-point reference. Cardinal Richelieu followed suit in 1634, adopting the westernmost Canary Island of Ferro (19° 55′ west of Paris) as his prime meridian. The Ferro meridian became a popular choice as prime meridian in the decades and centuries to come, particularly among French geographers: their tendency was to round off the longitudinal distance between Paris and Ferro to 20°, thus rendering it the 'meridian of Paris disguised' [25].

E.3 Establishing the Greenwich Meridian

John Seller (1630–1697) from London, instrument maker, surveyor, cartographer, artist, and Hydrographer to the King, published a map of the English county of Hertfordshire in 1676, which is the first map that used the London (*not* Greenwich) prime meridian.

It took until 1721, however, for Great Britain to formally establish its own national meridian. Yet, prior to 1721, the first Astronomer Royal, John Flamsteed, and his contemporaries already referred to the 'Meridian of Greenwich.' It is likely, although not unequivocally confirmed, that they adopted a line running through the centre of the Octagon Room in Flamsteed House (see figure 6.3) as their positional reference. After all, the Octagon Room was also used for measurements of the eclipse timings of Jupiter's satellites, referred to as time 'nominally' determined in the Octagon Room, which in turn allowed determinations of relative longitudes. (It is unlikely that Flamsteed used any of his mural arcs for this purpose, because they were poorly aligned with the actual meridian.) The 1721 meridian of choice passed through an early transit circle (telescope) acquired by Edmond Halley, the second Astronomer Royal, at the Royal Observatory at Greenwich. Halley's transit circle was erected in the northwestern corner of the Observatory, between Flamsteed House and the Western Summer House.

Transit circles are telescopes designed to make two measurements, that is, to determine the local sidereal time when a star crosses the local meridian owing to the Earth's rotation, and its elevation above the local horizon. These two measurements can be used to calculate the star's coordinates on the sky. For stars with accurately known positions, so-called 'clock stars,' their meridian transit times can be used to precisely calibrate and correct the local clock time, since their precisely known coordinates can be used, in tandem with the Earth's rotation rate, to accurately predict the local transit times. This thus enabled the Greenwich astronomers to set GMT to high accuracy. Clear instructions for transit circle operation were left by E. (Edward) Walter Maunder (1851–1928), at the time assistant at the Royal Observatory:

> *'The watcher who wishes to observe the passing of a star must note two things: he must know in what direction to point his telescope, and at what time to look for the star. Then, about two minutes before the appointed time, he takes his place at the eyepiece. As he looks in he sees a number of vertical lines across his field of view ... On comes the star, 'without haste, without rest,' till it reaches one of the gleaming threads. Tap! The watcher's finger falls sharply on the button. Some three or four seconds later and the star has reached another 'wire,' as the spider-threads are commonly called. Tap!'* [26]

At the time he was appointed to the role in 1742, the third Astronomer Royal, James Bradley (1693–1762), commissioned a new, improved transit circle to replace Halley's earlier instrument. Bradley's transit circle, located slightly less than 6 metres (or 0.02 seconds of time) to the west of the eventual Airy Transit Circle, defined the Greenwich Meridian between its inauguration in 1750 until 1816, when it was replaced by a larger instrument at the same position (although raised) by John Pond (1767–1836), the sixth Astronomer Royal. The Greenwich Meridian defined by the Bradley and Pond transit circles, now known as the Bradley Meridian, has been adopted as the definition of 0° longitude by the Ordnance Survey, the UK's national mapping agency, ever since the first Ordnance Survey map was published in 1801. It is still in use as the Ordnance Survey's prime meridian today.

Over the years, the Greenwich meridian was moved on a number of occasions to accommodate the operation of new and improved transit circles, travelling of order 10 metres on three different occasions to the next-generation transit circle nextdoors and covering a total distance of 43 metres (equivalent to 0.15 seconds of time) between Halley's first transit circle and that established in 1851 by Sir George Biddell Airy (1801–1892), seventh Astronomer Royal. The Airy Transit Circle was eventually adopted as the universal Greenwich Meridian [27] by the 1884 International Meridian Conference. Construction of the instrument turned out to be a major project. However, in 1847 Airy suggested to the Royal Observatory's Board of Visitors that they consider ...

'whether meridional instruments carrying larger telescopes should not be substituted for those which we possess. Whatever we do, we ought to do well. Our present instruments were, at the time of their erection, the best in the world; but they are not so now: and we actually feel this in our observations' [28].

The Halley, Bradley, Pond, and Airy transit circles defined both the Greenwich Meridian and (until 1927) GMT. Prior to Halley's installation of the Observatory's first transit telescope in 1721, the Observatory's clocks were calibrated using observations with a quadrant of the Sun at local noon based on the 'Equal Altitude Method.' Since recording the precise timing of the Sun's local meridian crossing is rather challenging, the Astronomer Royal or his assistant would instead record the time when the Sun reached a predefined altitude shortly before and again shortly after the actual meridian crossing. These measurements, corrected for the Sun's slightly changing position on the sky over the measurement period, could then be used to adjust the clock's hands for accurate local timekeeping.

The first printed map known to have adopted the Greenwich Meridian as its reference meridian was published in 1738. Contemporary maps published by Samuel Fearon and John Eyes, as well as those published by Thomas Jeffreys (c. 1719–1771), Geographer to King George III, are early examples of maps having adopted the Greenwich Meridian. Additionally, the cartographer Joseph Frederick Wallet DesBarres (1721–1824) adopted the Greenwich Meridian for his monumental and comprehensive four-volume collection of maps, charts, and views of the East Coast of North America, *Atlantic Neptune* (1777) [29].

However, throughout the 18th Century, English charts and maps did not systematically adopt the Greenwich Meridian as their reference line. Rather, they were randomly based on meridians passing through St Paul's Cathedral in London, Lizard Point (Cornwall), or sometimes on that running through the Canary Island of Ferro. The Lizard Point meridian was also popular with Dutch sailors travelling south through the English Channel along the coast to France, Spain, or Portugal.

The Greenwich Meridian was truly propelled to prominence once the *Nautical Almanac* began to be published regularly from 1767. The Greenwich Meridian formed the basis of the ephemeris tables in the 49 issues of the *Nautical Almanac* published by the Reverend Nevil Maskelyne (1732–1811), fifth Astronomer Royal, between 1767 and 1811. Summarized lucidly by Dava Sobel and William Andrewes,

'Maskelyne's tables not only made the lunar method practicable, they also made the Greenwich Meridian the universal reference point. Even the French trans- lations of the Nautical Almanac retained Maskelyne's calculations from Greenwich—in spite of the fact that every other table in the Connaissance des Temps considered the Paris meridian as the prime' [30].

Dominance in mapmaking shifted a few times during the 18th Century, first from Amsterdam to France, and roughly around the time when Captain Cook's charts were first published in 1773 it shifted once more, this time to England. Cook's charts used the Greenwich Meridian; he set the stage in adopting this reference line, which was closely followed by the British Admiralty charts subsequently published. The latter soon became the preferred navigational aids around the world.

Maskelyne's introduction of the *Nautical Almanac* in 1767 had an undesired side effect, since it required sailors to use astronomical instead of civil time. The astronomical day starts at noon, a full 12 hours after the start of the civil day. However, marine navigation at the time was based on the use of the civil day alongside the nautical day; the nautical day also starts at noon, like the astronomical day, but it starts 12 hours *before* the civil day. Clearly, this created significant potential for confusion because of the similarity between the astronomical and nautical days. Therefore, on 11 October 1805 the British Admiralty ordered that the nautical day no longer be used; the American authorities followed suit in 1848.

In this context, Resolution 6 issued by the 1884 International Meridian Conference unexpectedly led to much debate and disagreement:

6. That the Conference expresses the hope that as soon as may be practicable the astronomical and nautical days will be arranged everywhere to begin at mean midnight.

The resolution's driver was the practical desire to equalize the astronomical (and nautical) days with the civil day. Surprisingly, implementation of this resolution was problematic because of the implications of changing the astronomical day; the potential for confusion and misunderstandings was deemed too great by many (political) leaders. Nevertheless, although the political will to make this change was lacking, practical developments eventually drove that change[3]. Specifically, the invention and practical implementation of wireless telegraphy, and its potential to widely distribute time signals, greatly facilitated standardization.

The French established a wireless transmitter on the Eiffel Tower, thus assigning themselves a global leadership role. However, for this leadership role to become viable and successful, they found that they were forced to transmit time signals in GMT instead of Paris time... This eventually led to France's adoption of GMT as the basis of her national time system on 11 March 1911. However, the French still

---

[3] In 1918, the Council of the Royal Astronomical Society recommended that the astronomical day should be included in the *Nautical Almanac* from 1 January 1925, thus defusing the long-standing disagreement whether to use civil or astronomical time.

stuck with the Paris meridian as their preferred positional reference, instead of adopting the Greenwich Meridian alongside GMT. Change came from an unexpected angle, however.

In the spirit of Britain's *Nautical Almanac*, the French *Bureau de Longitude* also published a range of annual ephemeris tables. They were keen to include larger numbers of stars in their tables, but in a cost-effective manner. This led to the practical suggestion that observatories around the world should work together to compile a joint set of observations covering many more stars than each individual observatory could manage on its own; such a combination of scientific forces would simultaneously improve the existing situation in which significant duplication of time-consuming calculations was the norm rather than an exception.

To achieve this type of joint ownership of the ephemeris tables, the scientists would need to establish common standards, including a single reference meridian and a single time standard. At the International Congress on Astronomical Ephemerides of 23 October 1911, held in Paris, the astronomers in attendance unanimously agreed to adopt the Greenwich Meridian. However, this did not imply that the French authorities would follow suit. Eventually, however, the French were left to their own devices. Portugal adopted the Greenwich Meridian in 1913, and finally the French decreed that all of their nautical documents must adopt the Greenwich Meridian as positional reference as of 1 January 1914, thus completing the process in which all of Europe's sea-faring nations moved to adoption of a common reference meridian.

E.4 Airy's Transit Circle as precursor of the modern reference meridian

The modern Greenwich Meridian is, in essence, based on the location of the Airy Transit Circle in the Royal Observatory's Transit Circle Room. It was adopted as reference meridian by the 1884 International Meridian Conference—with the French delegates abstaining following their unsuccessful appeal to adopt the Paris meridian instead [31].

The first observation with the Airy Transit Circle (shown in figure E.6) was taken on 4 January 1851, delayed by three days owing to inclement weather conditions. It continued to be used regularly until 1938. Once a transit observation had been recorded, the telescope's tilt angle was measured using a system of high-precision microscopes. This allowed the observer to determine the star's elevation above the horizon, which in turn could be converted into its elevation angle (that is, its declination) on the sky to an accuracy of 0.01 second of arc.

All of the Greenwich Meridians, from Halley's to Airy's, were located using astronomical observations based on the direction of a plumb line tracing the local gravitational attraction. Similarly to other common transit telescopes, the Airy Transit Circle employs a mercury basin to align the telescope to the perpendicular direction, that is, aligned with the local 'normal' or plumb line. The location of the astronomical Greenwich Meridian was published widely, at first by reference to longitude differences determined on the basis of the lunar distance method, later by

Figure E.6. The Airy Transit Circle at the Royal Observatory at Greenwich. (© National Maritime Museum, Greenwich, London.)

means of chronometers onboard ships and through radio transmissions. The local plumb-line direction at Greenwich, however, deviates slightly from the normal direction with respect to the reference ellipsoid (in essence, the Earth's three-dimensional gravitational profile) used to define geodetic latitude and longitude (which take into account the Earth's full three-dimensional mass distribution rather than only the local gravitational direction) in the International Terrestrial Reference Frame, or ITRF, which is nearly the same as the 1984 World Geodetic System (the WGS-84 system) used by today's Global Positioning System.

Aware of the Earth's slightly flattened shape, around the time of the 1884 International Meridian Conference, scientists had in fact embarked on a large-scale project to determine these systematic deflections from the vertical direction at numerous locations worldwide [32]. In an attempt to reconcile their measurements, the scientific practitioners defined ellipsoids of revolution: a particular ellipsoid would be a reasonable approximation to the underlying gravitational mass distribution for measurements across carefully defined regions, such as countries or even entire continents.

In 1912, the Bureau International de l'Heure, or BIH, was established at Paris Observatory with as main aim to coordinate time referencing internationally and improve longitude determination anywhere on Earth [33]. To modernize time-keeping and global positional referencing, from 1929 Greenwich Observatory was no longer used as the single reference station for longitude and time referencing. Instead, a 'mean observatory' was established, to which Greenwich Observatory contributes alongside many other reference stations around the world [34]. This also led to replacement of GMT by Universal Time. In 1968, the BIH implemented statistically justified corrections for polar motion into its standard reference frame, the 1968 BIH System.

The Universal Time reference distributed by the BIH became known as UT1. It is based on contributions from many individual stations, referred to as UT0 for a given station. Throughout the changes implemented by the BIH, the overriding principle has always been that any new system should smoothly follow from its predecessor; in other words, the key condition that was imposed is that of continuity in astronomical time. Since 1984, the BIH Terrestrial System (BTS 84) is no longer based on optical observations of celestial objects; instead, it relies on weighted solutions based on very long baseline interferometry observations at radio frequencies, and on satellite and lunar laser ranging [35,36]. Since 1988, the International Earth Rotation Service (IERS) has been in charge of time and position referencing around the globe [37].

Although the local normal at the Airy Transit Circle still points to the modern celestial meridian, that is, the projection of the prime meridian onto the celestial sphere, its local plumb line does not pass through the Earth's axis of rotation. With the advent of satellite-based measurements, the Earth's reference moved from its surface to the planet's centre of mass, its geocentre, independently of any surface features. This caused an offset in the modern prime meridian, the ITRF reference meridian—defined by a plane passing through the Earth's rotation axis and its centre—of 5.3″ to the East compared with the astronomical Greenwich Meridian

Figure E.7. The Airy meridian (dotted line) compared with the International Terrestrial Reference Frame zero meridian (solid line). (Malys *et al* 2015.)

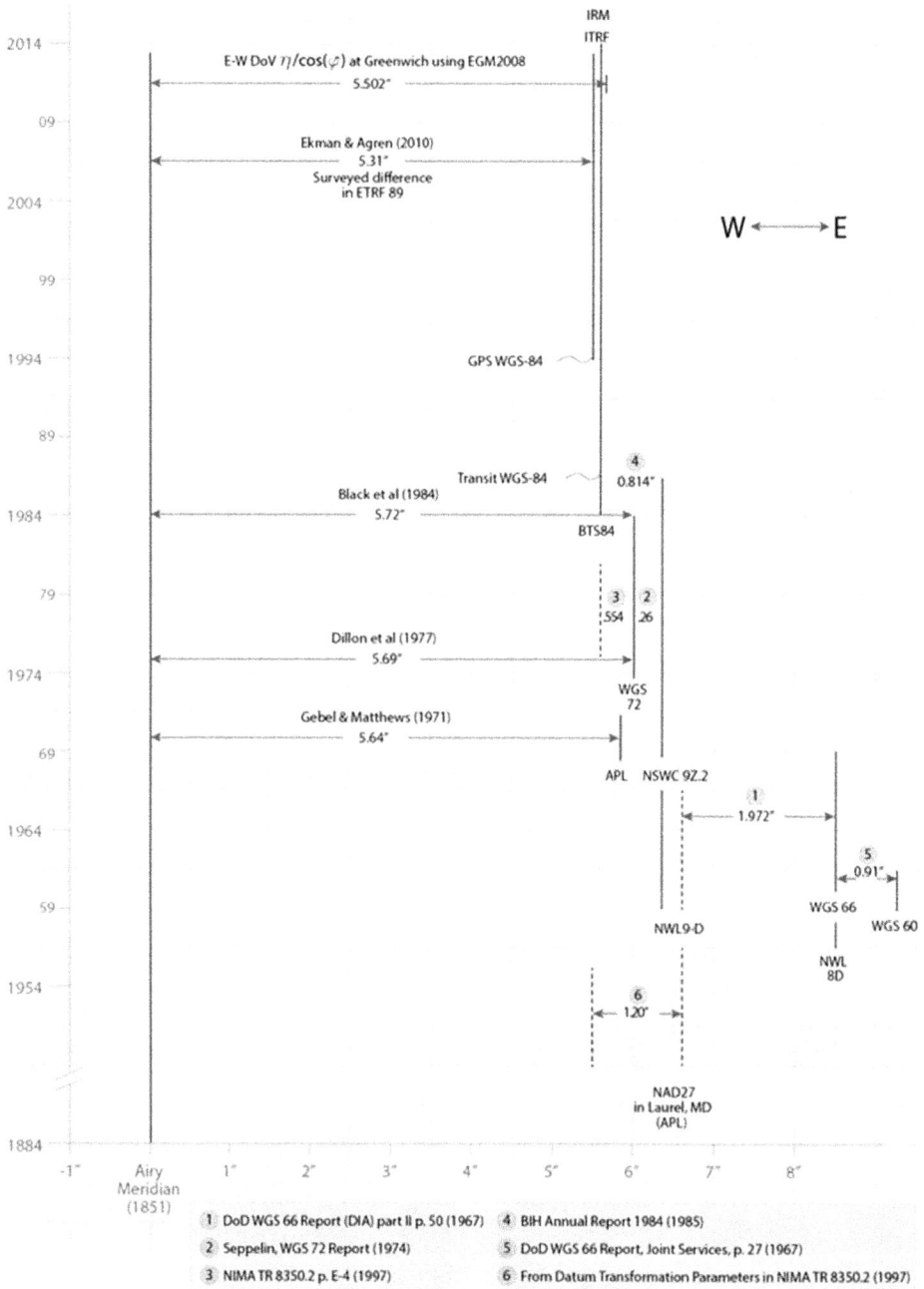

Figure E.8. Selected offsets (including references) among modern prime meridians from the Airy meridian. (Malys *et al* 2015.)

(which is now at longitude 00°00′05.3″ W), corresponding to a shift of 102.5 metres to the East compared with the prime meridian defined by the Airy Transit Circle [38,39]: see figure E.7. Figure E.8 shows a number of selected modern meridian determinations in relation to the Airy Meridian.

Natural causes, in particular the movement of the Earth's tectonic plates, now seem intent on slowly moving the 0° longitude line back northeastwards, back towards the Airy Transit Circle, by approximately 2.5 to 3 centimetres each year. The 5.3″ offset between the ITRF and Airy reference meridians therefore only strictly applies to measurements made in 1999. Meanwhile, satellite technology has enabled the construction of ever more accurate and detailed global maps, allowing the definition of a reference meridian which also takes into account tectonic plate movements and effects owing to variations in the Earth's rotation rate [40]. The ITRF zero meridian was thus defined by international agreement in 1984. This continues to be the basis of modern global position determinations.

References

[1] *Staatsblad* 1908/236, 1908/336, and 1937/82: 'De wettelijke tijd in Nederland is de middelbare zonnetijd van Amsterdam' (The legal time in the Netherlands is the mean solar time in Amsterdam), which took effect on 1 May 1909 at 0:00 am; the precise meridian was defined at 5° E only in 1937. This meridian is also known as the Gorinchem meridian or the Loenen meridian.

[2] *Explanatory Supplement to the Astronomical Ephemeris and The American Ephemeris and Nautical Almanac* 1961 (London: Her Majesty's Stationery Office)

[3] Washburn W 1982 The Canary Island and the question of the Prime Meridian: the search for precision in the measurement of the Earth *Coloquio de Historia Canario–Americano* **5**(4) 874–83 http://mdc.ulpgc.es/cdm/ref/collection/coloquios/id/420 [accessed 19 August 2017]

[4] The idea for this expedition, as well as Varin's name, appears first in a report of the *Académie's* mathematics section covering the period from August 1680 to 15 June 1681; Archives de l'Académie des Sciences (Paris) *Registres des Procès-Verbaux* **9 bis** f. 107 r-v

[5] Dew N 2010 Scientific travel in the Atlantic world: the French expedition to Gorée and the Antilles, 1681–1683 *Brit. J. Hist. Sci.* **43** 1–17

[6] Bartky I R 2007 *One Time Fits All: The Campaigns for Global Uniformity* (Stanford CA: Stanford University Press) p 40

[7] *International Conference Held at Washington for the Purpose of Fixing a Prime Meridian and a Universal Day*, October 1884, *Protocols of the Proceedings* (Washington DC: Gibson Bros); http://www.ucolick.org/~sla/leapsecs/imc1884.pdf [accessed 18 August 2017]

[8] Barrows A 2011 *The Cosmic Time of Empire: Modern Britain and World Literature* (Berkeley CA: University of California Press)

[9] *Ibid.*

[10] *Ibid.*

[11] *The Work of the Meridian Conference* 1884 *Science* **91** 414–5

[12] Fleming S 1879 *Time-Reckoning and the selection of a prime meridian to be common to all nations* (Toronto: Copp, Clark)

[13] *An act to authorize the President of the United States to call an international conference to fix on and recommend for universal adoption a common prime meridian, to be used in the*

reckoning of longitude and in the regulation of time throughout the world, 47th Congr. Sess. 1 Ch 380; https://www.loc.gov/law/help/statutes-at-large/47th-congress/session-1/c47s1ch380.pdf [accessed 18 August 2017]

[14] Dolan G 2009–2017 The adoption of a Prime Meridian and the International Meridian Conference of 1884 *The Greenwich Meridian ... where East meets West*; http://www.thegreenwichmeridian.org/tgm/articles.php?article=10 [accessed 18 August 2017]

[15] Holdich T H 1891 *Proc. Fifth Int. Geograph. Congr.*

[16] Struik D J 1987 *A Concise History of Mathematics* (New York: Dover) pp 32–3

[17] Norgate M and Norgate J 2006 *Prime Meridian*; http://www.geog.port.ac.uk/webmap/hantsmap/hantsmap/meridian.htm [accessed 18 August 2017]

[18] Swerdlow N 1973 A Lost Monument of Indian Astronomy: Das heliozentrische System in der griechischen, persischen und indischen Astronomie (B. L. van der Waerden) *Isis* **64** 239–43

[19] Hooker B 2006 *A multiplicity of prime meridians*; http://zeehaen.tripod.com/unpub_2/multitude_meridians.htm [accessed 18 August 2017]

[20] Koeman C 1980 The Astrolabium Catholicum *Revista da Universidade de Coimbra* **XXVIII** 65–76

[21] Crone E 1916 Het gebruik van het Astrolabium Catholicum *De Zee* March 1916 180–193

[22] Koeman C 1980 *op. cit.*

[23] Kanas N 2009 *Star Maps: History, Artistry, and Cartography* 2nd edn (Berlin: Springer) p 119

[24] *Nova verior et integra totivs orbis descriptio nvne primvm in lvcem edita per Alfonsvm de Santa Cruz Cæsaris Charoli V archicosmographvm, A.D. MDXLII* (A new, more true, and complete description of the whole world, first published by Alonzo de Santa Cruz, principal cosmographer to Emperor Charles V, 1542), Stockholm: National Library of Sweden, shelf mark KoB AB 50 St.f.

[25] *International Conference Held at Washington for the Purpose of Fixing a Prime Meridian and a Universal Day*, October 1884, *Protocols of the Proceedings* (Washington DC: Gibson Bros), p 57

[26] Maunder W E 1900 *The Royal Observatory Greenwich: A Glance at its History and Work* Cambridge Libr Coll (Cambridge: Cambridge University Press) p 156

[27] Meadows A J 1975 *Greenwich Observatory: the Royal Observatory at Greenwich and Herstmonceux, 1675–1975 2. Recent history (1836–1975)* (London: Taylor & Francis) p 10

[28] Airy G B 1847 Report of the Astronomer Royal to the Board of Visitors (5 June 1847) *Greenwich Astron Obs 1845*; in Stott C 1985 The Greenwich Meridional Instruments *Vistas Astron* **28** 133–45

[29] Lockett J 2011 *Captain James Cook in Atlantic Canada. The adventurer & Map Maker's Formative Years* (Halifax NS: Formac) p 187

[30] Sobel D and Andrewes W J H 1998 *The Illustrated Longitude* (London: Fourth Estate) 197–9

[31] McCarthy D and Seidelmann P K 2009 *Time: From Earth Rotation to Atomic Physics* (Weinheim: Wiley) pp 244–5

[32] Dracup J F 2006 Geodetic Surveys in the United States: The Beginning and the Next One Hundred Years *NOAA History: A Science Odyssey*; http://www.history.noaa.gov/stories_tales/geodetic1.html [accessed 20 August 2017]

[33] Guinot B 2000 *History of the Bureau International de l'Heure* in: Dick S, McCarthy D, and Luzum B (eds) Polar motion: Historical and scientific problems *Int. Astron. Union Colloq.* **178**, *Astron. Soc. Pac. Conf. Ser.* **208** 175–84

[34] Feissel M 1980 Determination of the Earth rotation parameters by the Bureau International de l'Heure, 1962–1979 *Bull. Géodet* **54** 81–102

[35] Boucher C and Altamimu Z 1985 *Towards an improved realization of the BIH Terrestrial Frame* in: Mueller II (ed) *Proc. Int. Conf. on Earth rotation and reference frames*, MERIT/COTES Rep. vol. **2** (Columbus, OH: Ohio State University)

[36] Boucher C and Altamimu Z 1986 Status of the realization of the BIH Terrestrial System ed A K Babcock and G A Wilkins *The Earth's rotation an reference frames for geodesy and geodynamics Int. Astron. Union Symp.* **128** (Dordrecht: Kluwer) pp 107ff

[37] Wilkins G A 2000 *Project MERIT and the formation of the International Earth Rotation Service* in: Dick S, McCarthy D, and Luzum B (eds) Polar motion: Historical and scientific problems *Int. Astron. Union Colloq.* **178**, *Astron. Soc. Pac. Conf. Ser.* **208** 187–200

[38] Malys S Seago J H, Pavlis N K, Seidelmann P K and Kaplan G H 2015 Why the Greenwich Meridian Moved *J. Geodes* **89** 1263–72

[39] Guinot B 2011 Solar time, legal time, time in use *Metrologica* **48** S181–5

[40] Malys *et al* 2015 *op. cit.*

www.ingramcontent.com/pod-product-compliance
Lightning Source LLC
Chambersburg PA
CBHW082133210326